Green Chemistry and Sustainable Technology

Series Editors

The series Green Chemistry and Sustainable Technology aims to present cutting-edge research and important advances in green chemistry, green chemical engineering and sustainable industrial technology. The scope of coverage includes (but is not limited to):

- Environmentally benign chemical synthesis and processes (green catalysis, green solvents and reagents, atom-economy synthetic methods etc.)
- Green chemicals and energy produced from renewable resources (biomass, carbon dioxide etc.)
- Novel materials and technologies for energy production and storage (bio-fuels and bioenergies, hydrogen, fuel cells, solar cells, lithium-ion batteries etc.)
- Green chemical engineering processes (process integration, materials diversity, energy saving, waste minimization, efficient separation processes etc.)
- Green technologies for environmental sustainability (carbon dioxide capture, waste and harmful chemicals treatment, pollution prevention, environmental redemption etc.)

The series Green Chemistry and Sustainable Technology is intended to provide an accessible reference resource for postgraduate students, academic researchers and industrial professionals who are interested in green chemistry and technologies for sustainable development.

More information about this series at http://www.springer.com/series/11661

Ying Li · Farid Chemat
Editors

Plant Based "Green Chemistry 2.0"

Moving from Evolutionary to Revolutionary

Springer

Editors
Ying Li
Department of Food Science and Engineering, College of Science and Engineering
Jinan University
Guangzhou, Guangdong, China

Farid Chemat
INRA, SQPOV, UMR408
Avignon University
Avignon, France

ISSN 2196-6982 ISSN 2196-6990 (electronic)
Green Chemistry and Sustainable Technology
ISBN 978-981-13-3809-0 ISBN 978-981-13-3810-6 (eBook)
https://doi.org/10.1007/978-981-13-3810-6

This Springer imprint is published by the registered company Springer Nature Singapore Pte Ltd.
The registered company address is: 152 Beach Road, #21-01/04 Gateway East, Singapore 189721, Singapore

Preface

Plant based chemistry have been used probably since the discovery of fire. Egyptians and Phoenicians, Jews and Arabs, Indians and Chinese, Greeks and Romans and even Mayas and Aztecs all possessed a culture of using plants as source of reagents for cosmetic, perfumery, medicine, food ingredients and products, colors and dyes, and building materials. Until the start of the petroleum era plant-derived biomass was the main source of reagents, ingredients and products for food and non-food applications. The spectacular growth of petroleum-based processes led to a withdrawal from those based on biomass. However, the depletion of fossil resources, upon which the current international industry and economy heavily depends, and environmental considerations force us towards a post-petroleum society. The challenges launched by the environment protection and competitiveness of the globalized world strongly require innovations that break away from the past rather than simple continuity. While green chemistry has given rules for modern chemistry, plant based chemistry could be one of the solutions from the past to the future of humanity as an ecologic and an economic chemistry, and turning to "Plant Based" Green Chemistry 2.0 in the twenty-first century.

This book was prepared by a team of chemists, biochemists, chemical engineers, physicians, and food technologists with an objective to provide an actual picture of current knowledge on "Plant Based" Green Chemistry used at laboratory and industrial scale. It is aimed for professional from industry, academicians engaged into plant-based chemistry and engineering or natural product chemistry researches and graduate level students. Each chapter would be complementary to other chapters and based on presentation of the reputed international researchers and professionals, and would address the latest efforts in the field.

We wish to thank sincerely all of our colleagues who have collaborated in the writing of this book. We hope to express them our scientific gratitude for agreeing to devote their competence and time to ensure the success of this book. We are totally convinced that this book is the starting point of new discipline and for future collaborations in "Plant Based" Green Chemistry between research, industry and education, covering large applications: perfume, cosmetic, pharmaceutical, food ingredients, nutraceuticals, biofuels, or fine chemicals industries.

Guangzhou, China Ying Li
Avignon, France Farid Chemat

Contents

Editors and Contributors

About the Editors

Ying Li is an Associate Professor of Chemistry from College of Science and Engineering in Jinan University. He received his Master degree from University of Toulouse (2010) and his Ph.D. degree in Avignon University (2014), France. After periods of working in the GREEN Extraction Team as a postdoctoral research fellow (2014–2015) and a leading high-tech company (2015–2017) in Kunming China, he moved to Department of Food Science and Engineering in Jinan University (Guangzhou) as an associate professor. His main research interests focus on the integration of innovative technologies and bio-based alternative solvents for total valorization of renewable indigenous bioresources, particularly for green extraction of bioactive compounds (e.g. essential oils, antioxidants, pigments, lipids, etc.), which leads to safe and cost-efficient processing procedures, and high value-added final products with great potential in cosmetics, nutraceuticals, beverage, feed and health industries. He has accomplished dozens of scientific peer-reviewed papers, one Springer book, and four book chapters as the first or corresponding author, in recent years, as well as a similar number of international conference communications to his publications.

Farid Chemat is a full Professor of Chemistry at Avignon University, Director of GREEN Extraction Team (innovative techniques, alternative solvents, and original procedures for green extraction of natural products), and co-director of ORTESA LabCom research unit Naturex-UAPV. He received his engineer diploma from University of Blida-Algeria (1990) and his Ph.D. degree (1994) in process engineering from the Institut National Polytechnique of Toulouse-France. After periods of postdoctoral research work with Prolabo-Merck, Rhone-Poulenc and Unilever (1995–1997), he spent two years (1997–1999) as senior researcher at University of Wageningen-The Netherlands. In 1999, he moved to the University of La Réunion-France DOM as assistant professor and since 2006 holds the position of Professor of Food Chemistry at the University of Avignon-France. His research activity is documented by more than 200 scientific peer-reviewed papers, and about the same number of conferences and communications to scientific and industrial meetings, 12 books, 40 book chapters and 14 patents. His main research interests have focused on innovative and sustainable extraction techniques (especially microwave, ultrasound and green solvents) for food, pharmaceutical, bio-energy and cosmetic applications. He is scientific coordinator of "PEEV: Industrial Platform for Green Extraction" and "France Eco-Extraction", an academic and industrial association, dealing with international dissemination of research and education on green extraction of natural products in academia and industries.

Contributors

Kamal Aberkani Department of Biology and Geology, Faculty Multidisciplinary of Nador, University Mohamed First, Selouane, Morocco

Maryline Abert-Vian GREEN Extraction Team, Avignon University, INRA, UMR408, Avignon, France

Noureddine Allouche Laboratory of Organic Chemistry, Natural Substances Team, Faculty of Sciences of Sfax, University of Sfax, Sfax, Tunisia

Hichem Ben Salah Laboratory of Organic Chemistry, Natural Substances Team, Faculty of Sciences of Sfax, University of Sfax, Sfax, Tunisia

Nourreddine Bouarourou Taourirt, Morocco

Abdelhak Boublenza Sarl Boublenza Agroalimentaire et Produits Agricoles, Tlemcen, Algeria

Ibtissam Boublenza Sarl Boublenza Agroalimentaire et Produits Agricoles, Tlemcen, Algeria

Ikram Boublenza Sarl Boublenza Agroalimentaire et Produits Agricoles, Tlemcen, Algeria

Meriem Boukroufa GREEN Extraction Team, Université d'Avignon et des Pays de Vaucluse, INRA, UMR408, Avignon, France

Chahrazed Boutekedjiret Laboratoire des Sciences et Techniques de l'Environnement (LSTE), Ecole Nationale Polytechnique, Alger, Algeria

Jie Cai College of Food Science and Engineering, Wuhan Polytechnic University, Wuhan, China

Farid Chemat GREEN Extraction Team, Université d'Avignon et des Pays de Vaucluse, INRA, UMR408, Avignon, France

Smain Chemat Research Centre in Analytical Chemistry and Physics (C.R.A.P.C), Bousmail, Tipaza, Algeria

Anne-Sylvie Fabiano-Tixier GREEN Extraction Team, Université d'Avignon et des Pays de Vaucluse, INRA, UMR408, Avignon, France

Frédéric Fine Terres Inovia, Pessac, France

Zahia Ghouila Research Centre in Analytical Chemistry and Physics (C.R.A.P.C), Bousmail, Tipaza, Algeria

Yongbin Han College of Food Science and Technology, Nanjing Agricultural University, Nanjing, China

Sadia Hassan Institute of Home and Food Sciences, Faculty of Life Sciences, Government College University, Faisalabad, Pakistan

Yi He College of Food Science and Engineering, Wuhan Polytechnic University, Wuhan, China

Muhammad Kamran Khan Institute of Home and Food Sciences, Faculty of Life Sciences, Government College University, Faisalabad, Pakistan

Werner Kunz Institute of Physical and Theoretical Chemistry, Universität Regensburg, Regensburg, Germany

Shuyi Li College of Food Science and Engineering, Wuhan Polytechnic University, Wuhan, China

Wanrong Li College of Food Science and Engineering, Wuhan Polytechnic University, Wuhan, China

Ying Li Department of Food Science and Engineering, College of Science and Engineering, Jinan University, Guangzhou, China

Gang Liu College of Food Science and Engineering, Wuhan Polytechnic University, Wuhan, China

Sara Madji GREEN Extraction Team, Université d'Avignon et des Pays de Vaucluse, INRA, UMR408, Avignon, France

Adyl Oussaid Department of Chemistry, Faculty Multidisciplinary of Nador, University Mohamed First, Selouane, Morocco

Larysa Paniwnyk School of Life Sciences, Faculty of Health and Life Sciences, Coventry University, Coventry, UK

Sandrine Perino Université d'Avignon et des Pays de Vaucluse, INRA, Avignon, France

Xinguang Qin College of Food Science and Engineering, Wuhan Polytechnic University, Wuhan, China

Alain Quinsac Terres Inovia, Pessac, France

Njara Rakotomanomana GREEN Extraction Team, Université d'Avignon et des Pays de Vaucluse, INRA, UMR408, Avignon, France

Moussa Sehailia Research Centre in Analytical Chemistry and Physics (C.R.A.P.C), Bousmail, Tipaza, Algeria

Yaning Shi College of Food Science and Technology, Nanjing Agricultural University, Nanjing, China

Anne-Gaëlle Sicaire Terres Inovia, Pessac, France

Yang Tao College of Food Science and Technology, Nanjing Agricultural University, Nanjing, China

Mohamed Touaibia Département de Chimie et Biochimie, Université de Moncton, Moncton, NB, Canada

Carlos Vaca-Garcia Laboratory of Agro-Industrial Chemistry, Institut National Polytechnique de Toulouse (France), Toulouse, France

Léa Vernès GREEN Extraction Team, Avignon University, INRA, UMR408, Avignon, France

Qi Wang College of Food Science and Engineering, Wuhan Polytechnic University, Wuhan, China

Yue Wu College of Food Science and Technology, Nanjing Agricultural University, Nanjing, China

Zhenzhou Zhu College of Food Science and Engineering, Wuhan Polytechnic University, Wuhan, China

Chapter 1
Plant-Based Green Chemistry: Moving Towards Petroleum-Free Chemistry

Including: Definition, Classification and Potential Applications of Plant-Based Chemicals

Carlos Vaca-Garcia

Abstract Once that primary needs are satisfied, mankind utilises plants for different applications. First, the precious molecules given by nature are extracted using methods such as distillation or maceration. They are used for health (pharmaceuticals or traditional medicine), well-being (essential oils), or for decorative purposes (pigments). This method is called the deconstruction approach. It can be completed by chemical modification, yielding hemisynthetic products such as modified starch or cellulose acetate. On the other hand, plant material can be cracked to obtain simple molecules. For instance, gasification of biomass leads to CO, H2, CH4, etc. These molecules can enter the traditional synthetic routes widely developed by the petrochemistry to form more complex compounds for a large variety of products. In any case, the deconstruction or the cracking approach should take into account the green chemistry principles and be the object of a life-cycle assessment. Bio-based products are not necessarily good for the environment, or more precisely, they might have a bigger impact on the environment than the fossil alternative. This chapter demonstrates through a few examples that smarter and more sustainable solutions always exist. The advantages of biomass and the big principles to a sustainable approach of the biomass valorisation are given.

Keyword Biomass utilisation · Biomass deconstruction · Biomass cracking · Environmental impact

1.1 Introduction and Definition

Mankind has always lived with and from plants. It is simply vital: We could never survive without them. That is a fact. Plants produce oxygen and nutrients for food,

---

C. Vaca-Garcia (✉)
Laboratory of Agro-Industrial Chemistry, Institut National Polytechnique de Toulouse (France), Toulouse, France
e-mail: Carlos.VacaGarcia@ensiacet.fr

Y. Li and F. Chemat (eds.), *Plant Based "Green Chemistry 2.0"*,
Green Chemistry and Sustainable Technology,
https://doi.org/10.1007/978-981-13-3810-6_1

and they allow the humankind to consume other animals' proteins. From plants, we used to, and we still do, build artefacts and construct shelter. Then, when our primary needs were satisfied, we explored the astonishing diversity of the vegetal kingdom to extract odorous and colouring substances. We discovered empirically that plants can heal disorders and diseases and many precursors of the pharmaceutical industry are, still in the twenty-first century, obtained from plants.

The reason for this panoply of applications is the rich variety in species and the easiness to find or produce them—at least for the moment. Mankind has almost always lived on a bio-economy basis, and the socle of all of it is the vegetal kingdom.

A radical change started in the beginning of the twentieth century, when we explored the potential of fossil resources for energy and chemicals. In the scale of the whole life of the humankind, this change can be qualified of "abrupt". In only 100 years, petroleum has become an abundant and inexpensive resource thanks to the progress of the extraction techniques.

However, if we compare petroleum and plant resources, we can affirm that both of them are abundant and inexpensive. Thus, what is the reason why, since the 1950s, we have come to the oligopoly of petroleum-derived commodities? One of the many possible answers for its hegemony is the fact that petroleum has a relatively simple and constant composition and does not possess hierarchical structures, and consequently, a simple cracking process allows obtaining a limited number of molecular bricks, almost constantly, independently of the provenance of the feedstock. All the opposite of the plant resources, in fact.

Plants are much more complex, and by the way, it would be injudicious to apply a cracking process to convert them into molecular bricks. Indeed, would it be wise to propose the thermal cracking of plants such as woad (*Isatis tinctoria*, Brassicaceae, which produces a dye precursor, much more appreciated than indigo) or Madagascar periwinkle (*Catharanthus roseus*, Apocynaceae, which produces a very successful anti-breast cancer precursor)? These examples are voluntarily extravagant, but they help us to realise that, contrarily to petroleum, plants bear extraordinary molecules built by the miracle of Nature. The global strategy for taking benefit from them is not thermal cracking but "deconstruction".

With the deconstruction of plant material, it is possible to separate the valuable molecules that are too difficult, if not impossible, to synthesise out of ex-petroleum molecules. The limit of this approach is the efficiency of separation techniques. It is common to obtain enriched fractions rather than pure molecules. But for many applications, an isolate is often enough.

Some fractions—or pure molecules—need to be modified chemically, or enzymatically, to attain interesting levels of worthiness. Thus, sugars can be transformed into succinic acid; pharmaceutical precursors are transformed into molecules with improved therapeutic activity; wood pulp can be transformed into cellophane[1] or insulation thermoplastics, etc. The list is extraordinarily long, but we can affirm that all the spectrum of organic chemistry can be applied to ex-plant material for the construction of new and enhanced substances, innovative polymers and convenient

[1]Cellophane is a registered mark in some countries.

types of fuels. The variety can theoretically be even wider than that obtained from the ex-petroleum molecules.

It is nevertheless important to consider that, regardless of their origin, petroleum and plant materials are transformed with practically the same chemical processes. Plant-based molecules are of “natural”[2] origin, but they and the processes used for their transformation are not necessarily inoffensive or without impact for humans or for the environment. Petroleum-based and plant-based molecules are subjected to the same considerations of green chemistry, as defined by Paul Anastas and John Warner [1]. Polluting processes are sadly sometimes used for the transformation of plant-based molecules, and low-impact processes are happily implemented for the transformation of petroleum.

This book will consider the chemical processes related to the transformation of plant material with regard to the principles of green chemistry. In the following paragraphs and chapters, when the term “bio-based” is used, it refers to “plant-based”, not necessarily including animal resources or microalgal biomass, but in most cases, they could also be considered under the same optic.

1.2 The Deconstruction Approach

Nature has created not only useful molecules for the mankind but also extraordinary construction patterns. Wood ultrastructure is a typical case. The traditional model of wood cell, that can be found in any textbook (Fig. 1.1), shows an organised distribution of the cellulose microfibrils in the different layers of the cell wall. The model makes us also understand that the different cells are linked together by the middle lamella (in black on the figure), with a neat frontier between the cell wall and the middle lamella. It is recognised that lignin and hemicelluloses are the main constituents of the middle lamella. The so-called lignin–carbohydrate complex (LCC) is sometimes taught as a “glue” binding the cell walls. The reality is more complex.

Some cellulose molecules in the cell wall are linked to hemicellulose molecules through hydrogen bonding. Then, some hemicelluloses molecules in the LCC are linked to lignin through covalent bonding, essentially ester and ether bonds (Fig. 1.2). There are no layers in the LCC structure. The primary and the secondary walls are penetrated by the LCC with a gradient profile. There is no neat border in fact. Demonstrating the entanglement of these three main molecules are quite difficult. The molecular modelling realised by a research team in the Massachusetts Institute of Technology [2] shows this intricacy to a certain extent. This intricacy phenomenon is observed not only in wood but in most of the natural ultrastructures (flesh and ligaments, plant cells, etc.).

[2]The dictionary definition of “natural” is «existing in or derived from nature, not made or caused by humankind». Petroleum is therefore included in this definition. Nevertheless, due to the recent negative image of fossil resources and to the numerous synthetic developments realised in the past

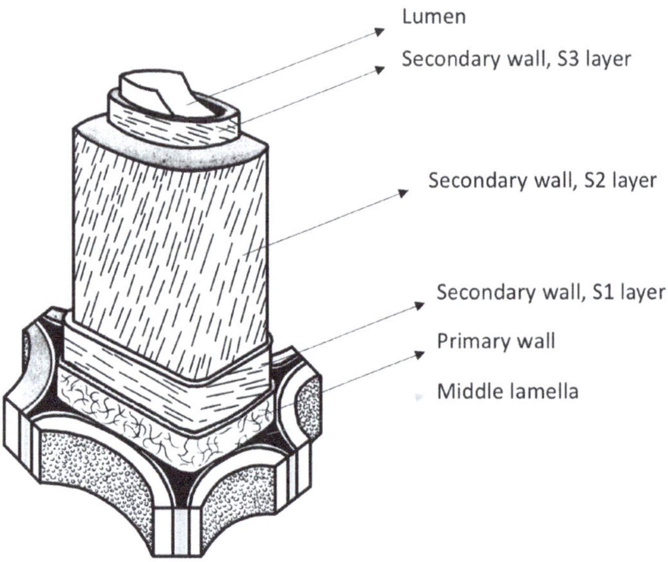

Fig. 1.1 Schematic representation of a wood cell with 3D representation of the cell wall layers

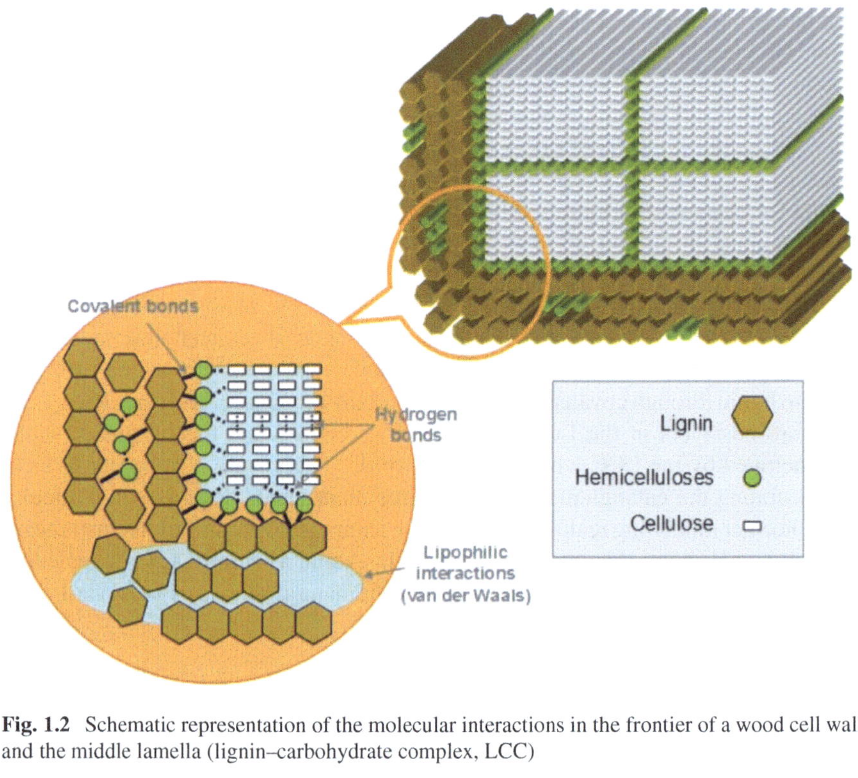

Fig. 1.2 Schematic representation of the molecular interactions in the frontier of a wood cell wall and the middle lamella (lignin–carbohydrate complex, LCC)

Fig. 1.3 A scanning electron microscope image of a carbon/carbon composite composed of fibres (in blue) and silicon carbide (in brown) Photograph credit: ©Eye of Science/Photo Researchers, Inc. www.pbs.org (MSC software)

Conversely, before the contribution of biomimetics in the inspiration of scientists and engineers, all of the human-made materials do not show complex molecular entanglement. Let us take, for instance, a composite material, consisting of a matrix and a carbon fibre reinforcement. When the material is broken by tensile effort, the fracture surface shows a perfect detachment of the fibre pulled out of the matrix (Fig. 1.3). A gap between the filler and the matrix can be observed as a consequence of lack of entanglement. This would never happen in a wood cell: primary and secondary layers pulling out of the middle lamella. The physical structure and the chemical bonds make the separation of lignin, hemicellulose and lignin extremely difficult. At least with the preservation of the main structure. That is one of the main principles of the deconstruction process.

Cellulose could be separated from hemicellulose through a dissolution process since they are linked only by low-energy hydrogen bonding. Indeed, hemicelluloses are soluble in certain solvents, whereas cellulose is insoluble in all the traditional solvents.

On the contrary, in order to separate hemicelluloses from lignin, it is necessary to break covalent bonding. Ether (C–O–C) bonds are difficult to break, while ester bonds (O–C(O)–) are easily broken with an alkali, but such chemical methods are aggressive to the structure of the three molecules. The challenge is still open to find effective deconstruction processes for separating the main components of lignocellulose.

This is only an example of the complexity of the plant material. It refers to the complex entanglement of biopolymers. But we must also consider the relative concentration of a certain molecule, or groups of molecules, in the plant.

Finally, from a more global perspective, let us consider the distribution of metabolites in a plant. Vegetable reserve oils, triacylglycerides, are concentrated sometimes

century from it, petroleum, as raw material, is erroneously not considered in common language as "natural" (by opposition to synthetic).

on the seeds (e.g. sunflower, rapeseed), but they can also be found in fruits (e.g. olive, coconut). Waxes are usually concentrated on the skin of fruits, and essential oils on the surface of leaves and fruits mainly encapsulated in fragile structures that allow the easy extraction of these terpeneous substances.

All these considerations of macroscopic and microscopic distribution, and easiness of separation/extraction, are to be taken into account when dealing with plant chemistry. If there is a viable path to obtain a desired fraction, a deconstruction process should be set up.

1.3 The Cracking Approach

When it is too expensive or too polluting to establish a deconstruction process, biomass can undergo a cracking process, based on thermal decomposition, to obtain energy and chemical compounds. Among the thermolytic processes, the combustion is the most evident and it has been known since immemorial times. As a general rule, any relatively dry biomass (usually under 20% of moisture content) can be burnt and fire is auto-maintained. But the process is not so simple. Combustion is based on exothermal chemical reactions that provide enough energy to have also undesired reactions, such as the conversion of sulphur (present in proteins for instance) into sulphur oxides, responsible for acid rain. Nitrogen from the air is also converted into undesired nitrogen oxides, toxic and with global warming effect. Halogen-containing biomass (in particular that growing in saline environment) can be responsible for toxic dioxins formation. Moreover, the amount of energy obtained per dry mass unit of biomass is always lower than that of fossil resources. There is a quasi-linear correlation between the energy content and the carbon content of the biomass.

$$Higher\ heating\ value\ (HHV)\ expressed\ in\ \frac{\mathrm{MJ}}{\mathrm{kg}} = 0.44 \times \%Carbon - 1.9$$

Cellulose has 44.4% C and provides 17.5 MJ/kg, whereas pine wood contains 51.8% C and provides 21.2 MJ/kg. We are still far from the 30 MJ/kg obtained from fossil resources such as coal. It is often said that the more oxygen a bioresource contains, the less energy we can obtain from it. It is because the carbon content diminishes. That is why the oxygenated molecules in bio-ethanol (CH_3CH_2OH) and biodiesel ($R–CO–O–CH_3$) are less efficient in terms of energy/mass with regard to gasoline and diesel, which are hydrocarbons (no oxygen).

From the biomass combustion, when it is smartly conducted to minimise pollution, we can obtain substantial energy, but the exhaust gases consist mainly in carbon dioxide. This gas does not have a significant application as raw material in organic chemistry, but combustion gas has now vocation to be employed to boost the microalgae growth by injection into photobioreactors [3]. This is known as biosequestration of carbon dioxide.

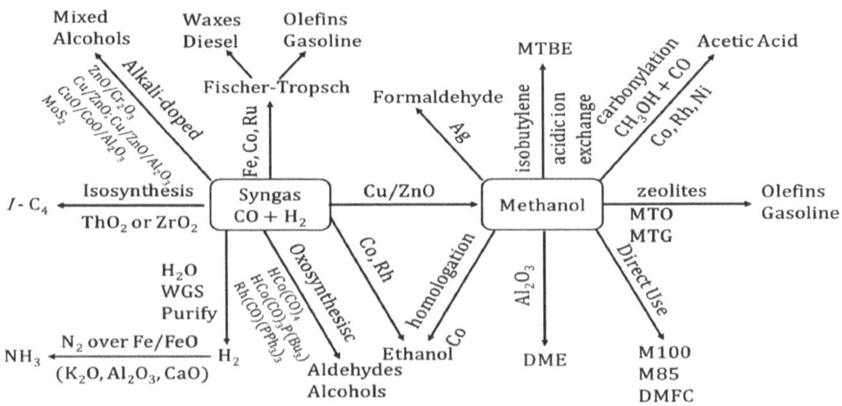

Fig. 1.4 Catalytic routes for the transformation of syngas into chemicals and fuels [4]

Other thermolytic processes are known as "pyrolytic", and they include pyrolysis, gasification and liquefaction. These subdivisions are established based on the amount of oxygen (from 10% to less than 100% of the stoichiometric amount needed for combustion) and also on the operation conditions favouring the formation of one main product: gas (called "syngas" from synthetic gas), liquid (called "bio-oil" from its physical—but not chemical—resemblance to petroleum) and solid (charcoal, obtained already from ancient civilisations by primitive pyrolysis).

All of them are interesting. Charcoal has become a popular solid fuel in barbeques. Bio-oil has a lot of potential for chemicals (acids and alcohols) but also still some technical challenges to overcome (corrosion, heterogeneity, and stability) to reach a wide viable application in industry. Syngas, which contains essentially carbon monoxide and dihydrogen, has been used in industry for decades as raw material for many synthetic pathways, including the Fischer–Tropsch (Fig. 1.4).

1.4 Bio-based, but not at Any Cost

Using biomass represents undeniably various advantages, either with the deconstruction approach or with the cracking approach. We spoke above about the "viability" of a process. It is time to better understand this axis of analysis. The term has always been used to refer to technical feasibility and benefit generation. It is time to include the idea of sustainability in this term. Life-cycle assessment or, even better, life-cycle sustainability assessment must be considered when qualifying a process as "viable". It seems so evident to say that what it is important is not to have a renewable raw material (the definition of bio-based), but to make the whole process of transformation viable. That is, sustainable in all the ways. That is the main challenge of Plant chemistry as a part of Green chemistry. In a whole concept: Plant Green Chemistry.

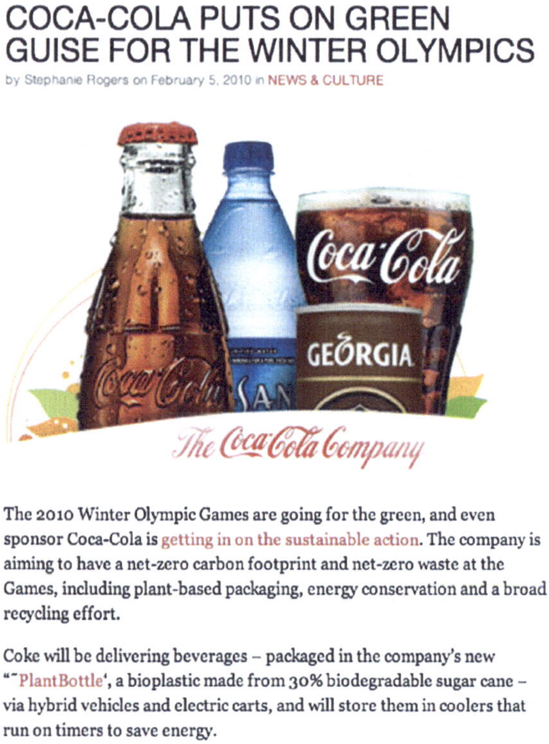

COCA-COLA PUTS ON GREEN GUISE FOR THE WINTER OLYMPICS

by Stephanie Rogers on February 5, 2010 in NEWS & CULTURE

The 2010 Winter Olympic Games are going for the green, and even sponsor Coca-Cola is getting in on the sustainable action. The company is aiming to have a net-zero carbon footprint and net-zero waste at the Games, including plant-based packaging, energy conservation and a broad recycling effort.

Coke will be delivering beverages – packaged in the company's new "PlantBottle', a bioplastic made from 30% biodegradable sugar cane – via hybrid vehicles and electric carts, and will store them in coolers that run on timers to save energy.

Fig. 1.5 Screenshot from www.thecoca-colacompany.com retrieved on 2010

1.5 The Case of Coca-Cola®'s PlantBottle™

In 2010, The Coca-Cola Company launched the PlantBottle on the occasion of the Winter Olympic Games in Vancouver. The characteristic of this PET bottle? Elaborated with 30% plant-based material. The green image of the brand was boosted by messages given in its official website (Fig. 1.5): "The company is aiming to have a net-zero carbon footprint and net-zero waste at the Games…". Although those objectives are laudable, some other parts of the advertising can be criticised: "… a bioplastic made from 30% biodegradable sugarcane". This phrase is not only approximate, but it deliberately misleads consumers who might associate the word "biodegradable" to "polyethylene terephthalate".

Beyond the marketing issues, questions have been raised: How is fabricated this bottle? How sugarcane can become part of the PET polymer? The answers were partially given in the website (Fig. 1.6). After careful observation, any chemist can determine that "Component B" corresponds to ethylene glycol, which represents 30% of the mass of the polymer. Thus, in this scheme, ethylene glycol is obtained from ethanol, which in turn is obtained from biomass by fermentation, as all the Brazilian *cachaça* fabricants know. Because, indeed, the claimed sourcing for sugarcane is Brazil, that is, 10 thousand kilometres away from Canada. Many scientists considered

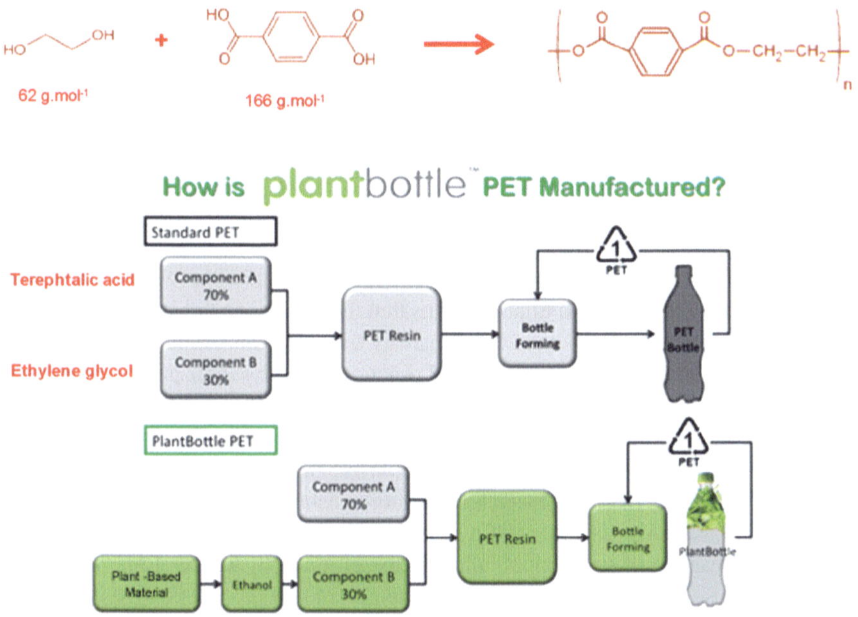

Fig. 1.6 Screenshot from www.thecoca-colacompany.com retrieved on 2010. The parts in red were added by the author of this chapter for a better comprehension

at that time, for right or wrong, that the environmental impact of the PlantBottle was higher that the fossil-based PET. And not only for transport reasons.

The Vice President of *Technology and Analytics Group* reported in 2014 [5] all the traditional and innovative ways to produce ethylene glycol, including from biomass. The conventional route converts ethylene to ethylene oxide, which in turn is hydrolysed to form ethylene glycol (but also di- and triethylene glycol). Ethylene is industrially obtained from ethane (natural gas liquids, NGLs) or from the cracking of naphtha. When bio-ethanol is involved, it can be dehydrated to form ethylene and then join the synthetic route to ethylene glycol.

Let us compare. On the one hand, the current industrial method to convert ethane to ethylene involves temperatures over 800 °C and creates 1.5–3.0 tons of CO_2 for every ton of ethylene made, even though new developments carry out this reaction at room temperature with an appropriate catalyst [6]. On the other hand, bio-ethanol dehydration is industrially conducted at 300–500 °C, at 1–2 bar, with alumina-based catalysts [7]. Which raw material should be used? NGLs or sugarcane? It is difficult to say without a life-cycle analysis (LCA) comparison. If deforestation occurs for planting sugarcane, and if intensive agricultural practices are employed, and if transport from Brazil to a chemical industry in a third country is done before the PET bottle reaches Vancouver, it is likely that the bio-based solution has a bigger impact to the environment than the fossil-based solution.

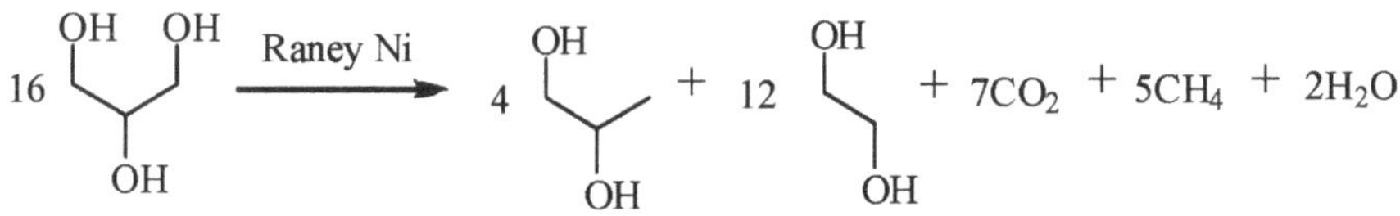

Fig. 1.7 Reaction approach to propylene glycol and ethylene glycol derived from glycerol reforming [9]

In a more recent version of the website dedicated to PlantBottle (www.plantbottle.info 2018), The Coca-Cola company affirms that the sugarcane fields (in general, not its own) are situated at more than 2000 km away from the Amazonian forest and that some examples of sustainable agricultural practices used by their providers include organic fertilisers and natural watering of the fields by the rain. Let us hope that they are not just "some examples" but that "all" of their farmers use these practices. Beyond these considerations, the real question is: Are there other possibilities to obtain this monomer from biomass that are more sustainable?

Option n°1: The development of a novel catalytic system, based on Ni/C catalysts in the presence of solid bases, allowed the efficient hydrogenolytic process of pentoses (xylose and xylitol) to yield ethylene glycol and propylene glycol [8]. The evident benefit of this process is the sourcing of the raw material. Xylose-containing hemicelluloses are of much lesser food interest than sucrose and are available almost worldwide. They can be obtained from trees and some annual plants' residues, for instance, corn cobs. By contrast, the reaction takes place at 200 °C and 40 bar of H_2 pressure. Temperature is lower than the ethanol dehydration, but pressure is much higher and the use of hydrogen gas makes the process riskier.

Option n°2: A new catalytic route was developed for the direct synthesis of propylene glycol and ethylene glycol from glycerol [9]. The reaction is conducted in a one-pot aqueous-phase process, using Raney Ni catalyst at 200 °C at ambient pressure. This option seems more convenient for the following reasons explained by Yin et al.:

> Firstly, the reaction conditions are much milder than those previously reported, since no hydrogen was needed and the reaction can be conducted under ambient pressure and the diluted aqueous glycerol can be directly used as the substrate without any concentration.
>
> Secondly, the catalyst is the commonly used commercial Raney Ni which is remarkably cheaper than the noble metal catalysts. It is interesting to find that the Raney Ni can be easily recovered from the reaction mixture by filtration or magnetic separation since Raney Ni is paramagnetic, and the catalyst can be conveniently reused without any further treatment.
>
> Thirdly, the two diol products can be easily separated from the reaction system because glycerol is totally converted and the catalyst can be easily removed.
>
> Finally, the gaseous by-products are very useful since they are mixture of CH_4 and CO_2; no CO was detected, which is very convenient for its further use as fuel or a hydrogen resource. (Fig. 1.7)

We can also add some other arguments to Option n°2, for instance, the high availability of glycerol worldwide thanks to the biodiesel production from vegetable oils. One ton of biodiesel is accompanied by 100 kg of glycerol generation. This raw material is therefore available in all the industrialised countries currently fabricating this first-generation biofuel. Its price is also low because the glycerol market was flooded when the biodiesel started to rise facing limited applications to growing supplies.

We hope that The Coca-Cola Company and other challengers such as PepsiCo do consider the life-cycle sustainability assessment when they choose new bio-based technologies that they qualify as "green". The trendy search for a 100% bio-based PET bottle or a 100% bio-based PEF (polyethylene furanoate) bottle is concerned by these considerations.

1.6 Towards a Petroleum-Free World?

From the previous paragraphs, we can outline that the most important thing is not to have a 100% bio-based product, but to have a 100% bio-based product obtained with smart technologies, involving physical, chemical, and biochemical tools, in such a way that the life-cycle sustainability assessment is always improved with regard to the substituted existing solution. This is part of the life-cycle thinking [10, 11].

From a global perspective, nevertheless, we can remind the thinking of Claude Roy, *Ingénieur général des ponts, des eaux et des forêts*, for his long collaboration with the French government. His ideas influenced the French policy of development in the sector of biofuels and of bio-economy in the past two decades. While he recognised that the biomass valorization was easy and necessary, he acknowledged that, at the same time, it was very difficult because it comprehends:

- Five large categories of feedstock: humid organic waste and residues, cellulosic waste and residues, wood, agriculture and silviculture, and specific cultures for grains and tubercules;
- Eight large types of applications: fertilization, food, traditional materials and products, polymers and composites, chemistry, biofuels, heat and power.

Consequently, dozens of different ministers, agencies, unions and clusters are concerned by the policies adopted. A global vision and a solid political will are therefore necessary.

At a time where the consequences of global warming are part of our reality, and where the vision of the depletion of petroleum, uranium, natural gas and coal feedstocks is more than clear, the wise realises that the soil, the Sun and the water will become, again, the primary fundamental factors on Earth for the production necessary to satisfy the needs of the growing population. The twenty-second century will depend on the quality of the soil, the availability of water and the biomass production for our survival. Farmers and foresters will be maybe the only ones capable to counterbalance naturally the overaccumulation of carbon dioxide in the atmosphere. The urban population will have to upgrade their consideration to them.

The mutation that has started, but is still to come, if we want to survive, implies that we need to produce more and better, while protecting the soils, the water resources, the ecosystems equilibria and the diversity of production systems. This will be more and more challenging with the growing pressure on resources and with the global warming. Strong-willed policies are needed. They have to be "renewable" as well.

1.7 Conclusions

As conclusion of this introductory chapter, we can put in a balance the pros and cons of the use of biomass:

Advantages of biomass:

- Renewable resource by photosynthesis;
- Present and potential stocks being huge, diversified, and rather well distributed in the five continents (compared to fossil feedstocks or uranium);
- High diversity of applications, including the residues of primary valorisation;
- High potential of local jobs creation, mainly because jobs cannot be relocated outside the production sites;
- Low pollution during exploitation and utilisation;
- (Almost) Carbon-neutral with regard to global warming;
- Rapid and easy biosequestration of carbon dioxide.

Present and future limits:

- Feedstock limited by the available arable land;
- Need of a solid and efficient water resource management plan;
- Need of a solid and efficient agronomic regeneration of soils plan;
- Real competition between the numerous utilisations of biomass, mainly food, energy and materials;
- Cost of bioproducts often higher than their counterparts from fossil resources[3];
- Seasonal availability of crops, logistics and storage.

Four great principles may guide a sustainable approach of the biomass valorisation:

(1) Valorise first what is the easiest and the less expensive to produce, to mobilise and to transform.
(2) Allocate first the bioresources, which are by nature limited even if they are huge, to the applications where biomass is not substitutable. At least, for the moment, allocate to the other uses the alternative resources that are more abundant, in particular sunlight or wind or tides for energy.
(3) Prefer, among the different applications of biomass, those that are at the same time more performant that do not threaten the pre-existing production channels (e.g. materials, food, fertilisers) and that maximise the implicit benefits (e.g. employments).

[3]This element must be reconsidered on a fair basis, because bioproducts include the cost of the regeneration of the resource, whereas fossil products do not.

(4) Use first the feedstock that are concentrated and in the end of chain (i.e. residues from other transformation activities, especially from food industry).

Thus, the major processing sectors can be presented below, taking in account that the agronomic regeneration of the soils and the food sector remain, of course, high priority:

- The traditional biomaterials: wood and timber, paper and pulp, particleboard. Let us remind that from them, wood is the only "renewable material". Its transformation is source of residues that generate other applications. Materials prolongate the CO_2 capture effect initiated by the plant growth. Recycling would multiply such effect.
- The new biomaterials: bioplastics, biocomposites, etc. They are called to replace eventually the "classic" materials such as plastics, aluminium, steel, mineral fibres and concrete.
- The "biomolecules": solvents, lubricants, surfactants, etc. They will widen the existing sectors where biomass is included (soap, starch, pharmaceuticals, fine chemistry, etc.).
- The biofuels: with priority to second and third generation to avoid competition with first-generation biofuels competing with food. The obtained molecules feed also the biomolecules sector, and they support the already established sectors such as E10, E85 and 100% alcohol.
- The industrial "bioheat": including agriculture and urban heat (heat networks). It must be always backed-up by fossil resources for evident reasons.
- The domestic "bioheat": more and more popular, it depends at present mainly on the wood residues feedstock but needs to be diversified.
- Combined heat and power from biomass origin: this can a by-product of the cracking approach technologies such as gasification and pyrolysis.

In the following chapters, we will focus essentially on the non-energy applications: chemistry in the first line.

References

1. Anastas P, Warner J (2000) Green chemistry: theory and practice. Oxford University Press, Oxford, New York
2. Jin K, Qin Z, Buehler MJ (2015) Molecular deformation mechanisms of the wood cell wall material. J Mech Behav Biomed Mater 42:198–206. https://doi.org/10.1016/j.jmbbm.2014.11.010
3. Pires JCM, Alvim-Ferraz MCM, Martins FG, Simões M (2012) Carbon dioxide capture from flue gases using microalgae: engineering aspects and biorefinery concept. Renew Sustain Energy Rev 16:3043–3053. https://doi.org/10.1016/j.rser.2012.02.055
4. Sikarwar VS, Zhao M, Fennell PS et al (2017) Progress in biofuel production from gasification. Prog Energy Combust Sci 61:189–248. https://doi.org/10.1016/j.pecs.2017.04.001
5. Bari D (2014) Alternative Routes to MEG. Paper presented at the IHS Chemical World Petrochemical Conference, Houston, TX USA, 27–28 March 2014

6. Cavaliere VN, Crestani MG, Pinter B et al (2011) Room temperature dehydrogenation of ethane to ethylene. J Am Chem Soc 133:10700–10703. https://doi.org/10.1021/ja202316m
7. Zhang M, Yu Y (2013) Dehydration of ethanol to ethylene. Ind Eng Chem Res 52:9505–9514. https://doi.org/10.1021/ie401157c
8. Sun J, Liu H (2014) Selective hydrogenolysis of biomass-derived xylitol to ethylene glycol and propylene glycol on Ni/C and basic oxide-promoted Ni/C catalysts. Catal Today 234:75–82. https://doi.org/10.1016/j.cattod.2013.12.040
9. Yin A-Y, Guo X-Y, Dai W-L, Fan K-N (2009) The synthesis of propylene glycol and ethylene glycol from glycerol using Raney Ni as a versatile catalyst. Green Chem 11:1514. https://doi.org/10.1039/b913395j
10. Julio R, Albet J, Vialle C et al (2017) Sustainable design of biorefinery processes: existing practices and new methodology. Biofuels Bioprod Bioref 11:373–395. https://doi.org/10.1002/bbb.1749
11. Sablayrolles C, Vaca-Garcia C (2016) Assessment of sustainability for the bioprocesses: does it (really) follow life cycle thinking? Biofuels Bioprod Biorefin 10:111–112. https://doi.org/10.1002/bbb.1642

Chapter 2
Biorefinery Concept as a Key for Sustainable Future to Green Chemistry—The Case of Microalgae

Léa Vernès, Ying Li, Farid Chemat and Maryline Abert-Vian

Abstract Biorefinery is a concept inspired from traditional oil refinery for petrol production, which enables to generate numerous bioproducts without any wastes from a unique source. This idea fits in with the actual context of sustainable development, where every part of the biomass treated is valorized through different unit operations in an economic and eco-friendly way to create high value-added compounds. Microalgae, a unique sustainable resource with a wide range of interesting biomolecules, could lead to a large quantity of biocompounds based on the biorefinery concept for cosmetic, pharmaceutical, food, feed, and health industries. This chapter provides a successful biorefinery case of microalgae from initial cultivation, harvesting, and green extraction of bioactive molecules to various potential applications in the end, which is heuristic for sustainable future to green chemistry.

Keywords Biorefinery · Microalgae · Green extraction · Biomolecules · Sustainable ingredient · Applications

2.1 Introduction

Microalgae are one of the most promising renewable resources that can strongly contribute to future sustainable development. Thanks to their extensive biodiversity and metabolism, microalgae can produce a wide range of interesting compounds (e.g., lipid, protein, pigment, vitamin, and minerals) with a high nutritional value. The production of microalgae, which is mainly located in Asia or Australia, has become more and more important all over the world. China accounts for over 85% of both microalgae and macroalgae biomass produced in the world [1], and some

L. Vernès · F. Chemat · M. Abert-Vian
GREEN Extraction Team, Avignon University, INRA, UMR408, 84000 Avignon, France

Y. Li (✉)
Department of Food Science and Engineering, College of Science and Engineering, Jinan University, Guangzhou 510632, China
e-mail: yingli@jnu.edu.cn

Y. Li and F. Chemat (eds.), *Plant Based "Green Chemistry 2.0"*,
Green Chemistry and Sustainable Technology,
https://doi.org/10.1007/978-981-13-3810-6_2

European firms have recently acquired several leading companies in Australia and in the USA like Earthrise. The production of microalgae-based food and feed products from European firms is currently estimated at around 5% of the global market [2].

Microalgae are considered as a sustainable resource since their cultivation requires very few water and no arable lands, which is more efficient and sustainable than that of cereals such as corn and wheat. For example, approximately 2000 L of water are needed to produce per gram of the *Spirulina* protein, which represents only 2% of the water needed for cattle farming [3]. What's more, the water could be waste, brackish or saline water and not necessarily drinkable.

Microalgae present also the advantage of being full of bioactive compounds with a broad variety of applications. Indeed, microalgae contain a large amount of useful proteins, lipids, and carbohydrates for food, feed, and biofuel industries. To take advantage of the potential of microalgae as a feedstock, it could be interesting to valorize the entire biomass through the concept of biorefinery, resulting in a wide range of final bioproducts. With the increased life expectancy, diet will play a key role in sustaining human health, which is a challenge for the food industry to supply tasty, healthy, and nutritious foods. For this, microalgae-based nutraceuticals are a coherent and attractive option for the food industry as there is a multitude of functional food ingredients. Thus, microalgae biorefinery could be a solution to produce high-valued compounds in an efficient and economically viable way.

This chapter provides a complete overview regarding microalgae biorefinery, including an introduction to biorefinery concepts, followed by the metabolic compounds with high nutritional value and health benefits found in microalgae as the microfactory. Last but not least, the successful valorization from microalgae biomass to various bioactive molecules is highlighted in terms of cultivation, harvesting, and representative green extraction techniques.

2.2 Biorefinery Concepts

Biorefinery is defined as a process consisting of the total valorization of a natural resource. The typical example is the oil refinery where numerous products are generated from fossil oils. Contrary to petrol refinery which involves fossil resources, the term "biorefinery" is related to natural resources such as vegetables or microalgae. With this new concept, every waste is considered as a new resource that can be extracted another time. For instance, a first extraction can target hydrophilic molecules, and the biomass then left containing hydrophobic compounds can be used to extract lipophilic compounds, which enable to drain all the biomass without producing any wastes or with very few wastes. The biorefinery enables to obtain a wide range of products from a unique raw material source. It consists of a succession of unit operations including pre-treatment, extraction, and separation of biomass component as shown in Fig. 2.1, in which all initial raw materials are exhausted through the step-wise operations, and wastes considered as by-products from the first step are then valorized to create new products with high added values. Hence, biorefinery

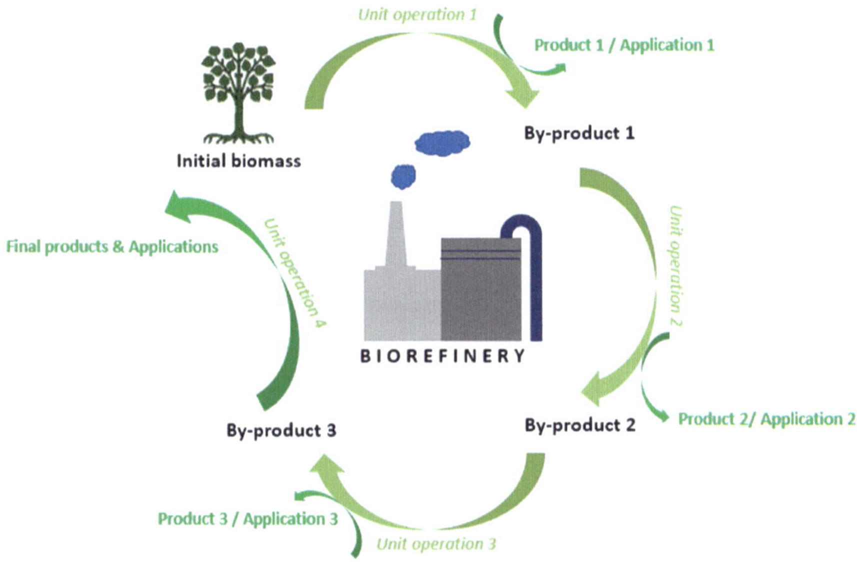

Fig. 2.1 Concept of bio-refinery

is indeed a profitable process with low environmental impacts since everything is recycled and used up for making new valuable products like food and feed products [4].

Microalgae are considered as one of the most promising feedstocks for the sustainable supply of commodities for both food and nonfood products. They are autotrophic microorganisms with different morphological, physiological, and genetic traits, which use light energy and inorganic nutrients (e.g., carbon dioxide, nitrogen, and phosphorus) to biologically synthesize active metabolites that have high nutritional value and desirable therapeutic functions, such as lipids, proteins, carbohydrates, pigments, and polymers (Fig. 2.2). Furthermore, microalgae can produce various biocompounds of commercial interest with different biological activities, such as carotenoids, phycobilin, polyunsaturated fatty acids (PUFA), proteins, polysaccharides, vitamins, and sterols.

Thanks to the biorefinery process, all microalgae components can be valorized in various sectors; e.g., lipids are extracted for biodiesel production, and their residues are used for the production of various biofuels, valuable chemicals, or other value-added products [5]. Microalgae can be considered as a microfactory of food ingredients, meaning that microalgae can provide many interesting target compounds that can be obtained by different extraction methods based on the developed biorefinery concept. Most of the time, the biomass is used once to extract one particular compound, and the biomass left after extraction is considered as a discarded waste. The potential of microalgae can be explored to produce new compounds such as nutrients mainly, as well as biofuels, biochemicals and other bioproducts, leading to various

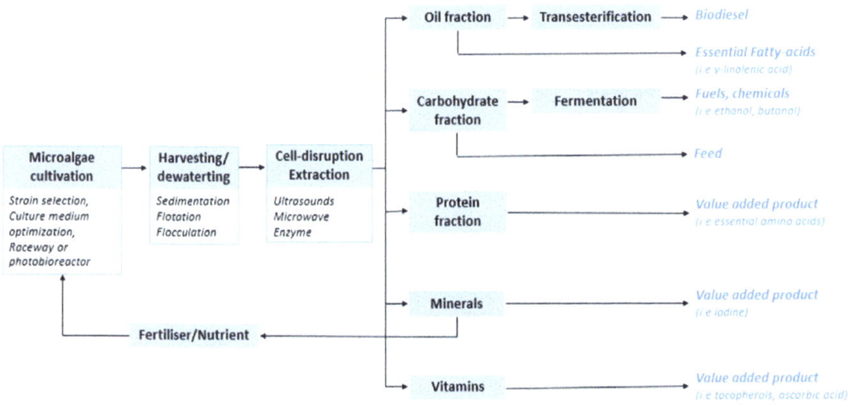

Fig. 2.2 Microalgae biorefinery and products obtained

products such as animal feed, bio-oil, fertilizer, plastics, and bio-ethanol used for numerous applications (Fig. 2.3) [5].

2.3 Microalgae as a Factory of Bioactive Compounds with Health Benefits

A nutraceutical or a functional food is a food or a part of a food for oral administration with demonstrated safety and health benefits in addition to its basic nutritional values, which can be used to promote general well-being, to prevent disease or to soften symptoms. It is presented in a nonfood matrix or nonconventional food formats, in a quantity higher than the one found in normal foods. The "nutraceutical" word combines "nutrient" (i.e., a nourishing food component) with "pharmaceutical" (i.e., a medical drug). This term was coined in 1989 by Stephen De Felice, Founder and Chairman of the Foundation for Innovation in Medicine, which is an American organization located in Cranford, New Jersey [6]. The philosophy behind nutraceuticals focuses on the prevention, where the relationship between nutrition and human health was emphasized by the principle of "let food be the medicine, and medicine be the food," advocated by Hippocrates (460–377 BC), the well-recognized father of modern medicine.

The global commercial production of microalgae for human nutrition has already been a reality. Nowadays, there is a marked trend in the food industry toward the development and manufacture of natural functional products. This new class of food products has found a great success in the market due to the growing interest of consumers for healthy food. A wide range of compounds are used or could be potentially employed as functional ingredients, e.g., carotenoids, polyphenols, and other antioxidative pigments, PUFA. Another important requirement for these functional

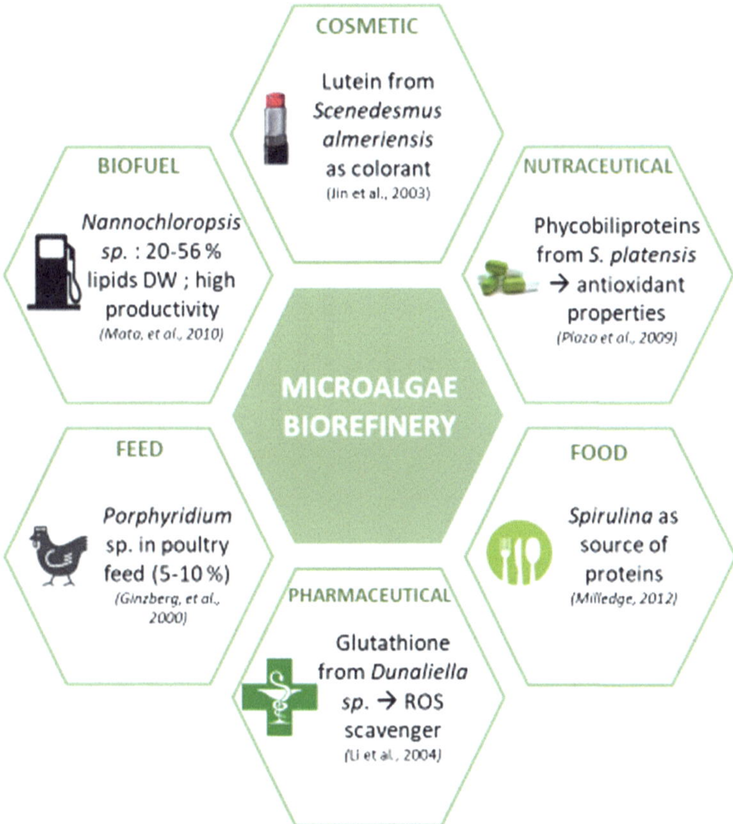

Fig. 2.3 Microalgal bioactive compounds and applications

ingredients is their natural origin, which is always preferred to the synthetic one. Functional foods supplemented with microalgae biomass are much more convenient and variable, thus combining health benefits with attractiveness to consumers. In some countries like Germany, France, Japan, USA, China, and Thailand, food production and distribution companies have already started serious activities to market functional foods with microalgae and cyanobacteria.

Microalgae are important sources of bioactive natural substances. Many metabolites isolated from these microorganisms possess biological activities and potential health benefits as described in Table 2.1. Microalgae accumulate specific secondary metabolites (e.g., pigments and vitamins) which are high-value products with applications in the cosmetic, food, or pharmaceutical industries. Bioactive compounds are physiologically active substances with functional properties for the human body so that there is great enthusiasm for the development and manufacture of various biocompounds, which can potentially be used as functional ingredients with different properties such as antioxidant, anti-inflammatory, antimicrobial, or antiviral.

Table 2.1 Classification of microalgal metabolites according to their biological activities

Function	Microalgae	Molecule	Extraction parameters	Specific use	References
Antioxidant	*C. vulgaris*	Canthaxanthin	Organic solvent (hexane/EtOH)	Immunomodulation and cancer prevention	[1–3]
	Nannochloropsis occulata	Tocopherol (α, β, γ)	Supercritical fluid extraction (40 °C, 3000 psi)	Protect membrane lipids from oxidative damage	[4, 5]
	Synechococcus sp.	Zeaxanthin	Supercritical EtOH extraction (200 bar)	Antioxidant activity	[2, 6]
	Arthrospira	Sinapic acid	Ultrasound and supercritical fluid extraction	Inhibitor of LDL cholesterol peroxidation	[7]
	Neochloris	Chlorogenic acid	Ultrasounds extraction	Scavengers of several types of ROS	[7, 8]
	Tetraselmis	Trans-cinnamic acid	Supercritical fluid extraction	Inhibitor of lipoxygenase	[7]
Anticancer	*Stigonema* sp.	Scytonemin	–	Protein serine/threonine kinase inhibitor, antiproliferative	[9]
	Lyngbya majuscule	Lagunamides A (1) and B (2)	–	Potent cytotoxic cyclic depsipeptides	[9]
	Aphanizomenon flos aquae	Aphanorphine, Siatoxin	–	Anticancer	[9]
Anti-inflammatory	*Scenedesmus almeriensis*	Lutein	Supercritical fluid extraction	Antiinflammatory activities	[10]
	Arthrospira platensis	C-Phycocyanin	Ultrasound extraction (PBS 0.1 M, pH 6.8, 10 °C)	Hepatoprotective effect and free radical scavenger	[6, 11, 12]

(continued)

Table 2.1 (continued)

Function	Microalgae	Molecule	Extraction parameters	Specific use	References
	A. flos aquae	α-Linolenic acid (C18:3, ω3)	Supercritical CO_2 EtOH extraction (400 bar, 60 °C)	Inhibit formation of inflammatory prostaglandins	[2, 13, 14]
	Phaeodactylum tricornutum	Fucoxanthin	Ultrasound extraction (70 kHz, 30 min, Room temperature)	Stimulation of cancer cells apoptosis	[2, 6, 15]
Anti-microbial	*A. platensis*	γ-Linolenic acid	Supercritical CO_2 EtOH extraction (400 bar, 60 °C)	Antimicrobial activity against human pathogens (*E. coli*, *P. aeruginosa*, *S. aureus*)	[2, 16]
	Nannochloropsis gaditana	Eicosapentanoic acid	Assisted solvent extraction ($MeOH/CHCl_3$)		[2, 17, 18]
	Crypthecodinium, Schizochytrium	Docosahexaenoic	Supercritical CO_2 extraction (55 Mpa, 55 °C)		[2, 17, 18]

2.3.1 Antioxidants in Microalgae

Oxidation is a chemical reaction that transfers electrons from a substance to an oxidizing agent, thus producing free radicals which facilitate chain reactions. The oxidative stress in human body would be created if free radicals reach high levels, leading to a variety of biochemical and physiological lesions resulting in metabolic impairment and cell death. On one hand, oxidative damage caused by reactive oxygen species to lipids, proteins, and nucleic acids can cause many chronic diseases such as heart disease, atherosclerosis, cancer, and aging. On the other hand, high level of active oxygen and free radicals could also cause lipid oxidation which led to a highly deteriorative process and unacceptable properties of foods as well as a loss in the nutritional value [7]. Antioxidants are capable of slowing or preventing the oxidation of other molecules, which can terminate chain reactions by removing free radical intermediates and inhibit other oxidation reactions by being oxidized themselves.

Microalgae are photoautotrophic organisms that are exposed to high oxygen and radical stresses. Consequently, they have developed several efficient protective systems against reactive oxygen species and free radicals. Hence, there is an increasing interest in using microalgae as natural antioxidant sources for cosmetics, functional foods/nutraceuticals, and medicine. Most commonly known antioxidants are vitamin A, carotenoids, vitamin C, vitamin E, polyphenols, and enzymes (superoxide dismutase, catalase, and glutathione peroxidase). Many compounds like pigments, lipids, and polysaccharides from microalgal biomass may have an antioxidant activity. The presence of antioxidants is essential not only for human health but also for the quality, retention, and safety of foods. In current years, intensive research on natural antioxidants has attained considerable attention because of that synthetic antioxidants such as butylated hydroxyanisole (BHA) and butylated hydroxytoluene (BHT) have been proved to cause harmful health effects due to their toxicity and involvement in carcinogenesis and liver damage.

A number of articles have publicized that microalgae possess potent antioxidants (Table 2.1) which act as inhibitors of lipid peroxidation and scavengers of free radicals in the form of phenolic compounds, vitamins, and pigments [8]. Therefore, the development and isolation of natural antioxidants from microalgae have become the research focus. Antioxidant activity is one of the important characteristics of carotenoids, and many of their biological effects are related to the ability to scavenge reactive oxygen species, which is one of the factors for their disease preventing effects.

Canthaxanthin (Fig. 2.4) is a carotenoid acting as a potent lipid-soluble antioxidant in animal tissues that is capable to scavenge free radicals [9].

Tocopherols (vitamin E) (Fig. 2.4) are lipid-soluble antioxidant molecules with the capacity to scavenge free radicals in the body. They are considered as essential nutrients because of their ability to protect membrane lipids from oxidative damages [10].

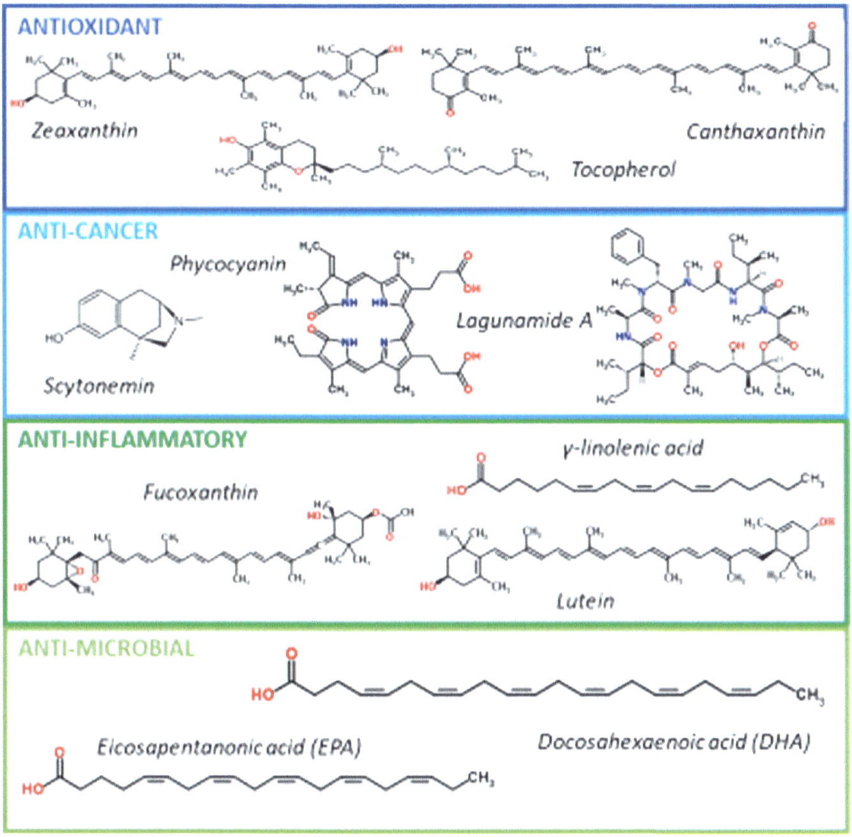

Fig. 2.4 Microalgal compounds with health benefits

Zeaxanthin (Fig. 2.4), belonging to the carotenoid compounds, is a molecule with antioxidant properties. In microalgal metabolism, it plays a role in protecting photosynthetic tissues against damage caused by light and oxygen [9].

Other antioxidant molecules such as phenols can also act as preventing agent against cardiovascular diseases. Indeed, chlorogenic, sinapic, and trans-cinnamic acid (Table 2.1) can slow the development of atherosclerotic plaques and inhibit LDL cholesterol peroxidation.

2.3.2 Anticancer Compounds in Microalgae

Cancer is one of the most dreadful human diseases and represents the largest cause of mortality in the world with over six million deaths annually [11]. Many synthetic drugs against cancer are available in the market; however, they are often related to side effects which can be very tough to overcome. Therefore, there is still a demand for potent drugs of natural origin. Microalgae show a wide species diversity and represent a promising source of natural anticancer molecules. The bioactive substances either kill cancer cells by inducing apoptotic death or stimulate the immune system by affecting cell signaling through activating the member of protein kinase C family of signaling enzymes.

Cyanobacteria produce diverse compounds with health benefits. For instance, cyanotoxins such as anatoxin-A and microcystins/nodularin could be used as anti-cancer compounds. Likewise, Lagunamides (Fig. 2.4) A (1) and B (2) are potent cytotoxic cyclic depsipeptides isolated from the filamentous marine cyanobacterium, *L. majuscule*, and are structurally related to the aurilide class of molecules, which have been reported to possess potent anti-proliferative activities against cancer cells [11].

2.3.3 Anti-inflammatory Compounds in Microalgae

Inflammation is an immediate self-defensive reaction to a cell or tissue injury caused by various noxious stimuli, such as toxins and pathogens. It is characterized by attracting large amounts of leukocytes to the inflamed area, in which these inflammatory cells are triggered by inflammation mediators and may become a harmful self-damaging process. The body recognizes the agents responsible for the attack and attempts to neutralize them as quickly as possible. Inflammation causes redness, swelling, heat, and pain, usually located at the site of infection. Anti-inflammatory agents should reduce the inflammatory response by suppressing the productions of inflammatory cytokines (interleukin-1β and interleukin-6) and inflammatory mediators (nitric oxide, prostaglandin E2). Ingestion of anti-inflammatory compounds thus enhances the body's immune response, helps to prevent disease, and aids the healing process. Microalgae produce several anti-inflammatory compounds (Table 2.1) which exert a protective function in the body when they are consumed as food or used as pharmaceuticals and cosmetics.

Many microalgal polysaccharides possess the ability to modulate the immune system through the activation of macrophage functions and the induction of reactive oxygen species (ROS), nitric oxide (NO), and various other types of cytokines/chemokines. Macrophages are able to regulate several innate responses and secrete cytokines and chemocytokines that serve as signals for immune and inflammatory molecular reactions. The PUFAs, especially 3 and 6 as eicosapentaenoic (EPA), docosahexaenoic (DHA), and arachidonic (AA) acids, have been applied in the treatment of chronic inflammation such as rheumatism and skin diseases. Among the pigments with anti-inflammatory activities, fucoxanthin found in diatoms is capable of stimulating apoptosis in human cancer cells [12].

Lutein (Fig. 2.4) is another important carotenoid naturally found in algae, which acts as an anti-inflammatory compound.

Gamma-linolenic acid (GLA) is an omega-6 essential fatty acid for human, which should be assimilated only through the food instead of self-production. There are several different types of omega-6 fatty acids that can be found in the diet coming from vegetable oils in the form of linoleic acids (LA), and from microalgae as well such as *Arthrospira.* All omega-3 and omega-6 fatty acids play a crucial role in brain function, normal growth, and development.

Fucoxanthin (Fig. 2.4) is a xanthophyll as one of the most abundant carotenoids and is an orange-colored molecule structurally similar to β-carotene and vitamin A. Fucoxanthin was the first pigment isolated in 1914 from most brown seaweeds like Fucus, Dictyota, and Laminaria, as well as a few other microalgae sources like *P. tricornutum* [13]. Fucoxanthin, stored in fat cells for a prolonged period of time, can induce fat loss while inhibiting fat cell differentiation and proliferation. Indeed, it inhibits inflammatory cytokines and mediators in lipopolysaccharide-stimulated RAW 264.7 macrophages, which can be served as an anti-inflammatory agent [13]. Moreover, it is able to inhibit the inducible nitric oxide synthase and cyclooxygenase 2 protein expressions and to reduce the level of cytokines and mediators (nitric oxide, prostaglandin E2, interleukin-1β). Furthermore, fucoxanthin suppressed the degranulation of mast cells by inhibiting antigen-induced aggregation of high-affinity IgE receptor followed by the activation of degranulating signals of mast cells, which played an important role in inflammation and immediate-type allergic reactions. Fucoxanthin significantly inhibited ear swelling and reduced the level of tumor necrosis factor-α and histamine, suggesting that fucoxanthin exerted an anti-inflammatory effect by suppressing mast cell degranulation in vivo. Besides, fucoxanthin is also used as anticancer and anti-obese agent.

Phycocyanin (Fig. 2.4) is a phycobiliprotein mainly found it the microalgae *A. platensis* (20% of its dry weight). Among different phycobiliproteins, phycocyanin is one of the greater importance compounds because of its various biological and pharmacological properties, its production, and stability as a promising and efficient pharmaceutical and biological reagent. The pharmacological property attributed to phycocyanin includes antioxidant (free radical scavenger), anti-inflammatory, neuroprotective, and hepatoprotective activity. The anti-inflammatory effect of phycocyanin is due to the inhibition of leukotriene formation, an inflammatory metabolite of arachidonic acid. Moreover, it helps to reduce edema, histamine (Hi) release, myeloperoxide (MPO) activity, and the levels of prostaglandin (PGE2) and leukotriene (LTB4) in the inflammed tissues [14]. As an anti-inflammatory agent, phycocyanin has been found to function as Cox-2 inhibitor that is effective in treating arthritis and other inflammatory conditions.

2.3.4 Antimicrobial Compounds in Microalgae

The importance of discovering new compounds with antimicrobial activity is driven by the development of antibiotic resistance in humans due to constant clinical use of antibiotics. Microalgae are an important source of antibiotics with a broad and efficient antibacterial activity. The antimicrobial activity of these microorganisms is attributed to their synthesized compounds such as fatty acids, terpenoids, sterols, phenols, and sulfur-containing heterocyclic compounds, especially for the lipid composition. Omega-3 fatty acids including DHA and EPA have gained considerable importance due to their association with the prevention and treatment of several diseases like atherosclerosis, thrombosis, arthritis, and cancers. The conventional source of EPA and DHA is marine fish oil; however, research studies have proved that a higher amount of EPA and some DHA can be produced by microalgae [15].

The antimicrobial activity of microalgae (Table 2.1) against human pathogens, such as *Escherichia coli, Pseudomonas aeruginosa, Staphylococcus aureus*, and *epidermidis*, has been attributed to -linolenic acid, EPA, DHA, palmitoleic acid, oleic acid, arachidonic acid, etc. [12]. The mechanism of these fatty acid actions affects various structures in microorganisms. The membrane damage most likely leads to a loss of internal substances from the cells, and the entry of harmful components reduces the nutrient absorption and inhibits the cellular respiration.

DHA (Fig. 2.4) is a long-chain polyunsaturated omega-3 fatty acid, which is important for brain, eye, and heart health throughout the lifecycle. Indeed, DHA is the primary compound of the human brain and retina, and it has several applications including infant formulas, products for pregnant and nursing women, food and beverage products and dietary supplements [12]. DHA also helps body to fight against diseases. Marine microalgae have significantly higher DHA contents compared to freshwater microalgae [16]. Most of the DHA in fish and complex organisms originates from photosynthetic and heterotrophic microalgae and becomes increasingly concentrated in organisms, as they move up the food chain. Nowadays, DHA is commercially produced from microalgae, e.g., Nutrinova in Germany produce DHA products from microalgal biotechnology for human and other applications.

EPA (Fig. 2.4) is a n-3 PUFA that acts as a precursor for prostaglandin-3, thromboxane-3, and leukotriene-5 groups. *Nannochloropsis* sp. has been proposed as a source of PUFAs due to its high contents of EPA [17]. EPA plays an important role in the regulation of biological functions and prevention and treatment of a number of human diseases such as heart and inflammatory diseases [16].

Linoleic acid (LA) is a polyunsaturated omega-6 fatty acid with an 18-carbon chain and two cis double bonds. As the first double bond is located at the sixth carbon, this carboxylic acid is also called "omega-6" and is considered as an essential fatty acid since it is not synthetized by human body. Linoleic acid has to be consumed by human for a proper health. LA is used in the synthesis of AA and thus some prostaglandins.

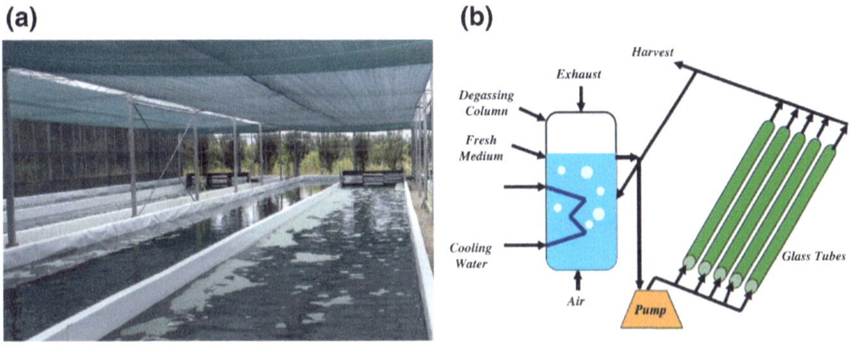

Fig. 2.5 *Arthrospira* cultivation in open pond (Château d'Isère, France) (**a**) and schematic tubular photobioreactor (**b**)

2.4 From Microalgal Biomass to Bioactive Molecules

2.4.1 Cultivation

Microalgae are a promising aquatic biomass source of sustainable bioproducts. They have significantly higher growth rates than terrestrial biomass and conventional crops, as most strains can double their cell mass within 24 h [18]. Other advantages of microalgae include the fact that they can be grown on nonarable land which can be arid or poor-quality land. Furthermore, many species of microalgae are able to grow in saline or brackish water. These features make microalgae a noncompetitive resource with agriculture, hence enhancing sustainability. Nowadays, algal biomass is produced at industrial scale for various applications such as human food supplements, animal feed, nutraceuticals, and pharmaceutical products.

The production of microalgae is increasing dramatically worldwide. The annual production is estimated at more than 6000 tons of dry matter [19]. The cultivation of microalgae on a large scale may be performed in two modes, either using high-yield open ponds with the raceway or in an enclosed transparent system using natural or artificial light, namely the photobioreactor.

Currently, microalgae are cultivated at large scale mostly in open ponds (or raceway) due to their low capital and operating costs (Fig. 2.5a). However, such systems provide little control on operational conditions, which can lead to low biomass productivities, and are constantly exposed to potential contaminants, which may increase the risk of culture collapse.

On the other hand, closed cultivation systems also called photobioreactors (PBRs) provide higher photosynthetic efficiency, enabling to achieve higher biomass productivity and concentration. Various photobioreactors have been proposed and developed during the last few years. These are closed systems which have been widely investigated since they can offer many advantages over open pond system, such as minimizing contamination, controllable cultivation condition and less CO_2 loss and water

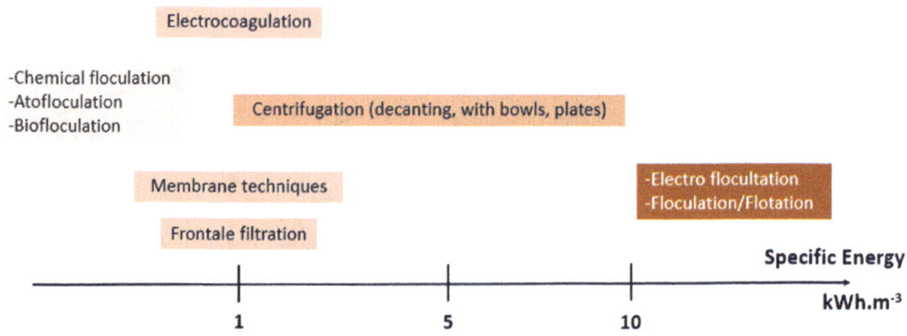

Fig. 2.6 Comparison of harvesting systems according to their energy consumption

evaporation. Right now, closed systems for mass microalgae cultivation include many types, such as flat plate, vertical column, stirred tank, plastic bag, tubular (Fig. 2.5b), and other configurations [20]. However, photobioreactor is expensive in terms of energy consumption, capital cost, and maintenance.

The conditions for microalgae cultivation are important factors that influence the metabolism of these microorganisms, thus directing the synthesis of specific compounds of interest. Several researchers have noted the influence of incubation temperature, pH in the medium, cultivation period on the synthesis of bioactive compounds, as well as salinity, light intensity, and medium constituents [21].

2.4.2 *Harvesting*

In the case of culture ponds, harvesting is carried out to maintain an algal concentration at a desired level, enabling an optimal production. In the absence of regular harvest, a high concentration is reached and engenders an important cell mortality. Conversely, a lower concentration less than 0.4 g/L destabilizes the culture and causes a reduction of bioactive compound contents in microalgae. Most of the microalgae like *A. platensis* are harvested in the morning because their protein content is generally higher than that in the evening.

The costly and difficult harvesting is a limiting step of many recovery methods for micron-sized microalgae, which represents an important economic role in an industrial-scale process [22]. Harvesting can be done by centrifugation, filtration, flocculation, sedimentation, or flotation, where corresponding energy consumption is compared in Fig. 2.6.

After the harvesting, the drying of microalgae provides a good preservation of vitamins and phytonutrients, which takes place at low temperatures under continuous air flow in ventilated rooms equipped with a dehumidifier [23]. Some industries have recourse to drying methods with atomization or spray drying, specifically to prevent the oxidation of vitamins and fatty acids.

2.4.3 *Green Extraction of Microalgal Bioactive Molecules*

Microalgae contain high amounts of lipids, proteins, and carbohydrates, which all can be extracted and used for different purposes. As most microalgae species have tough cell walls, a pre-treatment is often required to disrupt and compounds of interest release. Considering cost and time efficiency, the extraction technique adapted in this step must be optimized. Extraction is a process consisting of separating certain compounds from an organism (plant or animal). More precisely, the extraction consists of the releasing of one or various molecules of interest in the extraction solvent. The extraction can be achieved through various devices and using different types of solvents depending on their affinity to targeted molecules or even without any solvents. Thanks to extraction steps performed in succession, many different products can be obtained simultaneously to achieve microalgae biorefinery, which makes the market value higher than the production costs when exploiting the whole potential of microalgae ingredients.

Most popular extraction techniques used for natural compounds are maceration, percolation, and Soxhlet extraction, which require a large amount of solvents and a long treatment time to achieve good yields. Nowadays, green extraction techniques have developed to extract bioactive compounds from numerous bioresources, which present their advantages of being faster and more efficient with higher yields than conventional extraction techniques above.

Ultrasound-assisted extraction (UAE) The application of ultrasound to enhance extraction yields began in the fifties at the laboratory scale. Nowadays, UAE is a well-established technique commonly used to facilitate the extraction of any type of plant matrices, which has developed into a full-fledged extraction technique for microalgae [24]. Its mechanical waves are capable of traveling in an elastic medium at a frequency greater than the maximum limit of audibility of human ears (16 kHz).

Ultrasounds are mechanical vibrations characterized by their frequencies ranging from 20 kHz to 10 MHz, which can be differentiated into diagnostic and power ultrasound types depending on the frequency:

- Diagnostic ultrasound (low power and high frequency), which frequency is comprised between 1 and 10 MHz and ultrasonic intensity below 1 W cm^{-2}, has enormously applied in medical field as a diagnostic or a control tool.
- Power ultrasound (high power and low frequency), where frequency ranged from 20 kHz to 1 MHz and ultrasonic intensity above 1 W cm^{-2}, has applied to generate cavitation bubbles, leading to physical or chemical effects in the medium in the field of sonochemistry (chemical reaction acceleration), agriculture (water dispersion), and industry (cutting and plastic welding).

As the power ultrasound wave spreads through an elastic medium, it induces a longitudinal displacement of particles, acting as a piston on the medium surface, resulting in a succession of compression and rarefaction phases [25]. The ultrasound wave induces cavitation bubbles, which are able to grow during rarefaction cycles,

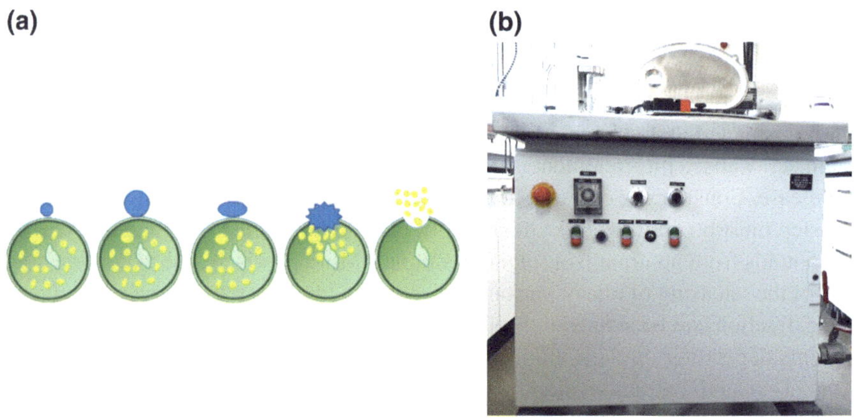

Fig. 2.7 Collapse of a cavitation bubble near a cell (**a**) and pilot ultrasonic reactor of 30 L (**b**)

and they decrease in size during compression phases. When the bubbles reach a critical diameter, they collapse during a compression cycle, inducing a release of huge amounts of energy. Temperature and pressure reached during bubble collapses have been estimated to be up to 5000 K and 1000 atm in a simple ultrasonic bath at room temperature [26]. The hot spots generated can significantly accelerate the chemical reactivity into a medium. If the cavitation bubbles are formed near to a solid surface, the implosion resulting generates microjets and shock waves directed towards the solid surface as shown in Fig. 2.7a [26]. In the case of biological compounds, when the cavitation occurs close to the material surface, its collapses create microjets directed toward the cell wall. The high pressure and temperature involved in this process will destroy microalgae cells, and the content inside will be released and solubilized into the medium [27].

Ultrasound could be generated by reactor (Fig. 2.7b), ultrasonic baths, or probes, where mechanical effect accelerates the release of organic compounds contained within the microalgae, enhances the mass transfer, and facilitates solvents or any kind of fluids access to the cell content. It is worth mentioning that UAE can be used with or without solvents. In fact, it is possible to extract the wet microalgal biomass directly after harvesting using the culture medium as the solvent.

Microwave-assisted extraction (MAE) Microwave energy is well known to have a significant effect on the rate of various processes in the chemical and food industry [28]. In the last decade, MAE has been successfully applied to various fields of chemistry since Ganzler et al. firstly reported this technique [29]. This technology was initially used in the extraction of various compounds from food products (citrus, aromatic plants, cereal, etc.). Microwave (MW) irradiation uses an electromagnetic field at a specific frequency, which ranges from 300 MHz to 300 GHz [30]. However, only a few frequencies are allowed for industrial, scientific, and medical uses (ISM frequencies), among which 0.915 and 2.45 GHz are mostly used worldwide. A typical MW generator for such frequencies can be found in the magnetron, the same device that equips domestic and laboratory MW ovens. Magnetrons for industrial applications can reach power ratings in tens of kilowatts (kW), while laboratory appliances usually use ratings below 1 kW. The fundamentals of the MAE process are different from those of conventional methods (solid–liquid or simple extraction) because the extraction occurs as the result of changes in the cell structure caused by electromagnetic waves. In MAE, the process acceleration and high extraction yield may be the result of a synergistic combination of two transport phenomena of heat and mass gradients working in the same direction [31].

Although heat is transferred from the heating medium to the interior of the sample in the conventional extraction, heat in the MAE is dissipated volumetrically inside the irradiated medium. Many forces such as the physicochemical interactions and relationships (e.g., dispersion forces, interstitial diffusion, driving forces, and chemical interactions) can be exposed during the extraction. The persistence and strength of these phenomena may be closely tied to solvent properties like solubilization power, solubility in water, purity, and polarity.

MAE can be an alternative to conventional techniques for many matrices. It performed under the action of microwave radiation that is believed to be affected in part by polarization and volumetric and selective heating. Microwave radiation is currently used for the rapid extraction of several classes of bioactive compounds, phytonutrients, functional food ingredients, and pharma-active substances from biomasses. The main advantages of MAE are minimal energy and solvent consumption, shorter operation times, good selectivity, recovery yields, and reproducibility. MAE causes the compounds of interest to desorb from the microalgal matrix because the free water molecules presenting inside the cells are heated. By this, localized heating and an expansion occur, during which microalgal cell walls are ruptured, allowing the extracted molecules to flow toward the solvent as illustrated in Fig. 2.8a. The effect of MW energy is strongly dependent on the dielectric susceptibility of both solvent and solid matrix. Owing to its high natural moisture content, matrix such as microalgae is highly susceptible to microwave irradiation. Rapid internal heating of these structures brings about effective cell ruptures, which facilitates releasing analytes into the cold solvent [11]. Figure 2.8b and c presents lab and pilot scale of microwave devices, respectively. The Mac-75 pilot enables to treat important quantities of biomass with microwave energy, which has successfully been used for diverse applications such as oil flavoring with aromatic plants.

Enzymatic-assisted extraction (EAE) Solvent extraction is one of the most frequently employed methods of extracting active components from plant materials. However, the rigidity of microalgae cell walls limits the efficiency of classic extraction methods. Enzymatic treatment has been proposed as an alternative to the solvent extraction process to improve the yield and quality of various natural products from algae.

Enzymes can be extracted from any living organisms such as bacteria, fungi, plants, or animals, and most enzymes used industrially come from fungi or yeasts. They are natural biological tools which allow accelerating metabolic reactions of a living organism. Most of the enzymes are proteins with the exception of ribozymes

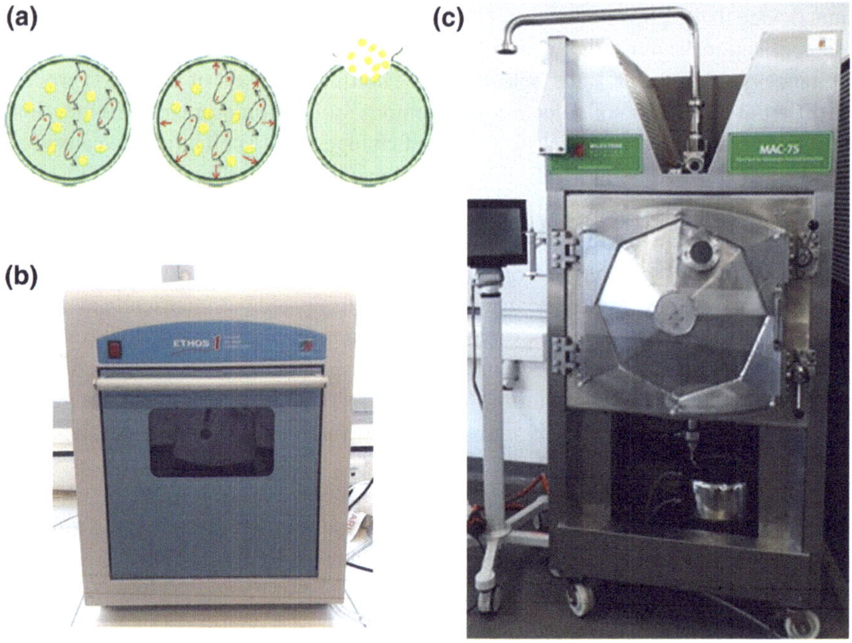

Fig. 2.8 Microwave mechanism on a cell (**a**), microwave device at lab scale (**b**) and pilot scale (**c**)

(RNA), and they exhibit a biological catalysis activity and selectivity toward substrates [32]. Similar to other catalysts, enzymes which are not involved in the reaction process stay intact at end of the reaction with the role of increasing the kinetics of chemical reactions and decreasing the free activation energy without changing the equilibrium constant. A greater specificity of reactions and a possible regulation of catalysis could be obtained by changing concentrations in products and substrates, pH, temperature, and metabolism as well. Enzymes are used to hydrolyze structural polysaccharides of the cell wall of biological materials such as microalgae so that the cell wall becomes permeable and the targeted compounds could be released into the extraction solvent. EAE is an efficient method employed to improve lipid extraction from several different biomasses such as soybean, sunflower, and microalgae [33, 34]. Enzyme-assisted aqueous extraction processing has been promoted for 40 years for fractionation of biological materials and extraction of interest molecules in an economical and safe manner.

In conclusion, hydrolytic enzymes work primarily by macerating the tissues of algae and breaking down the cell walls and complex interior storage materials in order to release interior compounds, which make enzymatic extraction a potentially suitable technique for the extraction of bioactive components such as pigments and peptides from microalgae.

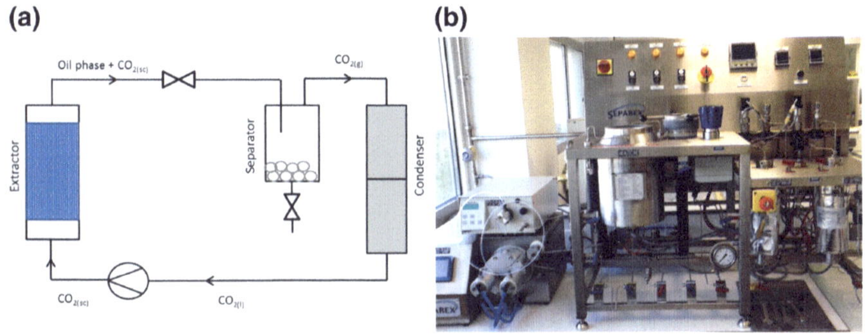

Fig. 2.9 Schematic supercritical fluid extraction (**a**) and supercritical fluid extraction device (**b**)

Supercritical fluid extraction (SFE) Fluid is considered as supercritical when its temperature and pressure are above its critical point, where distinction between liquid and gas phases does not exist [35]. Supercritical state, discovered by a French scientist Baron Charles Cagniard de la Tour in 1822, has been known for almost two centuries. Similar to the conventional solid-fluid extraction, SFE was emerged at the industrial stage in 1970s as an alternative process to plant distillation. The particularity of supercritical fluids is based on alternation of their physical properties, which can be modulated by an increase of pressure and/or temperature above their critical values (Fig. 2.9a). The supercritical fluid owns a density similar to liquids, which induces apparition of solvating power and a viscosity similar to gases, which diffusivity is intermediate between gas and liquid states, thus favoring the mass transfer between the solute and supercritical fluids in the extraction particularly. According to Taher's work [36], lipids are extracted from *Scenedesmus* thanks to supercritical carbon dioxide extraction at 53 °C and 500 bar.

Nowadays, the frequently used supercritical carbon dioxide (SC-CO_2) constituted more than 90% of industrial applications due to the fact that its low critical conditions ($T_c = 31$ °C and $P_c = 7.38$ MPa) are easily achievable in practice. Moreover, it has significant advantages such as inert, nonflammable, inexpensive, easily available, environment-friendly, and generally recognized as safe (GRAS) solvent [37]. As CO_2 is volatilized at atmospheric pressure, the extracts are free of solvents after depressurization [38]. This characteristic permits to remove evaporation unit operation after the extraction process, which is the most energy consuming in the industry. The different compartments of the device generating supercritical carbon dioxide are presented in Fig. 2.9b.

Pulsed electric field extraction (PEF) Since the destructive effect on microorganisms discovered by Doevenspeck in 1961 [39], PEF has become an attracting technique in the food processing such as preservation, transformation, and extraction because of its major advantage of saving in time and energy consumption. The principle of PEF technology is the application of short pulses of high electric fields, where the potential formation of pores inside the cell membranes occurs due to their exposure to low-moderate external electric fields of adequate strength (<10 kV/cm) during

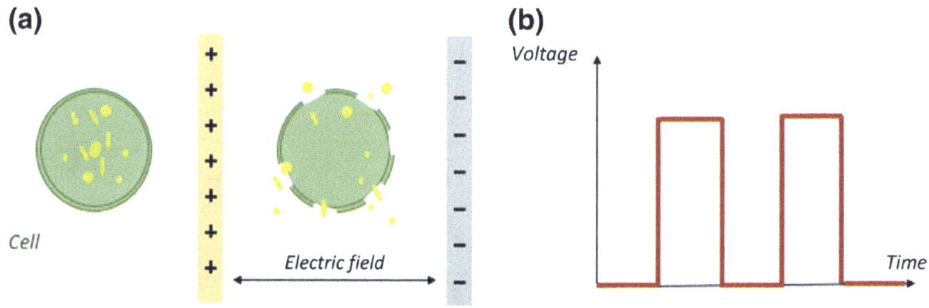

Fig. 2.10 Cell exposed to pulsed electric field (**a**) and voltage behavior over time (**b**)

a very short time ranging from around micro to milliseconds. This phenomenon is called "electroporation" or "electro-permeabilization" [40], which enable the release of compounds of interest (Fig. 2.10a). The processing time is calculated in a coupled device by multiplying the number of pulses times with effective pulse duration (Fig. 2.10b). Lebovka et al. [41] hypothesized that PEF treatment affected the cellular tissues as a correlated percolation was through two processes including the releasing of cells and moisture mass transfer inside the cellular structure, which is sensitive to PEF treatment repetitions. Thus, electroporation mechanism increases the permeability of microalgal cells, which enhance the solute diffusion through cell membranes [42].

Natural deep eutectic solvent (NADES) extraction Deep eutectic solvents (DESs) are mixtures of solids that have melting points far below the melting points of each compound, usually below 100 °C (Fig. 2.11). For instance, a mixture of choline chloride (melting point 302 °C) with urea (melting point 133 °C) at a molar ratio of 1:2 results in a DES with a melting point of only 12 °C. The most common DES is based on choline chloride, carboxylic acids, and other hydrogen-bond donors such as urea, citric acid, succinic acid, and glycerol [43]. NADES are DES composed of a natural substance, particularly primary metabolites such as organic acids, amino acids, and sugars. They possess better properties than DES for the extraction like liquid state below 0 °C, adjustable viscosity, sustainability, and a very high solubilization ability of both nonpolar and polar compounds. Since NADES can be used to extract metabolites from microalgae, primary metabolites such as protein, lipids, and carbohydrates are extracted from *Scenedesmus* with 1,2-propanediol-choline chloride-water at the molar ratio of 1:1:1 [44]. Furthermore, thanks to their H-bonds, NADES can interact with polar molecules such as polyphenols so as to extract them [45].

Accelerated solvent extraction (ASE) ASE is a patented technique for the extraction of solid and semisolid sample matrices using common solvents at elevated temperatures and pressures [46]. Many steps involved in the sample preparation can now be automated with ASE flow-through technology. On the contrary to other extraction techniques such as maceration or Soxhlet, filtration and clean-up of solid samples

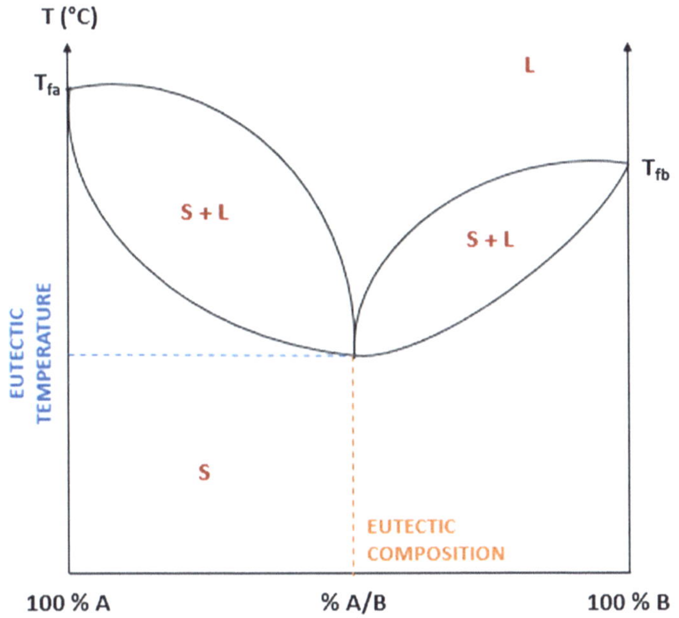

Fig. 2.11 Eutectic diagram

can be achieved as parts of the solvent extraction process in a single step [46]. This extraction technique is called "accelerated" since extraction time is reduced from hours to minutes. ASE presents the advantage of filtering the nonsoluble matter from the liquid extract in comparison with other extraction techniques where this step is an extra. This extraction technique is applied to microalgae to disrupt the cell for carotenoids extraction from *Haematococcus pluvialis* and other microalgae in a dedicated device as presented in Fig. 2.12 [47].

Bead milling This technique is a simple technique of cell disruption, which consists of an important stirring of the medium containing microalgae and solvent with quartz or metal beads. This process permits to separate or disperse very thin particles of microalgal biomass powder and to generate shearing effects due to the friction and collision between beads and cells. The dispersion system in a bead mill consists of a milling chamber and an agitator as shown in Fig. 2.13, where the milling chamber is filled with the grinding beads (e.g., glass, zircon oxide, and steel) and the product will be dispersed. The dispersion process takes place between the grinding beads sliding on each other and between the rotor and/or the vessel sides and the grinding beads [48]. This is an efficient technique which does consume few energy and can be transformed to the industrial scale. In the case of microalgae, beads of 0.3–0.4 mm are used to extract proteins and carbohydrates from *Tetraselmis suecica* and other two species of microalgae [49].

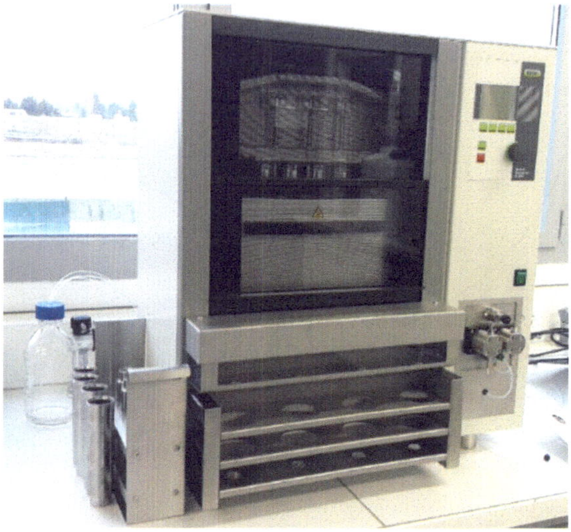

Fig. 2.12 Accelerated solvent extraction device for the lab scale

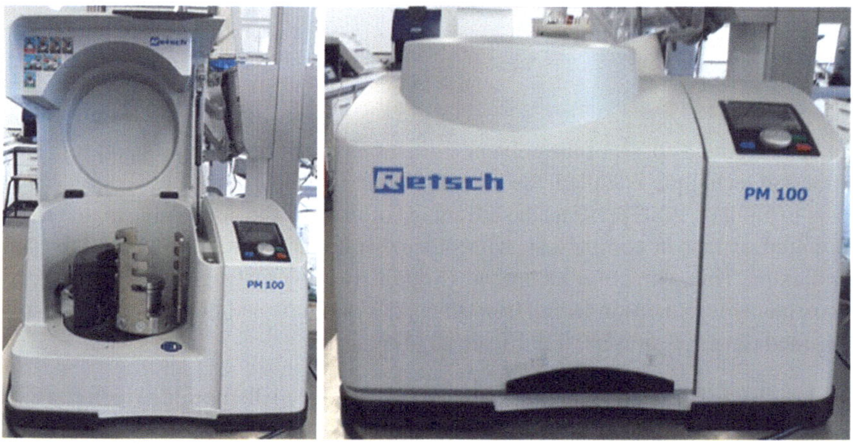

Fig. 2.13 Lab-scale bead milling equipment

2.5 Microalgae Biorefinery for Valuable Food Ingredients and Natural Products

Biorefinery techniques are necessary to exploit all products extracted from microalgae after cultivation. The main bottleneck is to separate the different fractions without damaging one or more of product fractions. It is thus conceivable to valorize proteins and enzymes, vitamins like B_{12}, minerals (e.g., Fe, Mg), pigments like β-carotene,

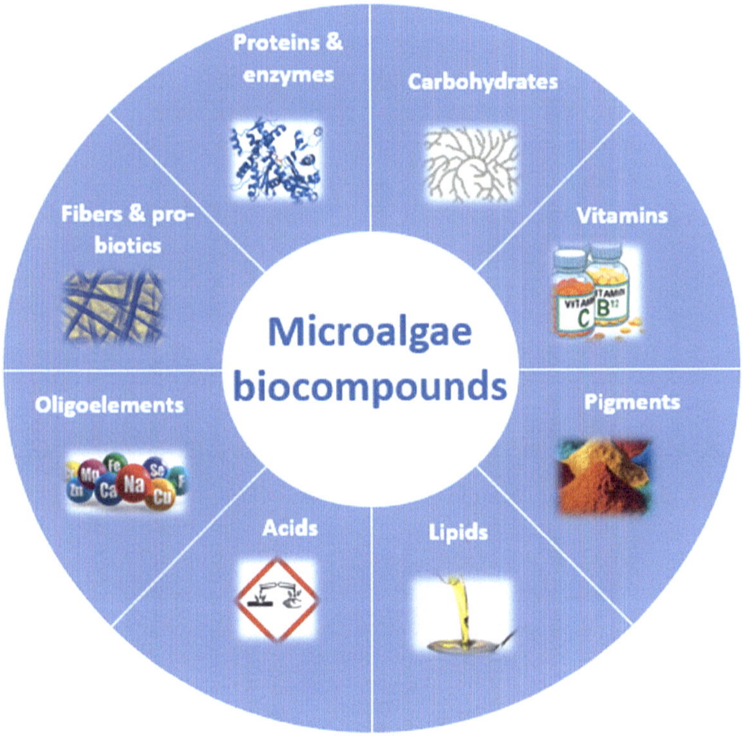

Fig. 2.14 Biocompounds of interest from microalgae biorefinery

lipids (e.g., EPA and DHA), polysaccharides like glycan, and acids like glucuronic acid from a unique microalgae thanks to a biorefinery process (Fig. 2.14).

2.5.1 Functional Ingredient

2.5.1.1 Coloring Agents

One of the most obvious and arresting characteristics of algae is their colors. In general, microalgae have their own particular combination of pigments and an individual color. Apart from chlorophylls as the primary photosynthetic pigment, microalgae also form various accessory or secondary pigments, such as phycobiliproteins and a wide range of carotenoids as listed in Table 2.2. These natural pigments can improve the efficiency of light energy utilization of the microalgae (phycobiliproteins) and protect them against solar radiation and related effects (carotenoids).

Beta carotene (Fig. 2.15) belongs to the group of pigments called carotenoids, which is a class of naturally occurring fat-soluble pigments responsible for the dif-

Table 2.2 Classification of microalgal metabolites according to their functionalities

Function	Microalgae	Molecule	Extraction parameters	Specific use	References
Colouring agent	*Dunaliella salina*	β-carotene	Supercritical CO_2 extraction (400 °C, 60 bars)	Food colorant (drink, cheese, butter)	[1, 19]
	Cylindrotheca closterium	Chlorophyll	Microwave extraction (50 W, 5 min)	Natural colorant, wound healing properties	[13]
	Porphyrum cruentum	Phycoerythrin	Solvent extraction (acetate buffer US 10 min)	Natural colorants for use in food, cosmetics, and pharmaceuticals	[20]
	A. platensis	Phycocyanin	Ultrasound extraction (PBS 0.1 M, pH 6.8, 10 °C)	Natural food colorant (candy, drink) and antioxidant	[10]
	H. pluvialis	Astaxanthin	Supercritical EtOH extraction (50% w/w EtOH), 7 MPa, 45 °C, 120 min)	High value colorant, antioxidant	[1]
Texturing agent	*Chlorella stigmatophora*	Carrageenan	Organic solvent extraction (Soxhlet)	Hydrocolloid properties of thickening and gelling in water	[21]
	Rhodophytes	Galactans	–	Gelling properties	[22]
	Porphyridium spp.	Guar gum	–	Thickener, gelling agent	[23]
	P. tricornutum, Chlorella pyrenoidosa	Mannose	Ultrasound extraction (300 W, 24 min, distilled water)	Improvement of texture (ice cream, salad dressings…)	[2, 24]
Flavouring agent	*Chlorella*	Arabinose	Ultrasound extraction (800 W, 40 min)	Flavor raw materials, dietary supplements, food ingredients	[25]
	C. pyrenoidosa	Xylose	Ultrasound extraction (300 W, 24 min, distilled water)	Diabetic sweetener in food and beverage	[26, 27]
	Gyrodinium impudium	Galactose	Ultrasound extraction (800 W, 40 min)	Food additive in gums	[25, 26]
	C. stigmatophora	Glucose	Ultrasound extraction (800 W, 40 min)	Sweetener	[25, 26]
	C. pyrenoidosa	Rhamnose	Ultrasound extraction (300 W, 24 min, distilled water)	Natural food and flavor ingredient	[27]

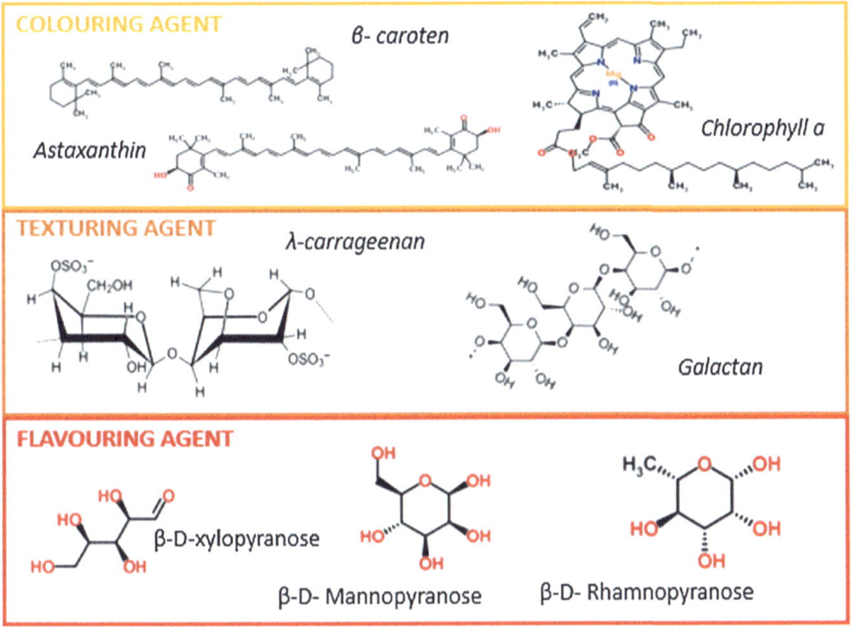

Fig. 2.15 Example of functional molecules from microalgae

ferent colors of fruits, vegetables, and other plants. There are more than 600 types of reported carotenoids in nature, and about 50 of them are provitamin A, such as α-carotene, β-carotene, and β-cryptoxanthin [50]. Carotenoids are the most important molecules extracted from microalgae regarding volumes [51], in which β-carotene is the only carotenoid with the potential of forming two molecules of vitamin A (retinol). It was the first carotenoid from algae to be commercialized, which serves itself as an essential nutrient, a natural food coloring agent, a provitamin A in food and animal feed, and an additive to cosmetics and multivitamin preparations in the market. Because of the limited supply from natural sources, a number of companies have introduced synthetically produced carotenoids to meet the high demand in the global market.

Natural β-carotene can be produced by the microalgae genus *Dunaliella*, which could produce and accumulate a significant amount of cellular β-carotene. The isomeric composition of β-carotene from the unicellular halotolerant alga *Dunaliella* contains more than 50% of 9-cis-β-carotene [52]. The average concentration of carotenoid in most algae is only 0.1–2%, but *D. salina* grown under the appropriate conditions of high salinity, light intensity, and temperature will produce up to 14% β-carotene of its dry weight [12]. As mentioned in Table 2.2, β-carotene can be efficiently extracted from *D. salina* thanks to SC-CO_2 extraction. Until 1980s, the production of β-carotene was synthetic because it is cheaper than the natural one

[53]. Nevertheless, the demand on natural β-carotene is increasing since it contains the natural isomer, which is more important than the trans-synthetic one.

Chlorophyll (Fig. 2.15), which amount is usually about 0.5–1.5% of microalgal dry weight, is a green fat-soluble pigment and an extremely important biomolecule. It is critical to photosynthesis, which allows plants to absorb solar energy for its conversion to chemical energy like oxygen and carbohydrates together with water and carbon dioxide [54]. Structurally, chlorophyll is essentially composed of two parts including a substituted porphyrin ring known as an excellent chelating ligand and a long carbon phytol chain. Chlorophyll molecules are arranged in and around the photosystems in the thylakoid membranes of chloroplasts. Chlorophyll selectively absorbs light in the red and blue regions and therefore emits a green color. *Chlorella* is usually called 'Emerald food' due to its high content of chlorophyll (7% of the biomass), which is 5 times higher than that in *Arthrospira*. All algae contain one or more type of chlorophyll: Chlorophyll a is the primary photosynthetic pigment in all algae, which is also the only chlorophyll in cyanobacteria and Rhodophyta. For instance, chlorophyll a is abundantly found in *Chlorella* and *Arthrospira*. As shown in Table 2.2, chlorophyll can also be extracted from *C. closterium* with microwave at 50 W. Like all higher plants, *Chlorophyta* and *Euglenophyta* contain chlorophyll b, while chlorophylls c, d, and e can be found in several marine algae and freshwater diatoms. The difference between chlorophyll a and b is that the methyl group in the ring II of chlorophyll a is replaced by a formyl group. This structural difference makes chlorophyll a be a blue/green pigment with a maximum absorbance from 660 to 665 nm and chlorophyll b be a green/yellow with a maximum absorbance from 642 to 652 nm [55].

In food industry, chlorophyll has been used as a natural pigment ingredient for many food processes and it has gain attention as an important food additive due to its strong green pigmenting and consumers' growing preferences for natural foods.

Phycobiliproteins could be found in cyanobacteria (blue-green algae), Rhodophyta (red algae), and Cryptomonads algae. Phycobiliproteins as deep-colored water-soluble fluorescent pigments are major components of a complex assemblage of photosynthetic light-harvesting antenna pigments, i.e., the phycobilisomes [56]. Phycobiliproteins are formed by a protein backbone covalently linked to tetrapyrrole chromophoric prosthetic groups, named phycobilins. The main natural resources of phycobiliproteins are the cyanobacterium *Arthrospira* for phycocyanin (blue) and the rhodophyte *Porphyridium* for phycoerythrin. From *A. platensis*, phycocyanin could be extracted at pH 6.8 in a phosphate buffer using an ultrasound-assisted extraction (Table 2.2). It is noted that phycocyanin (blue pigment), phycoerythrin (red pigment), and allophycocyanin (pale-blue pigment) can be differentiated according to their UV-visible absorption maxima. Phycobiliproteins have been commercially utilized as natural dyes with a variety of potential applications. For instance, phycocyanin is currently used in Japan and China as a natural coloring in food products like chewing gums, candies, dairy products, jellies, ice creams, and soft drinks as well.

Astaxanthin (Fig. 2.15) is a high-value molecule, mostly used for its colorant properties. The global astaxanthin market is estimated at about $257 million. Most astaxanthins are derived from the algae *H. pluvialis*, which are commonly consumed by fish and crustaceans like salmon and lobster, resulting in the pink coloration of their flesh. The value of 1 kg of natural astaxanthin has been estimated as high as $7150 [57]. Countries like USA, India, and Israel are the largest *Haematococcus* biomass producers. Astaxanthin is a naturally occurring high-value keto-carotenoid pigment with excellent antioxidant effects, which belongs to the xanthophyll group of carotenoids and also to the oxygenated carotenoids. It is composed of two asymmetric carbons located at 3 and 3′ positions of benzenoid rings on either end of the molecule. The hydroxyl and keto functional groups presenting in the ending ionone ring of astaxanthin are responsible for its potent antioxidant activity to penetrate the blood brain and retina barriers, which differs from other antioxidants. Therefore, it is believed to prevent brain and nervous systems from neurodegenerative diseases and aging. The most commonly used algae strain employed for astaxanthin production is *H. pluvialis*[38]. This microalga is believed to accumulate the highest level of astaxanthin in nature. Astaxanthin, which accumulates inside the thick, rigid cell walls of *Haematococcus* cysts during unfavorable growth conditions, can be recovered by cell-wall destruction and subsequent selective solvent extraction. Commercially grown *H. pluvialis* can accumulate more than 30 g of astaxanthin per kilogram of dry biomass, while other strains could also be used for astaxanthin production such as *Chlorella vulgaris, Chlorella zofingiensis, Coelastrella striolata var. multistriata* [9]. However, astaxanthin must be acquired from the diet due to the fact that mammals lack the ability to synthesize it or to convert dietary astaxanthin into vitamin A. Apart from conventional solvent extractions, astaxanthin could also be extracted from *H. pluvialis* by using supercritical fluids as described in Table 2.2.

2.5.1.2 Texturing Agents

Microalgae are a diverse group of photosynthetic organisms, in which polysaccharides as the main components of biomass are widely used in the food industry primarily as gelling and/or thickening agents. Many commercially used polysaccharides like agar, alginates, and carrageenans are extracted from macroalgae (e.g., *Laminaria, Gracilaria, Macrocystis*). Nevertheless, most microalgae produce polysaccharides and some of them could have industrial and commercial applications considering their fast growth rates and the possibility of controlling culture environmental conditions [58]. Gelling microalgal polysaccharides (phycocolloids), such as carrageenans presented in Table 2.2, are produced on a large scale with a wide range of applications in the food industry. The most promising microalgae for commercial purposes are the unicellular red alga *Porphyridium cruentum*, which produces a sulphated galactan exopolysaccharide that can replace carrageenans in many applications.

Carrageenan (Fig. 2.15) is a polysaccharide from algae and microalgae. According to Table 2.2, carrageenan can be extracted from *C. stigmatophora* in water with a high extraction temperature. Carrageenan has different types including iota, kappa, and lambda, in which iota and kappa carrageenans are widely used as gelling agents in food industry, cosmetics, and pharmaceutical formulations, whereas nongelling lambda carrageenan is served as a stabilizer due to the high viscosity of its solutions [11].

Galactan (Fig. 2.15) possessing extraordinary gelling properties is a polysaccharide consisting of polymerized galactoses, which is mostly localized in the cell wall of red algae like *Rhodophyta* [59].

Guar gum (Fig. 2.15), also called guaran, is a galactomannan. More precisely, it is a polysaccharide composed of galactose and mannose. The backbone is a linear chain of β-1,4-linked mannose residues to which galactose residues are 1,6-linked at every second mannose, forming short side branches. Guar gum is a potent thickener that can be also used in various multiphase formulations as an emulsifier or a stabilizer [53]. Moreover, guar gum has certified as a food additive (E412) in Europe, and the largest market for guar gum is the food industry.

2.5.1.3 Sweetening/Flavoring Agents

An important property of sugar is that it can strengthen tastes and aromas both above and below the sweetness threshold value. Microalgae contain a lot of carbohydrates with various properties, some of which belong to the class of sweeteners that can be used as food ingredients to improve the flavor of products (Table 2.2). For instance, monosaccharides such as xylose, arabinose, galactose, and glucose found in microalgae could be considered as sweetening agents [60], which can even replace molecules from vegetal sources like beetroot.

2.5.2 *Main Nutrients Contained in Microalgae*

2.5.2.1 Amino Acids

The quality of proteins can vary dramatically depending on digestibility and availability of essential amino acids, where content is the major quality factor of proteins. It exists 20 amino acids, among which eight are considered as essential (Table 2.3) [61]. Human requires these essential amino acids sufficiently in their nutrition since the human body is unable to biosynthesize them by itself.

Compared to plant-based proteins, animal-based proteins are generally considered as complete proteins as they contain essential amino acids, which are typically easier to digest due to the absence of cellulosic wall. Nevertheless, contrary to that of animal origin, plant-based proteins are more and more consumed since they are poor in saturated fats and cholesterol. Nutritionists and organizations like the Food and Agriculture Organization (FAO) advocate a more varied diet to avoid the development of diabetes and cardiovascular diseases [62]. Microalgae have been identified

Table 2.3 Main nutrients in microalgae

Compound	Molecule	Content (/100 g MS)	References
Essential amino acid	Histidine	1000 mg	[28]
	Isoleucine	3500 mg	
	Leucine	5380 mg	
	Lysine	2960 mg	
	Methionine	1170 mg	
	Phenylalanine	2750 mg	
	Threonine	2860 mg	
	Tryptophan	1090 mg	
	Valine	3940 mg	
Vitamin	Vitamin A	352 IU	[29]
	Vitamin K	1090 μg	
	Vitamin B1	0.5 mg	
	Vitamin B2	4.53 mg	
	Vitamin B3	14.9 mg	
	Vitamin B6	0.96 mg	
	Vitamin B12	162 μg	
Mineral	Calcium	468 mg	[29]
	Iron	87.4 mg	
	Phosphorus	961 μg	
	Iodine	142 μg	
	Magnesium	319 mg	
	Zinc	1.45 mg	
	Selenium	25.5 μg	
	Cooper	0.47 mg	
	Manganese	3.26 mg	
	Chromium	<400 μg	
	Potassium	1.660 mg	
	Sodium	641 mg	
Fatty acid	Myristic acid	0.46 mg	[30]
	Palmitic acid	40.65 mg	
	Palmitoleic acid	6.38 mg	
	Stearic acid	1.92 mg	
	Oleic acid	1.64 mg	
	Linoleic acid	17.95 mg	
	Gamma-linolenic acid	24.49 mg	
	Erucic acid	5.33 mg	
	Lignoceric acid	1.18 mg	

(continued)

Table 2.3 (continued)

Compound	Molecule	Content (/100 g MS)	References
Sugar	Rhamnose	4.46 g	[31]
	Ribbose	0.83 g	
	Xylose	0.36 g	
	Manose	0.17 g	
	Galactose	0.46 g	
	Glucose	1.20 g	

as excellent reservoirs of proteins and derivatives with potent biological properties, which play a major role in human and animal nutrition. For instance, *Arthrospira* contains all the essential amino acids good for human health [63]. The presence of microalgal proteins is in diverse forms and locations, e.g., cell-wall components, enzymes bounding to pigments, and/or carbohydrates. Amino acids as protein elements are of great importance in metabolism as they are building blocks of enzymes, peptides, and structural proteins. They are also precursors of numerous secondary metabolites (e.g., alkaloids, betalains, and polyphenols) [64].

2.5.2.2 Polyunsaturated Fatty Acids

Fatty acids are carboxylic acids with long aliphatic chains that may be straight or branched, saturated, or unsaturated. Some microalgae synthesize fatty acids of particular interests, especially for PUFAs such as γ-linolenic acid (GLA, 18:3w6) from *Arthrospira* and arachidonic acid (AA, 20:4w6) from *Porphyridium* (Table 2.3) [60]. It is worth mentioning that oleaginous eukaryotic microalgae can accumulate oils up to 50–70% [11]. Moreover, microalgae contain EPA (C20:5) and DHA (C22:6) essential fatty acids and other omega-3 fatty acids that most other crops do not contain. EPA from *Nannochloropsis, Phaeodactylum, Nitzschia, Isochrysis,* and *Diacronema* and DHA from *Crypthecodinium* and *Schizochytrium* are necessary for human and animal metabolism, and they cannot be synthesized by human body. Therefore, microalgae have become very important as food sources to obtain EPA and DHA [11].

Oleic acid (Fig. 2.16) is a monounsaturated fatty acid that is the most widely distributed and abundant fatty acid in nature. It belongs to the subgroup called long-chain fatty acids with 14–18 carbon atoms, and it is the precursor of most polyunsaturated fatty acids (PUFA). Microalgae produce both omega-3 and 6 polyunsaturated fatty acids from it, while animals can elongate and desaturated it to form a variety of omega-9 fatty acids. Regarding health benefits, the consumption of oleic acid enables to lower total low-density lipoprotein (LDL), while increasing high-density lipoprotein (HDL), also known as "good cholesterol", for retarding the development of heart diseases. Oleic acid can be extracted from *H. pluvialis* in DMF with ultrasound (Table 2.3).

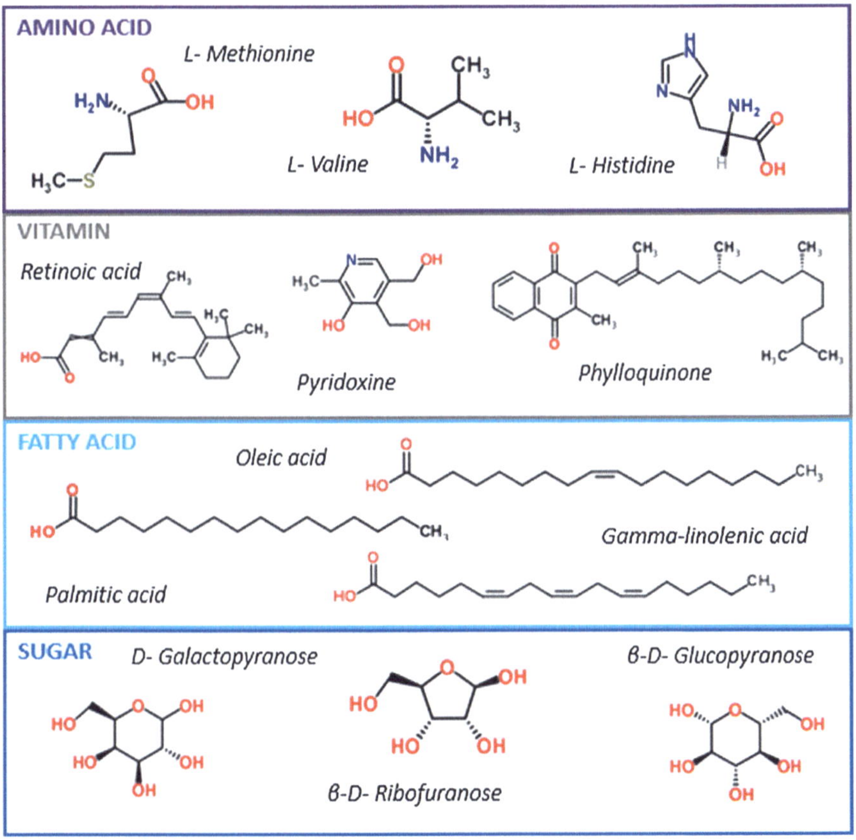

Fig. 2.16 Example of nutrients from *A. platensis*

2.5.2.3 Carbohydrates

In microalgae, most of the carbohydrates are stored in pyrenoids (starch grains) and cell walls as well. Depending on microalgal genera and species, different polysaccharides are found in the cell wall such as arabinose, xylose, mannose, galactose, glucose, and also less common sugars such as rhamnose, fucose, and uronic acids [65]. Polysaccharides are polymeric carbohydrate molecules composed of long chains of monosaccharide units bound together by glycosidic linkages. Many microalgae contain polysaccharide ranging in structure from linear to highly branched. Nutrition polysaccharides are common sources of energy, where nature and concentration depend on microalgae. Red microalgae, *Chlorella*, and cyanobacteria are particularly known for their high contents of polysaccharides, among which the most investigated representatives belong to the genus *Porphyridium*, *P. cruentum* for example [11]. The polysaccharide of *Porphyridium* sp. was shown to contain D-xylose, D- and L-galactose, D-glucose, D-glucuronic acid, several minor methylated sugars, and sulfate. Unicellular green algae belonging to the genus *Chlorella* is supposed to contain an acidic polysaccharide composed of rhamnose, arabinose, xylose, mannose, galactose, and glucuronic acid. *A. platensis* also contains the same monosaccharides, as listed in Table 2.3. Some cyanobacteria species can produce extracellular polysaccharides, which often have a very complex structure and contain unusual monosaccharide constituents. Thus, a sulfated polysaccharide named spirulan was isolated as Ca-salt from *A. platensis.*

Fucose is an aldohexose derived from galactose, which is attached to polyosides thanks to an osidic link. In the human body, fucose is metabolized by alpha fucosidase enzyme [58], or it can be extracted from the microalgae *C. stigmatophora.*

Mannose is an aldohexose composed of six carbons, which can prevent some bacterial diseases of the bladder [66]. It can be extracted from *Navicula salinarum* with a Soxhlet extraction.

2.5.2.4 Vitamins

Certain microalgae contain water and lipid-soluble vitamins that can be used as food supplements or food ingredients. A number of vitamins are presenting with higher concentrations in the microalgae than in conventional foods which are traditionally considered rich in vitamins. Ingestion of relatively small quantities of microalgae can cover the recommended daily amount of some vitamins in human nutrition. Microalgae can thus be considered to represent a nonconventional source of vitamins or a vitamin supplement for human nutrition. Indeed, almost all the essential vitamins (e.g., A, B_1, B_6, B_{12}, C, and E) can be found in microalgae in a variable amount depending on culture conditions and microalgae strains. For instance, *A. platensis* contains a lot of vitamins, mainly vitamins of B group (Table 2.3, Fig. 2.16).

2.5.2.5 Minerals

Minerals are structural components and significantly important elements that perform many necessary functions in the living body, including cell transport and a wide range of metabolic processes serving as various catalytic metalloenzyme cofactors. Minerals have an important role in human body since they participate in the formation of binding sites of metalloenzymes and play specific roles with beneficial functions in living systems [11]. Microalgae are known to be high in mineral content and are used as feed and food supplements to supply minerals. The mineral composition varies according to the strain as well as other various factors such as seasonal, environmental, geographical, and physiological variations. Mineral from Spirulina is listed in Table 2.3.

2.6 Conclusions and Future Prospects

Microalgae seem to be a wonderful resource with many applications in food and other sectors, which is very promising to replace the current vegetable oil and protein commodities in a sustainable way. However, the quality and the performance of the production and the marketing authorization are main brakes on the development of microalgae production and consumption.

One of the advantages of microalgae is the sustainable production without any competitions with food crops. Instead of growing in an arable land, they could be cultivated in the open pond under greenhouse, but its main drawback is the quality of biomass produced. Some other cultivation systems such as photobioreactors (PBR) develop little by little to avoid contamination. With these PBR, it is possible to control culture conditions to obtain microalgae with a higher quality. However, one major bottleneck which lies in the up-scaling production of PBR is the high cost resulting from the energy consumption and maintenance.

Furthermore, the consumption of microalgae is currently restricted to very few taxa, like *Spirulina* (*Arthrospira*), *Chlorella, Dunaliella, Haematococcus, Nostoc,* and *Aphanizomenon* as well, among which *Chlorella* and *Spirulina* dominate the microalgae market. The low exploitation rate of the biological diversity of microalgae is probably due to food safety regulations for human consumptions. However, the successful authorization (EC regulation 258/97) of the marine diatom *Odontella aurita* as a novel food by the French company INNOVALG in 2002 is a real plus for microalgal biotechnology [67]. Most of the time, the introduction of a novel food ingredient for human consumption in the market is not so simple and time-consuming so that an approval is required from the authorities to guarantee the safety of the new product. As microalgae are considered as a nontraditional source organism, some safety assessments are required. Thus, security threshold must be respected to market a food ingredient from microalgae such as toxins, heavy metals, and level of pathogenic microorganisms.

Many improvements should still be made in microalgae research to improve production and extraction efficiency and to generalize microalgae consumption. When more products can be obtained from one microalgae production process based on the biorefinery concept, its added value will increase significantly and will make microalgae production economically interesting for a sustainable future.

References

1. Lee Y-K (1997) Commercial production of microalgae in the Asia-Pacific rim. J Appl Phycol 9:403–411
2. Enzing C, Ploeg M, Barbosa M et al (2014) Microalgae-based products for the food and feed sector: an outlook for Europe. Publications Office
3. Habib MAB (2008) A review on culture, production and use of *Spirulina* as food for humans and feeds for domestic animals and fish. Food and Agriculture Organization of the United Nations, Rome

4. Zhu L (2015) Biorefinery as a promising approach to promote microalgae industry: an innovative framework. Renew Sustain Energy Rev 41:1376–1384. https://doi.org/10.1016/j.rser.2014.09.040
5. Rizwan M, Lee JH, Gani R (2015) Optimal design of microalgae-based biorefinery: economics, opportunities and challenges. Appl Energy 150:69–79. https://doi.org/10.1016/j.apenergy.2015.04.018
6. Bull E, Rapport L, Lockwood B (2018) 1. What is a nutraceutical? Stroke 13:57
7. Pan Y, Wang K, Huang S et al (2008) Antioxidant activity of microwave-assisted extract of longan (*Dimocarpus Longan* Lour.) peel. Food Chem 106:1264–1270. https://doi.org/10.1016/j.foodchem.2007.07.033
8. Sharma UK, Sharma K, Sharma N et al (2008) Microwave-assisted efficient extraction of different parts of *Hippophae rhamnoides* for the comparative evaluation of antioxidant activity and quantification of its phenolic constituents by reverse-phase high-performance liquid chromatography (RP-HPLC)[†]. J Agric Food Chem 56:374–379. https://doi.org/10.1021/jf072510j
9. Guerin M, Huntley ME, Olaizola M (2003) *Haematococcus* astaxanthin: applications for human health and nutrition. Trends Biotechnol 21:210–216. https://doi.org/10.1016/S0167-7799(03)00078-7
10. Plaza M, Herrero M, Cifuentes A, Ibáñez E (2009) Innovative natural functional ingredients from microalgae. J Agric Food Chem 57:7159–7170. https://doi.org/10.1021/jf901070g
11. Jacobsen C, Nielsen NS, Horn AF, Sørensen A-DM (2013) Food enrichment with omega-3 fatty acids. Woodhead Publishing, Oxford, Cambridge; Philadelphia; New Delhi
12. de Morais MG, Vaz BDS, de Morais EG, Costa JAV (2015) Biologically active metabolites synthesized by microalgae. Biomed Res Int 2015:1–15. https://doi.org/10.1155/2015/835761
13. Peng J, Yuan J-P, Wu C-F, Wang J-H (2011) Fucoxanthin, a marine carotenoid present in brown seaweeds and diatoms: metabolism and bioactivities relevant to human health. Mar Drugs 9:1806–1828. https://doi.org/10.3390/md9101806
14. Sekar S, Chandramohan M (2008) Phycobiliproteins as a commodity: trends in applied research, patents and commercialization. J Appl Phycol 20:113–136. https://doi.org/10.1007/s10811-007-9188-1
15. Winwood RJ (2013) Recent developments in the commercial production of DHA and EPA rich oils from micro-algae. OCL 20:D604. https://doi.org/10.1051/ocl/2013030
16. Yaakob Z, Ali E, Zainal A et al (2014) An overview: biomolecules from microalgae for animal feed and aquaculture. J Biol Res 21. https://doi.org/10.1186/2241-5793-21-6
17. Tang Y, Zhang Y, Rosenberg JN et al (2016) Efficient lipid extraction and quantification of fatty acids from algal biomass using accelerated solvent extraction (ASE). RSC Adv 6:29127–29134. https://doi.org/10.1039/C5RA23519G
18. Dogaris I, Welch M, Meiser A et al (2015) A novel horizontal photobioreactor for high-density cultivation of microalgae. Bioresour Technol 198:316–324. https://doi.org/10.1016/j.biortech.2015.09.030
19. Pulz O (2001) Photobioreactors: production systems for phototrophic microorganisms. Appl Microbiol Biotechnol 57:287–293. https://doi.org/10.1007/s002530100702
20. Liao Q, Li L, Chen R, Zhu X (2014) A novel photobioreactor generating the light/dark cycle to improve microalgae cultivation. Bioresour Technol 161:186–191. https://doi.org/10.1016/j.biortech.2014.02.119
21. Benavente-Valdés JR, Aguilar C, Contreras-Esquivel JC et al (2016) Strategies to enhance the production of photosynthetic pigments and lipids in chlorophycae species. Biotechnol Rep 10:117–125. https://doi.org/10.1016/j.btre.2016.04.001
22. Mata TM, Martins AA, Caetano NS (2010) Microalgae for biodiesel production and other applications: a review. Renew Sustain Energy Rev 14:217–232. https://doi.org/10.1016/j.rser.2009.07.020
23. Vernès L, Granvillain P, Chemat F, Vian M (2015) Phycocyanin from *Arthrospira platensis*. Production, extraction and analysis. Curr Biotechnol 4:481–491
24. King PM (2014) The use of ultrasound on the extraction of microalgal lipids. PhD thesis, Coventry University

25. Bendicho C, Lavilla I (2000) Ultrasound extractions. Encycl Sep Sci Acad Press Lond 1448
26. Suslick KS (1989) The chemical effects of ultrasound. Sci Am 260:80–86
27. Gerde JA, Montalbo-Lomboy M, Yao L et al (2012) Evaluation of microalgae cell disruption by ultrasonic treatment. Bioresour Technol 125:175–181. https://doi.org/10.1016/j.biortech.2012.08.110
28. Jambrak AR (2012) Application of high power ultrasound and microwave in food processing: extraction. J Food Process Technol 3. https://doi.org/10.4172/2157-7110.1000e111
29. Ganzler K, Salgo A (1987) Microwave-extraction-a new method superseding traditional Soxhlet extraction. Z Für Leb-Forsch A 184:274–276
30. Letellier M, Budzinski H (1999) Microwave assisted extraction of organic compounds. Analusis 27:259–270. https://doi.org/10.1051/analusis:1999116
31. Destandau E, Michel T, Elfakir C (2013) Chapter 4. Microwave-assisted extraction. In: Rostagno MA, Prado JM (eds) Green chemistry series. Royal Society of Chemistry, Cambridge, pp 113–156
32. Puri M, Sharma D, Barrow CJ (2012) Enzyme-assisted extraction of bioactives from plants. Trends Biotechnol 30:37–44. https://doi.org/10.1016/j.tibtech.2011.06.014
33. Shankar D, Agrawal YC, Sarkar BC, Singh BPN (1997) Enzymatic hydrolysis in conjunction with conventional pretreatments to soybean for enhanced oil availability and recovery. J Am Oil Chem Soc 74:1543–1547
34. Gerken HG, Donohoe B, Knoshaug EP (2013) Enzymatic cell wall degradation of *Chlorella vulgaris* and other microalgae for biofuels production. Planta 237:239–253. https://doi.org/10.1007/s00425-012-1765-0
35. Brunner G (2005) Supercritical fluids: technology and application to food processing. J Food Eng 67:21–33. https://doi.org/10.1016/j.jfoodeng.2004.05.060
36. Taher H, Al-Zuhair S, Al-Marzouqi AH et al (2014) Supercritical carbon dioxide extraction of microalgae lipid: process optimization and laboratory scale-up. J Supercrit Fluids 86:57–66. https://doi.org/10.1016/j.supflu.2013.11.020
37. Mukhopadhyay M (2009) Extraction and processing with supercritical fluids. J Chem Technol Biotechnol 84:6–12. https://doi.org/10.1002/jctb.2072
38. Rozzi NL, Singh RK (2002) Supercritical fluids and the food industry. Compr Rev Food Sci Food Saf 1:33–44. https://doi.org/10.1111/j.1541-4337.2002.tb00005.x
39. Jeyamkondan S, Jayas DS, Holley RA (1999) Pulsed electric field processing of foods: a review. J Food Prot 62:1088–1096
40. Weaver JC, Chizmadzhev YA (1996) Theory of electroporation: a review. Bioelectrochem Bioenerg 41:135–160
41. Lebovka NI, Bazhal MI, Vorobiev E (2001) Pulsed electric field breakage of cellular tissues: visualisation of percolative properties. Innov Food Sci Emerg Technol 2:113–125
42. ’t Lam GP, Postma PR, Fernandes DA et al (2017) Pulsed Electric Field for protein release of the microalgae *Chlorella vulgaris* and *Neochloris oleoabundans*. Algal Res 24:181–187. https://doi.org/10.1016/j.algal.2017.03.024
43. Hammond OS, Bowron DT, Edler KJ (2016) Liquid structure of the choline chloride-urea deep eutectic solvent (reline) from neutron diffraction and atomistic modelling. Green Chem 18:2736–2744. https://doi.org/10.1039/C5GC02914G
44. Cicci A, Sed G, Bravi M (2017) Potential of choline chloride–based natural deep eutectic solvents (NaDES) in the extraction of microalgal metabolites. Chem Eng 57
45. Faggian M, Sut S, Perissutti B et al (2016) Natural deep eutectic solvents (NADES) as a tool for bioavailability improvement: pharmacokinetics of rutin dissolved in proline/glycine after oral administration in rats: possible application in nutraceuticals. Molecules 21:1531. https://doi.org/10.3390/molecules21111531
46. Bucar F, Wube A, Schmid M (2013) Natural product isolation—how to get from biological material to pure compounds. Nat Prod Rep 30:525. https://doi.org/10.1039/c3np20106f
47. Poojary MM, Barba FJ, Aliakbarian B et al (2016) Innovative alternative technologies to extract carotenoids from microalgae and seaweeds. Mar Drugs 14. https://doi.org/10.3390/md14110214

48. Van Gaver D, Huyghebaert A (1991) Optimization of yeast cell disruption with a newly designed bead mill. Enzyme Microb Technol 13:665–671
49. Postma PR, Suarez-Garcia E, Safi C et al (2017) Energy efficient bead milling of microalgae: effect of bead size on disintegration and release of proteins and carbohydrates. Bioresour Technol 224:670–679. https://doi.org/10.1016/j.biortech.2016.11.071
50. Vershinin A (1999) Biological functions of carotenoids-diversity and evolution. BioFactors 10:99–104
51. Vigani M, Parisi C, Rodríguez-Cerezo E et al (2015) Food and feed products from micro-algae: market opportunities and challenges for the EU. Trends Food Sci Technol 42:81–92. https://doi.org/10.1016/j.tifs.2014.12.004
52. Orset SC, Young AJ (2000) Exposure to low irradiances favors the synthesis of 9-cis β, β-carotene in *Dunaliella salina* (Teod.). Plant Physiol 122:609–618
53. Borowitzka MA (2013) High-value products from microalgae—their development and commercialisation. J Appl Phycol 25:743–756. https://doi.org/10.1007/s10811-013-9983-9
54. D'Alessandro EB, Antoniosi Filho NR (2016) Concepts and studies on lipid and pigments of microalgae: a review. Renew Sustain Energy Rev 58:832–841. https://doi.org/10.1016/j.rser.2015.12.162
55. Hosikian A, Lim S, Halim R, Danquah MK (2010) Chlorophyll extraction from microalgae: a review on the process engineering aspects. Int J Chem Eng 2010:1–11. https://doi.org/10.1155/2010/391632
56. Christaki E, Bonos E, Giannenas I, Florou-Paneri P (2017) Phycobiliproteins: a new perspective in natural pigments derived from microalgae. J Oceanogr Mar Res 4. https://doi.org/10.4172/2572-3103.1000e114
57. Kim D-Y, Vijayan D, Praveenkumar R et al (2016) Cell-wall disruption and lipid/astaxanthin extraction from microalgae: *Chlorella* and Haematococcus. Bioresour Technol 199:300–310. https://doi.org/10.1016/j.biortech.2015.08.107
58. de Jesus Raposo MF, de Morais RMSC, de Morais AMMB (2013) Bioactivity and applications of sulphated polysaccharides from marine microalgae. Mar Drugs 11:233–252. https://doi.org/10.3390/md11010233
59. Pulz O, Gross W (2004) Valuable products from biotechnology of microalgae. Appl Microbiol Biotechnol 65:635–648. https://doi.org/10.1007/s00253-004-1647-x
60. Buono S, Langellotti AL, Martello A et al (2014) Functional ingredients from microalgae. Food Funct 5:1669. https://doi.org/10.1039/C4FO00125G
61. Vanthoor-Koopmans M, Wijffels RH, Barbosa MJ, Eppink MHM (2013) Biorefinery of microalgae for food and fuel. Bioresour Technol 135:142–149. https://doi.org/10.1016/j.biortech.2012.10.135
62. Bleakley S, Hayes M (2017) Algal proteins: extraction, application, and challenges concerning production. Foods 6:33. https://doi.org/10.3390/foods6050033
63. Gutiérrez-Salmeán G, Fabila-Castillo L, Chamorro-Cevallos G (2015) Nutritional and toxicological aspects of *Spirulina* (*Arthrospira*). Nutr Hosp 32:34–40
64. Reeds PJ (2000) Dispensable and indispensable amino acids for humans. J Nutr 130:1835S–1840S
65. Draaisma RB, Wijffels RH, (Ellen) Slegers P et al (2013) Food commodities from microalgae. Curr Opin Biotechnol 24:169–177. https://doi.org/10.1016/j.copbio.2012.09.012
66. Alassali A, Cybulska I (2015) Methods for upstream extraction and chemical characterization of secondary metabolites from algae biomass. Adv Tech Biol Med 4. https://doi.org/10.4172/2379-1764.1000163
67. Lordan S, Ross RP, Stanton C (2011) Marine bioactives as functional food ingredients: potential to reduce the incidence of chronic diseases. Mar Drugs 9:1056–1100. https://doi.org/10.3390/md9061056

Chapter 3
From Petroleum to Bio-Based Solvents: From Academia to Industry

Ying Li, Werner Kunz and Farid Chemat

Abstract The property and applications of bio-based solvents in place of petroleum-based ones for natural product extractions are discussed in this chapter including vegetable oils, terpenes, 2-methyltetrahydrofuran, bio-ethanol and natural deep eutectic solvents (NADES). Several commonly used prediction techniques such as Hansen solubility parameters (HSP), COnductor-like Screening MOdel for Real Solvents (COSMO-RS) and life-cycle assessment (LCA) are introduced to evaluate theoretical solubility and environmental impacts during all stages of a solvent's life. Besides, the revisit of polar paradox theory helps us to better understand the solubilization mechanism underpinning the extraction. This chapter is intended to describe not only past and current achievements and challenges, but also to give visions for more sustainable applications in this increasingly important field of chemistry. Finally, it is also intended to draw the way to more rigorous research going beyond empiricism, which is still the usual approach in plant extractions.

Keywords Bio-based alternative solvents · Natural product extraction · In silico evaluation · Polar paradox theory · Sustainable application · Solvent chemistry

---

Y. Li (✉)
Department of Food Science and Engineering, College of Science and Engineering, Jinan University, 510632 Guangzhou, China
e-mail: yingli@jnu.edu.cn

W. Kunz
Institute of Physical and Theoretical Chemistry, Universität Regensburg, 93040 Regensburg, Germany

F. Chemat
GREEN Extraction Team, Université d'Avignon et des Pays du Vaucluse, INRA, UMR 408, 84000 Avignon, France

Y. Li and F. Chemat (eds.), *Plant Based "Green Chemistry 2.0"*,
Green Chemistry and Sustainable Technology,
https://doi.org/10.1007/978-981-13-3810-6_3

3.1 Introduction

Sustainability is a compromise between environmental, social and economic demands. Chemists, chemical engineers and biotechnologists are challenged today all over the world to find such a compromise and to show innovative ways for minimizing negative effects on the environment, as they may arise from industry. The goals to achieve are usually summarized in terms of a set of fundamental principles, some of which emphasize the design and usage of safer solvents from renewable feedstocks, whenever technically and economically practicable [1]. Further, 12 principles of green engineering appeared to provide a feasible and economical guideline for modern sustainable processes, which matched quickly with the green chemistry principles [2]. Green chemistry has demonstrated and will, even more, demonstrate in the future its contributions to manifold aspects of our daily life, such as energy, the responsible use of resources including recycling and the attenuation of climate change.

The global market of plant extracts has enormous growth potential with current market size at more than billion dollars per year. Using plant extracts provides an invaluable solution for industries (e.g. pharmaceuticals, nutraceuticals, cosmetics, food and beverages, etc.) and developing safe and green processes to get them is one aspect of this global effort, and also using new green alternative solvents is a particular challenge in this context [3]. Over the course of the past decades, an ever-growing demand for using greener, biorenewable solvents has been identified to replace petroleum-based solvents in chemical processes (Fig. 3.1). The ideal alternative solvents suitable for green chemistry are non-volatile organic compounds (VOCs) that have high solvency, high flash points with low toxicity and low environmental impacts, being easily biodegradable [4]. They should be obtained from renewable (non-petrochemical) resources at a reasonable price and should be easy to recycle. Of course, there is no perfect solvent that can meet all above-mentioned requirements and apply to all situations, thus the decision for the optimum solvent will always be a compromise for a special process and, as far as plant extraction is concerned, for a special plant and the targeted extract.

Since solvents of petroleum origin are now strictly regulated worldwide, it is indispensable to encourage the development and implementation of greener solvents (Fig. 3.2) [5]. However, it must be admitted that the origin of a solvent does not necessarily define its green nature. For example, MMB (3-methoxy-3-methyl-butan-1ol), which is synthesized from iso-butylene (C4 fraction of cracked naphtha) is of very low toxicity and has a low vapour pressure and therefore can be reasonably called a green solvent, whereas the bio-based furfuraldehyde is toxic and carcinogenic and as such certainly not a green solvent. Nevertheless, these are exceptions and solvents with synthons of biogenic origin are at least in general "greener" than petrol-based ones.

Bio-based solvents can be obtained in bio-refineries. Just like petroleum refineries, nowadays, the objective of bio-refinery is to operate integrated processes to produce multiple products with different uses (e.g. solvent, surfactants, polymers, biofuels

Fig. 3.1 Development from conventional to alternative solvents based on the green principles

Fig. 3.2 Classification of alternative solvents for natural product extractions

Table 3.1 Energy comparison of solvents

Solvent	Heat capacity (J/mol*K)	Heat of vapourization (kJ/mol)	Energy for 1 kg solvent evaporation (kWh)	CO_2 emissions (g)
n-Hexane	195.6	28.9	0.121	96
Ethanol	112.3	38.6	0.265	212
Methanol	81.1	35.2	0.321	257
Ethyl lactate	254	45.6	0.185	148
Ethyl acetate	170.7	31.9	0.127	102
Water	75.3	40.7	0.715	572

and biomaterials, etc.). Further to water, the most natural solvent, solvents from bio-refineries can be both environmentally friendly and ecologically interesting, provided the bio-refinery provides a multitude of products out of the plants so to maximize their value. Let us see in the following if they are really superior to petroleum-based solvents and other more recently discussed ones such as supercritical CO_2 and ionic liquids (ILs).

3.2 Bio-Based Solvents as Alternative Solvents for Natural Product Extractions

It is well known that water has been recognized as the most available and the cheapest solvent for the green extraction of natural products. However, with respect to energy consumption and greenhouse gas emissions, water is not a truly green solvent due to its mournful evaporation so that several traditional solvents are still used in industries and some other polar solvents are proposed as alternatives. As can be seen in Table 3.1, water appears less green than other solvents due to its relatively high energy consumption required for evaporation and thus enhanced CO_2 emissions. Therefore, it is always favourable to limit the amount of water to evaporate as much as possible or to look for other ways to concentrate or to separate the extract from the aqueous solution. Further, the remaining aqueous solutions may be a serious threat to the environment and may be thermally or biologically treated before their release. It is promising to discover that water in its subcritical region, i.e. between boiling point (100 °C) and critical point (374 °C), has been proved to be efficiency in extracting both apolar and moderately polar compounds, which polarity change was mainly dependent on the temperature than the pressure [6–11].

On the other hand, various organic solvents used in current industrial vegetal extractions have aroused concerns in terms of economy, environment and competitiveness. In other words, appropriate alternative solvents should be used to deal with the rigorous set of intra-industry regulations, food and cosmetics in particular. For instance, n-hexane as a common organic solvent for many years in the industrial extraction of vegetable oils due to its apolar property and simple recovery. With regard to the quality and yield of oils, the use of n-hexane in solvent extraction outperforms mechanical extraction, where higher temperature is specifically required to maximize the yield of extracted oils. Nonetheless, the inevitable leakage of n-hexane during extraction and recovery has been considered as a harmful substance due to its negative influence on human health and atmospheric environment. Hence, it is compulsory to establish special precautions for operational personnel in industry, especially for the working areas with explosion risks. Although extraction with n-hexane has been considered as a Best Available Technology Not Entailing Excessive Costs (BATNEEC) by numerous environment regulation authorities, concerns with health, safety and environment have, nowadays, led to a new hotspot in exploring alternative solvents with the help of green chemistry development.

Once we talk about n-hexane alternatives in natural product extraction, on no account can we ignore the significance of some petroleum-based alternative solvents such as n-butane, which liquefied gas is a low-toxic, non-polar, linear hydrocarbon solvent with similar solvation properties to those of hexane and supercritical CO_2. The best extraction performances are obtained with a finely divided and dry matrix in the lipid or fat extraction, the interesting product could be either the extract itself, precipitated solid or liquid solute, or the purified matrix free of undesired molecules. Although the use of n-butane as solvent is controversial, it has been used in food and perfumery industries for decades, because it can be easily recovered and thoroughly removed by evaporation without any traces in the final extracts. Thanks to its low boiling point, n-butane can dissolve and extract waxes, oils and very delicate aromatics very effectively at very low temperatures and pressures while preserving the integrity of the highly evaporative volatile compounds (i.e. aroma, flavour and taste) in plant materials without reacting and creating other unwanted products [12]. In order to be versatile to various plant types and applications, the addition of polar or mid-polar solvents may be favourable as well as the proper tuning of other parameters to optimize the process in terms of efficiency and other economical requirements. The industrial equipment manufacturer CELSIUS has developed a one-litre pilot-scale extractor to make extractions of bioactive compounds under pressure using liquid butane with optionally added co-solvents. This process involves a solid–liquid extraction and a subsequent solute–solvent separation at low temperature, which is a particular advantage for thermo-sensitive or oxidable molecules. Butane used as the solvent is totally chemically inert to the extracts and to the environment. Further, a relatively low pressure (up to 8 bar) allows one to scale up to an industrial level. This procedure has currently been optimized (e.g. solid–liquid ratio, time, temperature, pressure, type and concentration of co-solvents added, etc.) and scaled up to 500 L. Moreover, the solvent recycling is in progress also in industrial exploitation. Nevertheless, organic solvents like n-hexane and n-butane derived from unrenewable petroleum resources are encouraged to be gradually substituted by bio-based solvents from agricultural resources that are economically viable and environmentally safe.

3.3 Vegetable Oils as Natural Agro-Solvents

Vegetable oils, generally produced by cold pressing from plant seeds that are abundant in lipids, play an important role in our diet and can be consumed directly in the form of refined or virgin oils, or through manifold products in food industry. Vegetable oils are non-polar and lipophilic systems with compositions that vary significantly depending on their origin, quality and the methods through which they have been obtained. They are non-allergenic, non-genetically modified, and normally food grade with relatively high flash point and cheap price (1.3–10.5 €/kg). Also, they have considerable and selective dissolution power in certain extractions. Furthermore, they are biodegradable and not necessary to separate afterwards. The good consistency of properties between vegetable oils and ideal solvents shows that such natural oils may become alternatives to conventional organic solvents. As the application of vegetable oils as alternative solvents has been emerged in many areas, a rigorous investigation of the solubilization mechanisms and capacities of vegetable oils can help developing novel green strategies for extraction techniques.

3.3.1 The Composition of Vegetable Oils

Regardless of vegetable oils as solvents or solutes, its composition should be considered as a priority for targeted extraction. The main building blocks of vegetable oils are triglycerides, which are composed of three fatty acid molecules esterified to one glycerol molecule. The characteristics of triglycerides are determined by the type, proportion and position of fatty acids on the glycerol backbone. The fatty acid composition of triglycerides in vegetable oils depends on the plant species, but also on the cultivation, and agronomic and climatic conditions. Typically, even-numbered carbon atoms dominate the fatty acid chain lengths in vegetable oils due to their biosynthesis. According to the degree of saturation of the fatty acid, they can be generally classified into saturated, mono- and poly-unsaturated fatty acids, which may have different configurations resulting in different physical and chemical properties.

In addition to triglycerides, the presence of minor components (2–5%) in vegetable oils cannot be ignored due to their significant characteristics and nutritional values, virgin oils in particular. They can be divided into two types: glycerolipids such as mono- and di-glycerides, phospholipids and non-glycerolipids including sterols, tecopherols/tocotrienols, free fatty acids, vitamins, pigments, proteins, phenolic compounds, water, etc. Although some undesirable minor components are removed during the refining process by most consumers and food manufacturers, there is no completely efficient and selective process to remove them entirely. As a consequence, usually a small part of such molecules acting as emulsifier remains inside the oil. For example, partial glycerides (mono- and di-glycerides) presenting naturally in virgin oils are well known for their emulsifying properties, which are authorized food additives (E471). Monoglycerides and di-glycerides, respectively, have a hydrophilic-lipophilic balance (HLB) of approximately 3.4–3.8 and 1.8, and thus are hydrophobic [13], which total content accounts for 1–6 wt% depending on the oil type.

Phospholipids are natural polar lipids in plant cell membranes, which could be co-extracted with other neutral triglycerides during squeezing or solvent extraction of virgin oils using hexane. Most phospholipids are removed in the degumming step of the oil-refining process in order to avoid foaming or browning during technological processing operations or edible oil use. All phospholipids consist of a glycerol backbone, two fatty acids and one polar phosphate group associated itself with an amino alcohol (choline ethanolamine, serine) or a sugar (inositol) as shown in Fig. 3.3. Lecithins (phosphatidyl choline) are authorized food additives (E322) with wide applications, especially in baking industry because of their significant surfactant effects (HLB $\approx$ 8). Moreover, lamellar structures can be formed in the presence of other phospholipids and surface-active compounds such as sterols.

Free fatty acids are unesterified fatty acids generated from the hydrolysis of triglycerides or phospholipids. They are usually eliminated in deodorization and physical refining processes due to their impact on foaming and on the smoke point of oils. Normally, virgin and refined vegetable oils or fats have free fatty acid content less than 5 and 0.1%, respectively [14, 15]. However, their contents are higher when oils have high mono- and di-glyceride levels. Free fatty acids with a low HLB value of approximately 1.0 can facilitate the formation of reverse micelles so as to stabilize water-in-oil emulsions.

Plant sterols, namely phytosterols are one of the important minor compounds with considerable biological activities [16, 17], which are also the predominant compounds in the unsaponifiable fraction of vegetable oils. In general, β-sitosterol is the most abundant phytosterol in vegetable oils accompanying with variable amounts of other sterols (e.g. stigmasterol, campesterol, brassicasterol, etc.) depending on the oil type [18]. Phytosterols are affected by several oil-refining processes, in which deodorization can significantly reduce total sterols due to distillation and esterification of free sterols [19]. Since phytosterols are barely hydrophobic and poorly soluble in fats, they are usually esterified to make them liposoluble for easily processed in food industries.

Tocopherols and tocotrienols with four common types (α, β, γ and δ) are well-known vitamin E compounds, which are vital for enhancing the oxidative stability of vegetable oils because of their strong inhibitory effects against lipid oxidation [20, 21]. Naturally, tocopherols represent as free alcohols only, whereas tocotrienols are in esterified forms [22]. The content of different types of tocopherols and tocotrienols is dependent on the oil type as well, which may also be influenced by the unintentional removal during oil refining, especially in deodorization [23].

The water content of less than 0.05% is generally controlled in commercial vegetable oils, which may alter during preservation periods after opening as the moisture absorption or loss can occur depending on the storage environment [24]. Since water is known for its immiscibility with oils, it can provide oil–water interfaces in vegetable oils, to which endogenous or exogenous surface-active compounds with low HLB values migrate and preferably concentrate to form and stabilize association

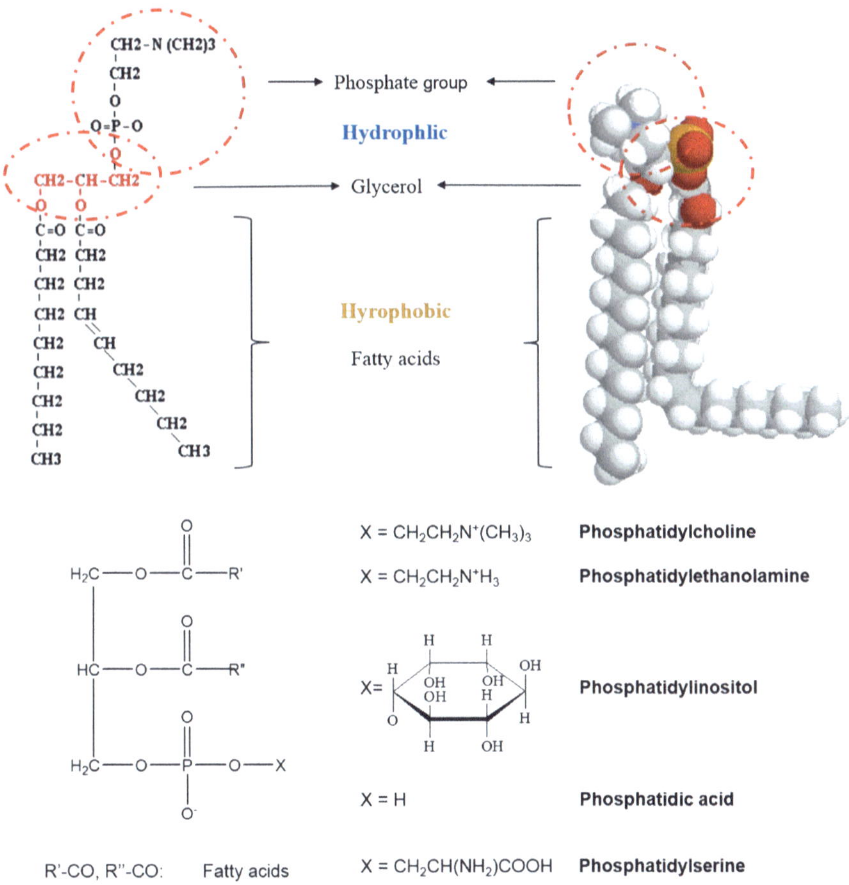

Fig. 3.3 Structures of phospholipids: glycerol phosphatides structure with hydrocarbon chains (R′, R″) and phosphate group with corresponding 'X' structures

colloids such as reverse micelles and lamellar structures [18]. These energetically more favoured structures enable the coexistence of both hydrophilic and lipophilic compounds, which has inspired a novel solution for the extraction of water-soluble compounds in vegetable oils.

A wide range of phenolic compounds was reported at varying levels in vegetable oils (particularly in olive oils) depending on the oil type, ripeness, process and storage conditions [25]. These polar compounds having free radical scavenging activity are important for the oxidative stability of the polyunsaturated fatty acids in vegetable oils. They tend to stay in the polar region of association colloids. Squalene is a triterpenoid hydrocarbon of 30 carbons in vegetable oils at a very low level [26], which can be used in functional food and its hydrogenated form can be applied as a moisturizing agent in cosmetics. Pigments such as carotenoids in most virgin oils are a conjugated system with alternated double bonds in a long chain of forty carbons, which can be classified into non-polar (e.g. α- or β-carotene and lycopene) and more polar carotenoids with oxygen at one or both end groups. Chlorophyll pigments often inadvertently mask the colour of carotenoids in virgin oils, and the variable contents of them are determined by the oilseed maturity and the oil extraction method. Furthermore, the presence of proteins and peptides has been investigated as minor components in virgin oils, which attracted increasing concerns on the stability and the allergenicity of oils [27]. Besides, trace quantities of other components such as metals, hydrocarbons and amphiphilic lipid oxidation products are deleterious to oil quality, and hence should be reduced to a minimum [18].

3.3.2 Applications of Vegetable Oils as Solvents in Natural Product Extractions

The first historical record of using vegetable oils as solvents could trace back to ancient civilizations. Ancient Egyptians originally infused aromatic plants into vegetable oils for aesthetic, spiritual and religious rituals during pharaonic eras. Furthermore, they invented frying in oils around 2500 BC in order to preserve meat and vegetables [28]. Meanwhile, ancient Chinese has experienced infused oils with various indigenous medicinal plants for therapeutic and nutritional purposes, Chinese massage in particular. The empirical application of such infused oils has gone through thousand years' development and evolvement until the rise of using petroleum-based solvents (e.g. hexane, benzene, etc.) in the nineteenth century. Nonetheless, the increasing health, safety and environmental concerns drive the reuse of vegetable oils as agro-based solvents for natural product extractions.

In recent years, edible vegetable oils in commercial markets have now been successfully enriched or aromatized with bioactive compounds from herbs, spices or other materials in order to improve their nutritional value and organoleptic quality and to increase shelf-life as well. The resulting oils can be defined as "flavoured oils" or "gourmet oils" after the name of plants from which additives are extracted. As the consumption of such kind of products attracts more interest due to their particular capacities to prevent diseases through a healthy diet, their versatility, conveniences, and wide range of tastes have made them rapidly diffused among traditional and non-traditional consumers across many countries in the world. Therefore, it is worth exploring more in detail their characteristics, which can help establishing a sound-labelling standard for strict regulations [29]. Nowadays, for consumers, lots of various types of enriched or aromatized oils as seasonings are available on the market. Moreover, due to the fact that several special phytochemicals are either destroyed or lost during distillation or very difficult to extract resulting in a low yield of essential oils, infused vegetable oils are indispensable for obtaining such compounds. Therefore, the solvent property of vegetable oils and their applications in conjunction with innovative techniques (e.g. supercritical fluids, ultrasound, etc.) have been explored in extractions of bioactive phytochemicals from natural biosources, which helps one to design greener extraction procedures and novel functional end-products with great potential in food, nutraceutical and cosmetics industries.

Li et al. summarized most works in the past few years that proved the good solvent power of vegetable oils. Vegetable oils enriched or flavoured with antioxidants or aroma extracts from plants or their by-products showed better organoleptic quality and oxidative stability, but the use of organic solvents has caused an increasing concern on their safety in human body contact applications and environment though they have been evaporated afterwards in enriched oils. The virgin vegetable oils have successfully proven their effectiveness in retaining high concentration of carotenoids from shrimp by-products and microalgae under optimal conditions without any loss and degradation of carotenoids or changes in their fatty acid profiles [30]. Inspired by this, Li et al. have developed direct green extractions of a series of bioactive compounds with different polarities, which showed comparable performance to that extracted by petroleum-based solvents [31–33]. Moreover, the integration of innovative techniques could obtain more compact process and final enriched oils without solvent evaporation and separation steps, which offers significant benefits concerning cost, time, energy and environment for potential industrial-scale applications in food, cosmetic and nutraceutical industries. For instance, the beneficial role of adding vegetable oils as co-solvents of supercritical CO_2 extraction has been demonstrated to enhance the carotenoid yield and prevents the degradation of the target carotenoids during extraction [34–36]. Microwave and ultrasound could improve the dissolving power of oils to ensure a good reproducibility of produced extracts with high yields and optimized organoleptic properties, and without lipid degradation and product contaminations [37, 38]. These newly developed techniques have already been put into real production on an industrial scale with the consideration of a range of technoeconomic and environmental factors (Fig. 3.4). Recently, we used vegetable oils as solvents to develop another greener procedure for concrete and absolute production in perfumery, flavour and fragrance fields. The integration of innovative techniques and vegetable oils as alternative solvents could give a more efficient and eco-friendly process for future industrial application.

3.4 Other Bio-Based Solvents and Alternatives

Biomass, as well as its by-/co-products or wastes, is a huge set of resources that can be recycled and transformed to alternative eco-friendly solvents by means of physical, chemical or biological methods. Although these bio-based derivatives have been widely used and studied for organic catalysis and synthesis, such solvents have not been thoroughly investigated for other applications such as natural product extractions [39]. Four representatives of such bio-based solvents will be elaborated in following subsections.

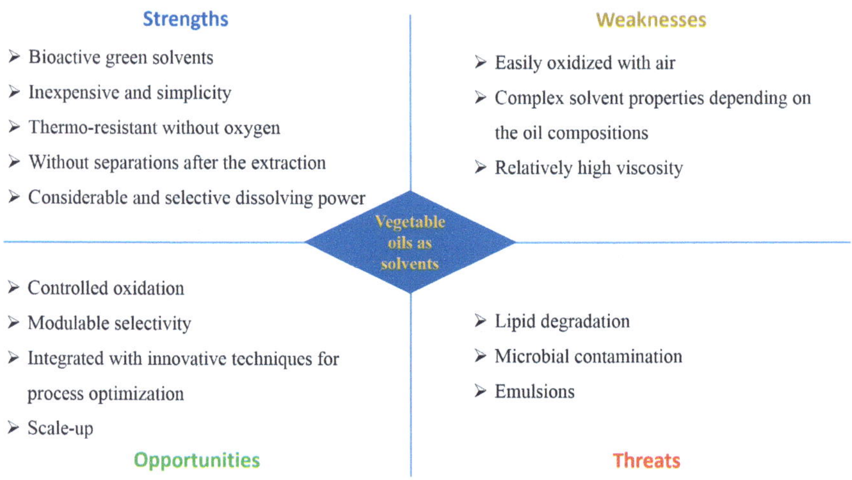

Fig. 3.4 SWOT analysis for extraction using vegetable oils as alternative solvents

3.4.1 Terpenes

Terpenes are one of the typical compounds in essential oils and can be isolated from raw plant materials by physical methods such as hydro-distillation or cold pressing from citrus peel oils. Monoterpenes representing more than 90% of the chemical composition in essential oils have shown their potency of being alternatives to petroleum-based solvents in various industrial applications. The relevant physiochemical properties of terpenes and classical solvents are predicted in Table 3.2, in which all terpenes are similar in molecular weight and chemical structures and boiling points, but different from classic solvents commonly used in extractions. Moreover, terpenes also have similar values of Hansen solubility parameters (HSP), confirming their potential of substituting petroleum-based solvents. Moreover, terpenes have higher dielectric constants, so a higher polarity and dissociating power than *n*-hexane, which is a frequently used solvent in oil extractions. Besides, terpenes with higher flash points are less flammable and less hazardous. The major drawback of using terpenes is their slightly higher viscosity and density as well as the higher energy consumption during recovery owing to their high boiling points. However, the extra energy and cost required could be compensated by the relatively high yield of bioactive compounds extracted, carotenoids in particular. It is worth mentioning that terpenes are generally more stable at higher temperatures than alcohols, which have less environmental impacts. As a consequence of all these properties, terpenes may be considered as promising bio-based alternatives to classic solvents in some cases, provided that a careful investigation of the extraction power and of both economic and environmental aspects is made. This has been experimentally confirmed for various purposeful applications, where successful scaling-ups have been demonstrated leading to greener processes, in which petrol-based solvents could be replaced

(Table 3.3). Moreover, using terpenes as alternative solvents in Dean-Stark distillations has successfully expanded the routine quality control analysis of moisture in food products [54, 55]. In addition, terpenes could be transformed as agrochemical building blocks for the production of other bio-based solvents or value-added compounds with industrial interest [56, 57]. For instance, α/β pinenes or turpentine oils from forestry by-products could be as synthons for pinnae as bio-based solvents through a solvent-free catalytic hydrogenation.

3.4.2 *MeTHF (2-Methyltetrahydrofuran)*

MeTHF is a commercially available solvent obtained by hydrogenation of hemicellulose derivatives from various feedstocks such as corn cobs or sugarcane bagasse (Fig. 3.5). Unlike *d*-limonene which is considered as a generally recognized as safe (GRAS) material by the Food and Drug Administration (FDA), MeTHF has not yet been approved by FDA. Furthermore, it requires special safety measures for explosion and can never replace hexane at the same quantitative level. However, it shows an excellent health safety environment (HSE) profile and has various other benefits, e.g. it has a considerable dissolving power, it is biodegradable, easy to recycle, non-carcinogens, nonhazardous to ozone and air, etc. As a result, MeTHF is not only a good solvent for the design of eco-efficient processes in catalysis and organic chemistry, it has currently also been considered as a promising alternative to petroleum-based organic solvents in the rarely studied field of natural product extraction. Lignin could be directly extracted into MeTHF solvents from pulp and soluble carbohydrates at high temperatures [58]. An efficient and highly selective extraction using MeTHF as the alternative solvent to chloroform has been attempted for membrane protein FhuA, which yielded 70 mg FhuA per litre fermenter broth with a purity of 95% [59]. The solvency of MeTHF has been studied as well, both theoretically and experimentally, for the extraction of aroma compounds in blackcurrant buds and caraway seeds, respectively [52, 53]. The former showed that MeTHF performed the best among all solvents (isopropanol, butanol, ethanol, ethyl lactate, ethyl acetate, α-pinene and supercritical CO_2 tested, whereas the latter showed that other bio-based solvents like ethyl acetate and dimethylcarbonate appeared better than MeTHF to replace *n*-hexane. The relative proportions of total oxygenated (17.06%) and non-oxygenated (42.88%) compounds in final MeTHF extracts is similar to that (25.26 and 47.73%, respectively) in *n*-hexane extracts, whereas a higher extraction yield (7.10%) was obtained by MeTHF than with *n*-hexane (3.87%), probably due to fact that MeTHF allows extraction of other concomitants such as amino acids, flavonoids and phospholipids. In addition, the extraction of carotenoids and rapeseed oils by MeTHF has been achieved in our laboratory as well. The starting accessibility and effective diffusivity for the MeTHF extraction of carotenoids from carrots appeared to be faster than that with *n*-hexane, yielding around 22% more β-carotene than with *n*-hexane in the same time interval [60, 61]. The higher yield was also found in rapeseed oil extraction by MeTHF accompanying with a low residual oil content at

Table 3.2 Physiochemical properties of petroleum and bio-based solvents predicted by HSPiP software

Sources	Biomass			Petroleum		
Solvent	*d*-Limonene	α-Pinene	ρ-Cymene	*n*-Hexane	Toluene	Dichloromethane
Chemical structure	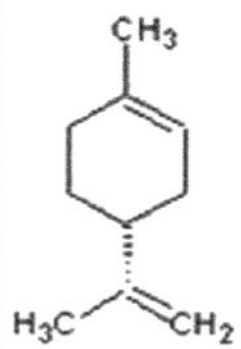	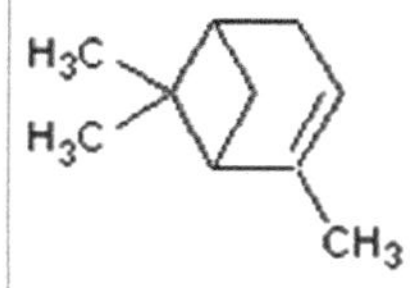	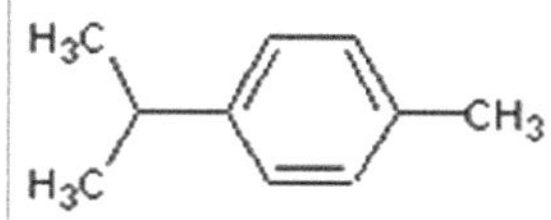		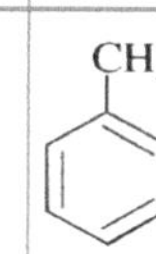	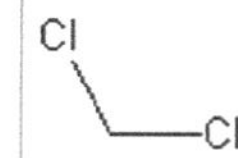
Molecular formula	$C_{10}H_{16}$	$C_{10}H_{16}$	$C_{10}H_{14}$	C_6H_{14}	C_7H_8	CH_2Cl_2
Physiochemical properties						
Molecular weight (g mol^{-1})	136.2	136.2	134.2	86.2	92.1	84.9
Molar volume (cm^3)	161.6	154.3	156.5	127.6	106.1	59.7
Density 25 °C (g cm^{-3})	0.834	0.879	0.861	0.675	0.868	1.325
Flash point (°C)	42.8	32.2	47.2	−23.3	19.9	−14
Boiling point (°C)	176	155	176	69	120.1	40
Viscosity, 25 °C (C_p)	0.83	1.32	0.83	0.31	0.59	0.413
Dielectric constant, 20 °C	2.44	2.58	2.34	1.89	2.38	8.93

(continued)

Table 3.2 (continued)

Sources	Biomass			Petroleum		
Solvent	*d*-Limonene	α-Pinene	ρ-Cymene	*n*-Hexane	Toluene	Dichloromethane
Enthalpy of vapourization (kJ/mol)	39.49	37.83	39.34	28.85	33.5	28.6
Surface tension (dyne/cm)	25.8	25.3	28.5	20.3	28.8	26.52
Log P	4.45	4.37	4.02	3.25	2.39	1.03
Hansen Solubility Parameters (δ_{total}, $MPa^{1/2}$)	17.1	16.6	17.6	15.0	18.4	21.3
Environmental aspects						
Environmental impact	Low			High		
Toxic	No			Yes		
Renewable	Yes			No		

Table 3.3 Terpenes as alternative solvents in natural product extractions

Raw materials	Extracts	Compared solvents	Experimental remarks	Refs.
Rice bran	Lipids	*d*-Limonene and *n*-hexane	The optimized conditions are solid to liquid ratio of 1:5 and 1 h, which gave the comparable results hexane extract in free fatty acid, oil colour and phospholipids	[40]
Rice bran	Lipids	*d*-Limonene and *n*-hexane	The optimized conditions are solid to liquid ratio of 1:5 and 30 min, which gave the best quality with considerable yield	[41]
Oilseeds of peanuts, soya and sunflower, Olives	Lipids	α-Pinene and *n*-hexane	The Soxhlet extraction using α-pinene from renewable materials showed a higher efficiency than conventional extraction using *n*-hexane. High recycling rate of α-pinene let the novel process more eco-friendly and safer	[42]
Aglandau olive	Lipids	*d*-Limonene and *n*-hexane	*d*-Limonene as an alternative solvent to hexane in Soxhlet extraction is more competitive in term of efficiency, energy and time saving	[43]
Aglandau olive seed	Lipids	*d*-Limonene and *n*-hexane	No significant difference in qualitative and quantitative analysis between extracts using both solvents; waste reduction, short operating time and energy saving have been achieved thanks to the microwave and Clevenger	[44]

(continued)

Table 3.3 (continued)

Raw materials	Extracts	Compared solvents	Experimental remarks	Refs.
Rapeseeds	Lipids	*d*-Limonene, α-pinene, *p*-cymene, *n*-hexane, ethanol, isopropanol and *n*-butanol	Terpene solvents are better alternatives than alcoholic solvents for *n*-hexane substitution, among which *p*-cymene is the most promising solvent for oil extraction in term of yield and selectivity	[45]
Dried microalgae (*Chlorella vulgaris*)	Lipids	*d*-Limonene, α-pinene, *p*-cymene and *n*-hexane	No significant difference in both qualitative and quantitative analysis between extracts using different solvents	[46]
Wet microalgae (*Nanochloropsis oculata*, *Dunaliella salina*)	Lipids	*d*-Limonene, α-pinene, *p*-cymene and *n*-hexane	No significant differences in terms of distribution of lipid classes and fatty acid composition have been obtained for different extraction techniques; Substantial saving in the cost and energy	[47]
Microalge (*Spirulina*, *Phormidium*, *Anabaena* and *Stigeoclonium*)	Lipids	Mixture of *d*-limonene and ethanol	The optimized condition are the mixture of limonene and ethanol (1:1, v/v) as solvent, 200 °C, 20.7 MPa and 15 min, which gave best yield for *Spirulina* and the highest amount of ω-3 fatty acids for *Stigeoclonium*	[48]
Rapeseed	Lipids	*d*-limonene, α-pinene, *p*-cymene, *n*-hexane, MeTHF, DMC, IPA, ethanol and ethyl acetate	Terpene solvents significantly gave a worst yield among all solvents though their theoretical solvation abilities looked better than hexane for the extraction of most components in rapeseed oils	[49]

(continued)

Table 3.3 (continued)

Raw materials	Extracts	Compared solvents	Experimental remarks	Refs.
Tomato fruits	Lycopene	*d*-limonene and dichloromethane	No significant difference in both qualitative and quantitative analysis between two extracts using different solvents	[50]
Microalgae (*Neochloris oleoabundans*)	Carotenoids	Food-grade d-limonene and ethanol	The predicted optimum conditions for pressurized liquid extraction were 112 °C and 100% ethanol as the solvent	[51]
Blackcurrant buds (*Ribes nigrum* L.)	Aroma	α-pinene, MeTHF, ethyl acetate, methyl acetate, ethyl lactate, butanol, isopropanol, ethanol, supercritical CO_2 and *n*-hexane	The simulation of Hansen solubility parameters described α-pinene as a good alternative for aroma extraction, but the flavour profile is experimentally very different from that of the hexane extract, which may be due to the difficulty in solvent evaporation	[52]
Caraway seeds	Aroma	α-pinene, MeTHF, ethyl acetate, ethyl lactate, butanol, isopropanol, ethanol, dimethyl carbonate and *n*-hexane	Alpha pinene appears as alternative solvent and most promising for limonene extraction, which was contrary to experimental results	[53]

the pilot scale. Using MeTHF could accelerate the lipid extraction with a factor of 3.5 compared to *n*-hexane, which has a positive impact from economic and energy aspects [62]. All these above further proved the potency of MeTHF to be a promising alternative solvent to *n*-hexane for natural product extractions.

3.4.3 Natural Deep Eutectic Solvents (NADES)

NADES have been discovered and explored as green and biogenetic alternatives to ILs and deep eutectic solvents (DESs) in the extraction of secondary metabo-

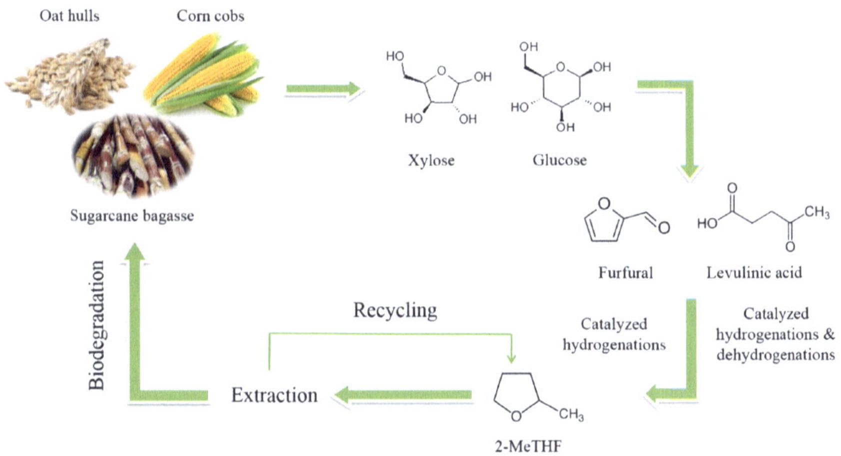

Fig. 3.5 Life cycle of 2-MeTHF as the solvent in natural product extractions

lites from natural plants, which are composed of primary metabolites in living cells with adequate molar ratios [63]. NADES are cheap, safe and sustainable liquids even below 0 °C with a tunable viscosity, in which the individual components (e.g. sugars, amino acids, etc.) are easily available in our daily diet. Most interestingly, they have high dissolving power for a wide range of solutes ranging from polar to non-polar because of their supramolecular structure and broad polarity range. Macromolecules such as DNA, proteins and polysaccharides are also soluble in NADES. In recent years, NADES as green solvents have been studied with a fast-growing trend in terms of their efficiency, optimization and recovery from their extracts [64]. The higher extractability of phenolic compounds was proved in certain NADES under optimized conditions with an efficient recovery and higher stabilizing ability compared to that in conventional organic solvents [65, 66]. Compared to water or 40% ethanol solution, natural pigments from safflower were more stable in sugar-based NADES and the stability of carthamin in NADES under various conditions (e.g. high temperature, light and storage time) could be improved [67]. It is noted that the water content in NADES proved to have the greatest influence on the extraction efficiency, which determined their physicochemical and solvent properties [68]. The high extractability and efficiency with NADES may be attributed to not only H-bonding interactions between molecules of NADES and bioactive compounds, but also their viscosity rather than polarity, which should be kept in mind when designing NADES applications. H-bonding interactions will be weakened with water dilution and even disappear as the water content increased. The physicochemical properties like viscosity can be reduced significantly with the addition of small amounts of water, which can drop down almost to that of water with the addition of 25% water (v/v) [69]. Notably, the stabilization capacity of NADES can be adjusted by decreasing water content with increasing viscosity. Therefore, the intentional adjustment of water content helps to develop tailor-made NADES with controllable structure and characteristics for various applications in future. Apart from fine-tuning the physicochemical characteristics of solvents, NADES could also be used to enhance the biological activities of plant extracts, such as antioxidant and antiproliferative activity of phenolic grape skin extracts using choline-chloride-based NADES containing malic acid [70].

The existence of NADES in plant cells plays a role in protecting organisms from extreme conditions, which indicates that NADES of biological origin might be solvents for various water- and non-water-soluble metabolites. Moreover, the proved desirable properties of NADES, such as sustainability, biodegradability, acceptable toxicity and high extractability of compounds with diverse polarity, make them suitable and attractive for food, cosmetic, agrochemical and pharmaceutical applications. Since NADES have been successfully used in certain phenolic extracts at laboratory scale with good compatibility with green analytical chemistry, the industrial application can be expected soon. Nevertheless, NADES in extractions of other natural compounds need to be further investigated and comparatively large number of eluants used and their post-processing may be a drawback for scaling up. Furthermore, questionable controversies arise [71], i.e. NADES is appropriate for a "real" DESs mixed with/solubilized in water or "natural hydrotropic mixtures"? NADES are not real "natural" due to their chemically synthesized components in most real cases. "Natural" *versus* "bio-based", etc.

3.4.4 Bio-Ethanol

Finally, we consider bio-ethanol, a readily available, clean fuel made for combustion engines, which can be produced from renewable feedstocks (e.g. corn, wheat, waste straw, sawdust, grass, etc.) by hydrolysis and sugar fermentation processes by microorganisms. Due to its origin, it produces considerably lower greenhouse gas emissions during combustion than conventional fuels do. Moreover, it is also biodegradable and far less toxic than fossil fuel and petroleum-based solvents. Although the production of bio-ethanol has been extensively studied, this non-synthetic ethanol has recently been explored to replace synthetic ethanol as the substituted solvent in several types of laboratory applications as well as industrial manufacturing, which can preserve efficacy of function while reducing the impact on environment [72–75]. Since ethanol is a well-known solvent in plant extractions, bio-ethanol can be immediately used for these processes.

3.5 Computer-Aided Tools for Solvent Evaluation

Regarding the increasing number of novel sustainable solvents discovered, how to evaluate them in an efficient and accurate way instead of the classic "trial and error" method before scaling up to real industrial practice? Although there are by far multiple choices including popular and niche computer-aided tools for solvent evaluation, several pragmatic ones are here recommended.

3.5.1 *In Silico Solvent Screening*

Usually and even today, the choice of the most appropriate solvent for plant extraction is still made empirically or even intuitively. This is a typical sign of the lack of scientific basis in this field. Today, powerful tools exist to characterize the properties of solvents and their suitability for a particular process. Prior to getting down to business, in silico design for bio-based solvents like glycerol and itaconic acid based solvents from Prof. Jean-Marie Aubry's team should be highlighted using different computer programs such as GeneratoR of Agro-based Sustainable Solvents (GRASS), InBioSynSolv (IBSS), Hansen Solubility Parameters in Practice (HSPiP) and COnductor-like Screening MOdel for Real Solvents (COSMO-RS) [76–79], where the latter two allowing for a rapid screening and thus avoiding time-consuming experiments will be shortly presented in what follows. The first one is the old and widespread concept, known as Hildebrand and later on Hansen solubility parameter model, the other one is the modern and much more powerful COSMO-RS approach. Finally, we will mention a more general mathematical tool—the principle component analysis (PCA)—that is useful for data reduction and solvent classification.

3.5.1.1 Hansen Solubility Parameters (HSP)

The well-known "like dissolves like" rule of chemistry intrigued that solvents can be considered as good solvents for corresponding solutes if the intermolecular forces between solvent and solute molecules with similar strength. The "Hansen solubility parameters" is an important concept which is still widely used for rational selection of appropriate solvents for given solutes through theoretical descriptors based on a given experimentally obtained data set. It is based on the aforementioned classic rule, and it derives from the theory of Hildebrand's total solubility parameter which presents a limited predictive value. Only one parameter describes a solvent in Hildebrand's theory, which in most cases is obviously not sufficient for a proper description of solubilities.

Hansen divided the total cohesion energy of a solvent, E, as defined by Hildebrand, into three separate parts. These three kinds of energies are attributed to dispersion (atomic), E_{D}, polar (molecule) E_{P} and hydrogen bonding (molecule), E_{H}, interactions, see Eq. (3.1).

$$E = E_{\mathrm{D}} + E_{\mathrm{P}} + E_{\mathrm{H}} \tag{3.1}$$

E is the required energy of liquid evaporization which is necessary to break all of the cohesion bonds in the liquid. D indicates dispersion forces which are the general van der Waals interactions between the atoms of the two molecules, P stands for polar forces which are the electrical attractions arising from dipole moments. H is a type of polar force which can generally be considered as a form of electron exchange and in most cases describes the power to form hydration bonds. E can be experimentally measured and is given by Eq. (3.2).

$$E = \Delta H - RT \tag{3.2}$$

where ΔH is the measured or predicted latent heat of vapourization, R is the universal gas constant and T is the absolute temperature.

The Hildebrand total solubility parameter (δ) equals the square root of total cohesion energy density, which can be obtained by dividing the total cohesion energy (E) by the molar volume (V) of the solvent or solute. Similarly, the respective Hansen solubility parameters (HSP) can be determined experimentally and then calculated by dividing by the molar volume. Thus δ_{total}, the Hansen total solubility parameter, which now consists of three HSPs in terms of dispersion (δ_{d}), polar (δ_{p}) and hydrogen-bonding (δ_{h}) interaction energies, is equivalent to the Hildebrand total solubility parameter. The unit of HSPs is $\text{MPa}^{1/2}$ or $(\text{J/cm}^3)^{1/2}$ in the SI system of units, which can be converted to the frequently used unit in older papers, $(\text{cal/cm}^3)^{1/2}$, by multiplying by a factor of 2.046. HSPs can be simply determined experimentally for solvents, polymers, C60 fullerene, carbon dioxide, DNA, and almost any other chemical component [80], whereas their determination for ILs is much subtler.

Solvent solubility can be readily visualized through a three-dimensional graph with axes δ_{d}, δ_{p} and δ_{h}, which can be easily made using HSP in Practice (HSPiP) programming (Fig. 3.6). This program proposes also a sphere around the parameters of a considered solute, based on experimental determinations, with the parameters of potentially good solvents inside and bad solvents outside. The radius of the sphere (R_{o}) allows us to know the probable maximum difference in the parameters of the solute and of different solvents so as to find the ideal solvent which is the closest to the sphere centre. To perform the HSP solvent optimization, a simple composite affinity parameter, the relative energy difference (RED) number, has been introduced to discriminate good and inappropriate solvents.

Since 3D graphs are difficult to understand when they are shown statically in documents, 2D graphs have been commonly used in case of diminishing the significance of one HSP that is very close among various solvents. Most people associate HSP plots with δ_{p} versus δ_{h} graphs as an instance. The HSPiP software offers four group contribution methods to determine the HSP values, which fragment the chemical structure of a compound according to its functional groups. However, it is common to find varying HSP values for the same molecule due to the different algorithms in light of each method. Therefore, the method involved in HSP prediction should be specified before the solvent selection. The Stefani-Panayiotou method breaks down a molecule into first order (molecular structure of the compound) and second

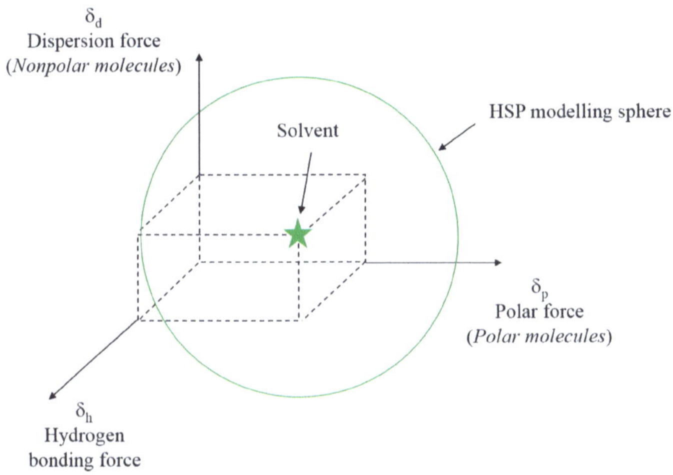

Fig. 3.6 Representation of a solvent centred in the modelling of a Hansen solubility parameter (HSP) sphere

order (conjugation system) functional groups which have their own parameters [81]. However, this method is only suitable for molecules with a simple structure. Nevertheless, this method provides partial solubility parameters (PSP) concerning the molecule polarity and is able to calculate solubility parameters for many molecules in a relatively easy way by coupling to the COSMOtherm software, see next subsection. The particular contribution of the Van Krevelen method is to introduce a "symmetry" option, which allows one to calculate HSPs corresponding to the plane of symmetry in molecules. The Hoy method includes options similar to the secondary group in the Stefani-Panayiotou method with the consideration of various x-membered rings and some isomeric forms. It also helps with molecular input to the numerical and Van Krevelen calculations by producing an approximate value for the molar volume. However, none of above-described group methods could give a satisfying HSP prediction for complex inter-group interactions. By now, only the Yamamoto-Molecule Breaking (Y-MB) method provides an automatic way of creating groups (first group only) according to the simplified molecular input line entry system (SMILES) or international chemical identifier (InChI) of individual molecules. In fact, it allows the automatic molecule breaking that creates subgroups from any molecules. This program combines full HSP data with neutral network methodology in such a way that the interaction between groups automatically gets fitted by the relative strengths of the neural interconnections so as to give the best predictive power for HSP.

Although the HSP database has been continuously updated to its newest versions, the predictions for complex solvents with large molecular weights are still unsatisfactory. Besides the HSP values, Y-MB can in principle also estimate many other important physiochemical parameters such as melting point, boiling point, vapour pressure, density, viscosity and HSE values. So, it seems that using Y-MB for HSP predictions of solubility in non-conventional solvents is probably the best way in the framework of this concept. This method has been used to predict HSPs as descriptors for an efficient and appropriate selection of green alternative solvents from a large solvent database with statistical methods [82]. HSPiP has also been used in the theoretical solvent screening for artemisinin solubility improvement, which isopropanol was found the best substitute of *n*-hexane with 65% of yield and >67% of purity for the potential large-scale crystallization [83]. Furthermore, safer bio-based alternatives to regulated solvents in amidation reactions were also successfully modelled by coupling HSPiP-to-Yalkowsky approximation, which led to the best performance of *p*-cymene that could be set up in a continuous flow reactor for several direct amidations [84]. However, as showed and discussed in other papers [85, 86], the HSP approach is by its nature a very crude and approximate method on a case-by-case basis. It should be improved or better to combined, whenever possible, with a more physical one, such as the COSMO-RS model, discussed in the next subsection.

3.5.1.2 COnductor-like Screening MOdel for Real Solvents (COSMO-RS)

COSMO-RS is one of the most powerful solvation models based on monomolecular quantum chemical computations that generate a priori descriptors deriving from molecular structure information [87]. The COSMO-RS allows the transfer from the molecular state of solutes that is considered to be embedded in a cavity with a virtual surrounding conductor to real solvents (Fig. 3.7). The generated σ-surface represents the repartition of charge density on the molecular surface (i.e. green-to-yellow: weakly polar surfaces, blue: electron-deficient regions and red: electron-rich regions), which can be reduced to a histogram called σ-profile. The σ-profiles and σ-potentials in the COSMO-RS framework, as well as other thermodynamic parameters, can help pre-screening and classifying various solvent candidates with the help of multivariate statistical analyses so as to select appropriate solvents for different applications [76]. The coupling approach of HSP and COSMO-RS for *n*-hexane substitution showed that the distribution of *n*-hexane and its close-by alternatives in the HSP 2D space is in good accordance with the close positioning in the COSMO-RS diagram [88]. Compared to conventional group contribution methods, this modern approach can well distinguish between stereoisomers though their real molecular structures. Moreover, it can properly describe interactions of complex fluid systems at a desired temperature or pressure, including multiphase mixture and supramolecules. The performance of both COSMO-RS and HSP predictions were compared by the solubilization investigation of avobenzone in cosmetic solvents. The COSMO-RS values of good solvents such as dimethyl isosorbide gave more reliable and accurate solubility values than those predicted by the HSP approach [86].

Recently, Prof. Farid Chemat' GREEN team has conducted a number of studies on bio-based solvents for green extraction of natural products using COMSO-RS along with HSPiP. For alternatives to *n*-hexane, the extraction of food aromas from caraway seeds firstly showed that the similarity of σ-potential between solvents and extracts

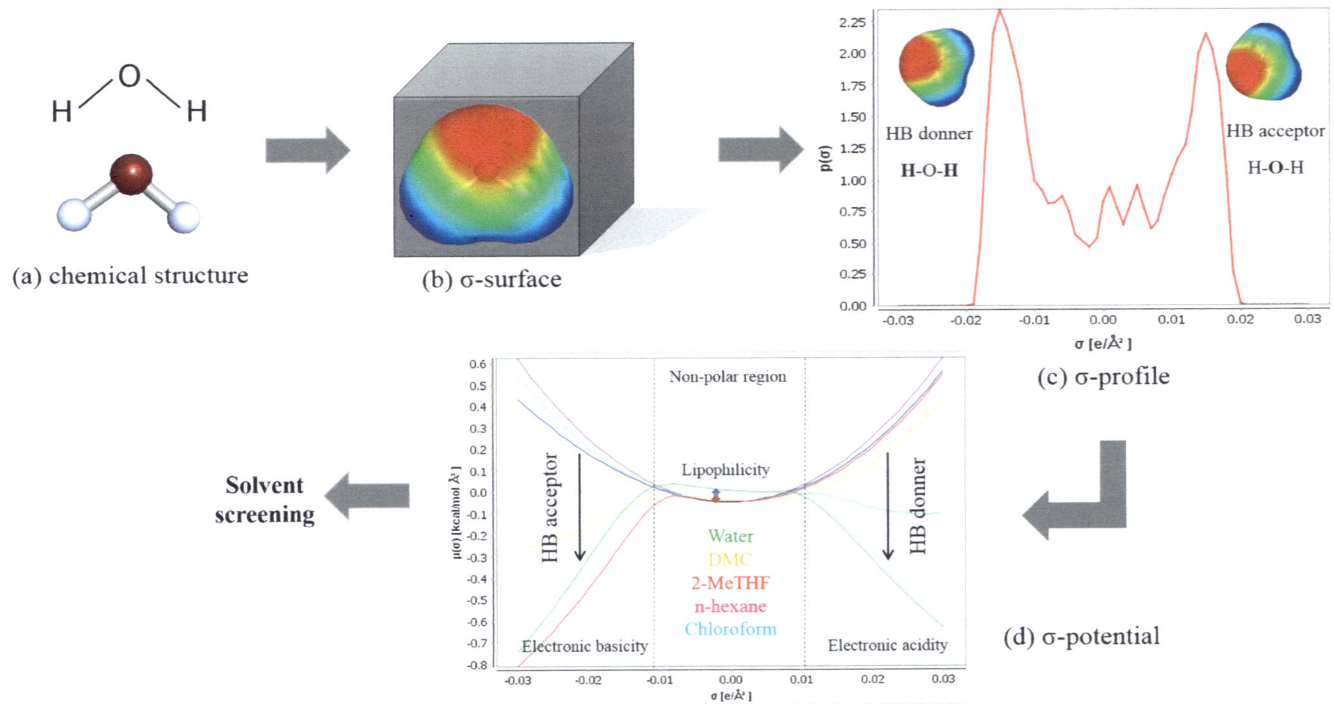

Fig. 3.7 Schematic diagram of COSMO-RS methods: the transformation of the **a** 2D molecular structure with **b** its σ-surfaces followed by the conversion into its **c** σ-profile and **d** σ-potentials of five representative solvents: non-polar (n-hexane), highly polar HB acceptor and donor (water), HB acceptor (2-Methyltetrahydrofuran, Dimethyl carbonate), and HB donor (chloroform)

was compared at the boiling point of each solvent. The values predicted by COSMO-RS for carvone were more or less consistent with the experimental values, whereas there was a discrepancy for limonene [53]. Further, the consistency of COSMO-RS and experimental results demonstrated that a higher solubility for natural carotenoids in bio-based solvents as compared to *n*-hexane with the highest carotenoid content observed in cyclopentyl methyl ether (CPME) where 66% was represented by β-carotene and 34% was α-carotene [61]. COSMO-RS has also succeeded in applying for lipid extraction from rapeseed, salmon fish and *Yarrowia lipolytica* yeast as well, which helped for a better comprehension of dissolving mechanisms towards substitution with bio-based solvents [49, 89, 90]. Similarly, in cooperation with CELSIUS company, a substantial progress has made in using liquefied gases as alternative solvents for extracting various natural products with the help of COSMO-RS modelling [12, 91, 92], where liquefied gases showed good consistency in both theoretical and experimental results and desirable "eco-footprint" based on the principles of green extraction. Besides this, COSMO-RS application for the extraction of other bioactive compounds from indigenous plants has been successfully expanded as well with some international partners [93, 94]. All above confirm that COSMO-RS could provide the theoretical basis for real practices, which facilitates the way of selecting bio-based alternatives to solvents of petroleum origin for developing a green strategy for natural product extractions and their industrialization.

3.5.1.3 Multivariate Statistical Analysis

Considering a database consisting of various solvents as individuals described by several characteristics as variables, multivariate statistical analyses are usually applied for solvent comparison in order to classify and to select the optimal solvent in a visualized and convenient way. The purpose of the principal component analysis (PCA) introduced by Pearson and developed by Hotelling is to reduce the dimension of the database through condensing the original variables into principle components while retaining the information better with a viewable representation. The principal components are linear combinations of the original variables. They are ordered and hierarchized in terms of the percentage of variance contributed by each principle component. The first principle component, specifically, the axis of greatest inertia of the scatter plot, contains information with the maximum percentage of variance. The second principle component, an axis perpendicular to the former which reflects the second largest inertia, contains information relating to the following percentage of variance. The rest principle components can be determined by such analogy until the cumulative variability of principle components is 100%. Generally, the PCA provides a faithful representation for above 90% of the original information by corresponding principle components on which individuals are projected orthogonally. It is possible to directly visualize the similarities or dissimilarities between individuals and to evaluate the position of their representative point graphically. The PCA is a well-known statistical analysis method for analysing the structure of multidimensional databases.

Hierarchical clustering is useful to be considered after PCA application on a dataset, which can bring together all individuals observed in groups of solvents having similar properties [95]. This is a method of iterative classification, which considers each individual to start as a single group. Firstly, the dissimilarity is calculated between all individuals studied, e.g. using the Euclidean distance to measure how individuals are dissimilar. Subsequently, the two objects whose combination minimizes a given aggregation criterion, are combined in the same group. Once it is achieved, this operation is repeated by calculating the dissimilarity between the formed group and the remaining individuals. The objects or formed groups continue to group according to the same aggregation criterion until the total regrouping of all individuals is completed, which can be represented by a classification binary tree called dendrogram that allows one to view the hierarchy of partitions. There are several aggregation criteria for regrouping individuals, in which the best known one is proposed by Ward [96]. This involves an aggregation of two groups finally so that the classification must be as compact as possible.

Let us look through several cases. The above-described multivariate statistical methods partitioned the green solvents into clusters together with other toxic solvents, which suggests favourable substitutions in potential applications [97]. The physicochemical properties of potentially green solvents were determined as variables in PCA, which can easily provide a visualized three-dimensional space with 90.9% of cumulative variability. All solvents and physiochemical properties distributing in the PCA space could be well discriminated by the similarity of each physiochemical property so that an efficient solvent screening could be achieved. Agglomerative hierarchical clustering (AHC) has been performed afterwards to subdivide all solvents into clusters. In general, the AHC results are in good accordance with PCA observations thus confirming them. Furthermore, a pseudo-3D-representation of potential solvents modelled by COSMO-RS was also illustrated by PCA, where the values of σ-potential predicted were successfully used as variables in PCA to classify classic organic solvents into ten families in which the green solvents have been positioned [76]. Last but not least, PCA and AHC can be also applied to investigate the dissolving power or sensorial quality within the same solvent family. The different types of vegetable oils were grouped into three classes according to their real experimental dissolving capacity for six main volatile aroma compounds as variables in PCA plotting. This was further confirmed by the AHC dendrogram where the dissimilarity comparison could help to determine the optimal solvent based on the overall performance [32]. Similarly, the difference of obscure olfactory perception between extracts by conventional hexane and its alternatives could be processed and discriminated in a representational way thanks to PCA and AHC [12].

3.6 Life-Cycle Assessment (LCA)

Since the alternative solvents discussed in the preceding sections gradually come to the pilot or even industrial scale, the whole process including several green chemistry metrics (e.g. atom economy, environmental factor, transport costs, etc.) must be considered instead of their solubilization and extraction power only. Therefore, a life-cycle assessment should be sequentially conducted mainly in terms of manufacture, distribution, use and disposal, which are primary life-cycle stages in solvent usages [98]. Most alternative solvents can be produced from renewable resources, in which water and CO_2 require purification and condensation rather than manufacture. Since ILs, especially the popular imidazolium-based ones, require multistage syntheses, this major disadvantage must be carefully compared to their benefits. It can be expected that these ILs are then rapidly excluded from any further use in the context of green processes and in particular for plant extraction. Moreover, it should be paying attention that in many cases some alternative solvents are chemically synthesized or energy-consuming recovery even though they are considered "natural" or "green", such as choline in NADES or terpenes with high boiling points [99]. Distribution after manufacture is an often-overlooked area that may induce cost and environmental concern due to long-distance transportation, thereby a local supplier is advised when choosing the solvents. As previously discussed about the properties and evaluations of all alternative solvents, appropriate solvent selections for specific applications must include the issues of recovery, reuse and recycling. However, one should keep in mind that these pros and cons are quite general and valid for any solvent in its corresponding application rather than other applications. So, the best solvent is still simply no solvent and consequently solvent-free processes are the "greenest" processes, at least in this respect [100]. Finally, compared to the common incineration of volatile organic solvent, the disposal route for most bio-based alternative solvents seems to be easier thanks to their low environmental impact.

We should also mention that even water is not a perfectly green extraction solvent in all respects. Especially, when it must be distilled, the energy consumption is very high leading to a negative impact on the environment. Let us consider in more details the particular, but important case of the general LCA of MeTHF and *n*-hexane as an example. MeTHF can be manufactured from renewable resources so that it should be considered as a possible alternative solvent to *n*-hexane, which is usually produced from petrochemical feedstocks. Provided that they are both supplied locally for natural product extractions, MeTHF is more stable but less volatile than *n*-hexane. They have nearly the same molar mass but different physical characteristics. MeTHF has a slightly higher boiling point (80.3 °C versus 69.1 °C for *n*-hexane) that leads to slightly higher energy consumption (0.126 kWh versus 0.121 kWh for *n*-hexane) and CO_2 emissions (101 g versus 96 g for *n*-hexane) for evaporation of 1 kg solvent. Nonetheless, for steam production, the amount of steam (0.379 kg/kg MeTHF versus 0.521 kg/kg *n*-hexane) leads to a less energy requirement for MeTHF (0.271 kWh versus 0.373 kWh for *n*-hexane) corresponding to a lower CO_2 emission (217 g for MeTHF versus 298 g for *n*-hexane). Besides, MeTHF shows a better HSE profile like lower acute and long-term toxicity (CMR) compared to *n*-hexane. From the above, we can broadly conclude that MeTHF is a good candidate to replace *n*-hexane as the extraction solvent when regarding the whole cradle-to-grave process.

3.7 Polar Paradox Theory

The polar paradox begins as a hypothesis described by Porter et al., which illustrates the paradoxical behaviour of antioxidants in different media and rationalizes the fact that polar or amphiphilic primary antioxidants with high HLB values are more active in non-polar media such as oils with a low surface/volume (LSV) ratio, whereas non-polar or amphiphilic antioxidants with low HLB values tend to be more active in more polar media such as oil-in-water emulsions with a high surface/volume (HSV) ratio or in polar lipids [99]. This empirical observation turned into a putative theory when it was found that the partition of the antioxidants in the medium as well as their interfacial behaviour are crucial for their effectiveness in addition to their inherent potency and the previously believed assumption of oxidation in vegetable oils as homogeneous media was questioned. In fact, oils contain always and in particular after partial oxidation different types of colloidal associations which are considered now as the sites where oxidation is initiated and propagate into the bulk of the vegetable oils. The polar antioxidants are preferentially partitioned into the oil–water interface (oils always contain a tiny amount of water in equilibrium and this water is supposed to be trapped into the colloidal associates) and also to the oil–air interface, hence are more effective in oxidative inhibition than non-polar ones that are preferentially partitioned into the lipid bulk phase (Fig. 3.8). This was supported by the fact that the interfacial tension could be decreased by polar antioxidants, whereas a comparable decrease was not observed for the surface tension [18].

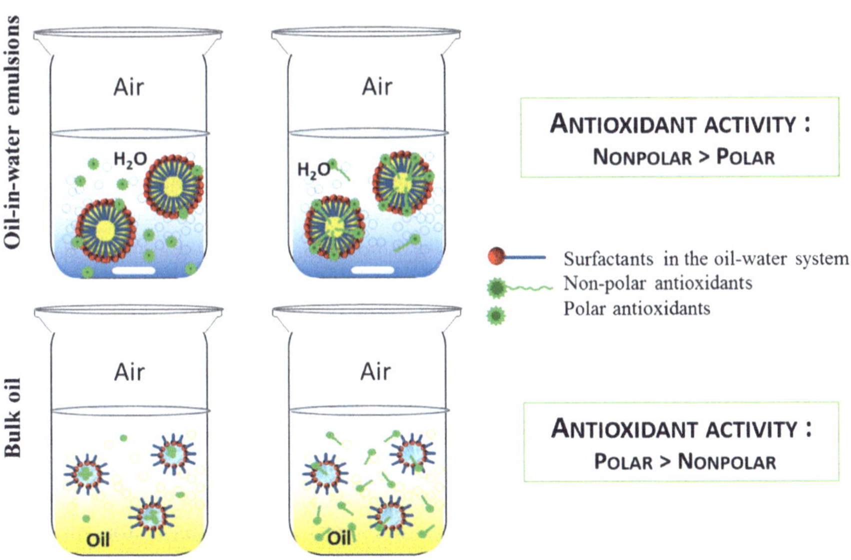

Fig. 3.8 Putative theory of polar paradox in foods

To resume the polar paradox theory, we can simply state that oil-in-water emulsions are better protected against oxidation by non-polar antioxidants than by polar antioxidants. The effectiveness of non-polar antioxidants in oil–water mainly depends on their affinity to the oil–water interface. Non-polar or amphiphilic antioxidants with a low HLB are mainly concentrated at the oil–water interface forming a protective membrane around the lipid droplet while polar antioxidants with high HLB are substantially dissolved in the aqueous phase. The amount and the type (cationic, anionic or non-ionic) of emulsifier in a heterogeneous system can dramatically change the relationship between the polarity and the effectiveness of antioxidants described by the polar paradox theory. For example, non-polar antioxidants can be transported by the emulsifiers to the polar phase, where they are useless or they can be entrapped into the interface.

In the case of vegetable oils, the dissolving mechanism underpinning the seemingly paradoxical extraction of hydrophilic compounds using hydrophobic solvents can be well interpreted by the polar paradox theory. Although the effectiveness of polar antioxidants in lipid medium is not determined, the colloidal structure where antioxidants concentrate has been confirmed by small angle X-ray scattering (SAXS) so as to further support the polar paradox theory (Fig. 3.9). Moreover, highest extractability was obtained in the extraction of natural phenolic compounds with the help of surfactant (soja lecithin) using refined sunflower oils as lipidic media. The structural transformation from anisotropic (elongated, rods or ellipse) to spherical aggregates was discovered by means of SAXS before and after the extraction, which indicates the formation of swollen reverse micelles with water droplets of at least 10 nm dispersed in the oil phase [33]. This observed shape transition of reverse micelles is presumably controlled by the presence of the extracted phenolic compounds that participated in the formation of the aggregates as co-surfactant.

Of course, polarity or HLB values only are not sufficient to predict the effectiveness of an antioxidant. Another obvious important parameter is its concentration. Non-polar antioxidants are more efficient in vegetable oils at low concentrations, while the opposite effect was observed at high concentrations. It turned out that the polar paradox theory can only be applied beyond a certain concentration threshold. The antioxidant activity increases as a function of the antioxidant concentration to a value where the optimal activity is achieved. After this critical value, the activity decreases with further increasing concentration. Below the critical concentration, the non-polar antioxidants are more effective than the polar antioxidants because the solubility outweighs the interfacial phenomena. The reversed trend is found above the value of the critical concentration. This nonlinear relationship (cut-off effect) between polarity and efficacy of antioxidants in multiphase system can be probably explained by their location, distribution and mobility, which are influenced by both polarity and molecular size (e.g. chain length) of the antioxidants [101–104]. In summary, the polar paradox theory is a valuable first approach, but other factors must not be neglected for a proper description or even prediction of antioxidant effectiveness. Even though this theory is not an evaluation tool for solvents, a revisit of it put forward new concepts, hypotheses and theories, which indeed help us to develop new extraction strategies (including solvents to used, antioxidant efficiency, etc.) through

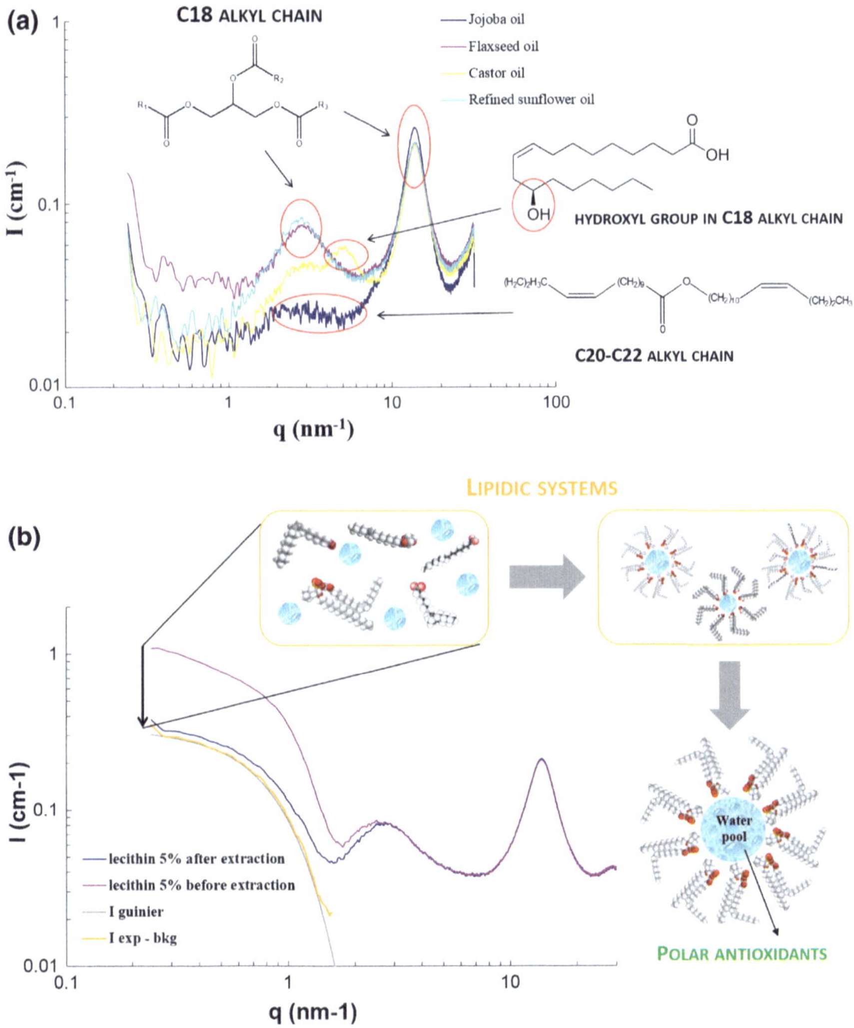

Fig. 3.9 Small angle X-ray scattering spectra of oil structures: **a** jojoba, castor, flaxseed and refined sunflower oil; **b** the structural transition during the extraction using refined sunflower oils and lecithin (5% w/w); I guinier: extrapolated scattering intensity using the Guinier's model for validation of non-interacting spherical aggregates in the oil sample with 5% of lecithin addition after extraction; I exp-bkg: experimental scattered intensity for the oil sample with 5% of lecithin addition after extraction

a better understanding of the underlying solubilization mechanism, in particular in coacervate-based extraction systems using supramolecular bio-based solvents [105].

3.8 Conclusions and Perspectives

The current knowledge of bio-based alternative solvents represents a promising basis for the future development of novel and improved technologies for natural product extractions and other applications. Current extraction techniques based on the above-mentioned bio-based solvents have been successfully performed mainly on a laboratory scale waiting now for their industrial scaling up. But also, on the laboratory scale, more efforts must be made towards the design and development of more bio-based solvents that meet the gradually more and more stringent standards or regulations. This means also that a solid database of the bio-based solvents' properties and for their usage for target extracts and plants is desirable together with information of costs and economical and ecological constraints. It is evident that each target extract and maybe even each plant from which it is obtained is a special case. But this does not mean that for each case empirical trial and error studies are the best or only approach. It is the duty of the scientists to find and to propose general rules that make the proper choice easier and less time- and cost-consuming. This scientific approach, however, is still at the very beginning, as far as plant extraction is concerned.

Although the property and adoption of bio-based solvents for chemical processes gain incremental acceptance, the replacement of petroleum-based solvents by them still needs a relatively long transitional period due to the balance of environmental sustainability and economic pressures. Once we choose bio-based solvents, the important point to think about is whether the solvent options are truly bio-based alternatives. In general, most industries still prefer economics to environmental factors now but reckoning without the hidden economic burden resulting from long-term environmental problems, which may be weighed against the gains. Therefore, the systematic development of application-oriented bio-based solvents for distinct processes involving LCA is of the highest priority. Furthermore, the combined use of computer-aided programs may have a synergistic effect on the design of tailor-made bio-based solvents for target molecules from natural biomass. In addition, the ethos of neoteric solvent forms may be emerged soon such as binary or ternary mixtures, switchable solvents, or even endogenous water and exudates of plants, which may be competitors or solubilizers of bio-based solvents.

References

1. Erythropel HC, Zimmerman JB, de Winter TM et al (2018) The Green ChemisTREE: 20 years after taking root with the 12 principles. Green Chem 20:1929–1961
2. Anastas PT, Zimmerman JB (2003) Design through the 12 principles of green engineering. Environ Sci Technol 37:94A–101A
3. Chemat F, Abert-Vian M, Cravotto G (2012) Green extraction of natural products: concept and principles. Int J Mol Sci 13:8615–8627
4. Capello C, Fischer U, Hungerbühler K (2007) What is a green solvent? A comprehensive framework for the environmental assessment of solvents. Green Chem 9:927–934
5. Kerton FM (2009) Renewable solvents. In: Kerton FM (ed) Alternative solvents for green chemistry. Royal Society of Chemistry, Cambridge, pp 97–116
6. Srinivas K, King JW, Monrad JK, Howard LR, Hansen CM (2009) Optimization of subcritical fluid extraction of bioactive compounds using Hansen solubility parameters. J Food Sci 74:E342–E354
7. Barbero GF, Palma M, Barroso CG (2006) Pressurized liquid extraction of capsaicinoids from peppers. J Agric Food Chem 54:3231–3236
8. Benito-Román Ó, Alonso E, Cocero MJ (2013) Pressurized hot water extraction of β-glucans from waxy barley. J Supercrit Fluid 73:120–125
9. Cacace JE, Mazza GJ (2006) Pressurized low polarity water extraction of lignans from whole flaxseed. J Food Eng 77:1087–1095
10. Gil-Ramírez A, Mendiola JA, Arranz E et al (2012) Highly isoxanthohumol enriched hop extract obtained by pressurized hot water extraction (PHWE). Chemical and functional characterization. Innovat Food Sci Emerg Tech 16:54–60
11. Deng C, Yao N, Wang A, Zhang X (2005) Determination of essential oil in a traditional Chinese medicine, Fructus amomi by pressurized hot water extraction followed by liquid-phase microextraction and gas chromatography–mass spectrometry. Anal Chim Acta 536:237–244
12. Rapinel V, Santerre C, Hanaei F et al (2018) Potentialities of using liquefied gases as alternative solvents to substitute hexane for the extraction of aromas from fresh and dry natural products. C R Chim 21:590–605
13. McClements DJ (2004) Food emulsions: Principles, practice and techniques, 2nd edn. CRC Press, Florida, USA
14. Gan HL, Che Man YB, Tan CP, Noraini I, Nazimah SAH (2005) Characterisation of vegetable oils by surface acoustic wave sensing electronic nose. Food Chem 89:507–518
15. Tan CP, Che Man YB, Selamat J, Yusoff MSA (2002) Comparative studies of oxidative stability of edible oils by differential scanning calorimetry and oxidative stability index methods. Food Chem 76:385–389
16. Hendriks HFJ, Weststrate JA, van Vliet T, Meijer GW (1999) Spreads enriched with three different levels of vegetable oil sterols and the degree of cholesterol lowering in normocholesterolaemic and mildly hypercholesterolaemic subjects. Eur J Clin Nutr 53:319–327
17. Reische DW, Lillard DA, Eitenmiller RR (2002) Antioxidants. In: Akoh CC, Min DB (eds) Food lipids: chemistry, nutrition, and biotechnology, 2nd edn. CRC Press, New York, USA, pp 488–511
18. Chaiyasit W, Elias RJ, McClements DJ, Decker EA (2007) Role of physical structures in bulk oils on lipid oxidation. Crit Rev Food Sci Nutr 47:299–317
19. Kamal-Eldin A (2005) Minor components of fats and oils. In: Shahidi F (eds) Bailey's industrial oil and fat products. Wiley, Canada, pp 319–347
20. Burton GW, Traber MG (1990) Vitamin E: antioxidant activity, biokinetics and bioavailability. Ann Rev Nutr 10:357–380
21. Van Acker SABE, Koymans LMH, Bast A (1993) Molecular pharmacology of vitamin E: structural aspects of antioxidant activity. Free Rad Biol Med 15:311–328
22. Combs, GF (2007) The vitamins. Fundamental aspects in nutrition and health, 3rd edn. Academic Press, USA

23. Johnson LA (2002) Recovery, refining, converting, and satbilizing edible fats and oils. In: Akoh CC, Min DB (eds) Food lipids: chemistry, nutrition, and biotechnology, 2nd edn. CRC Press, New York, pp 222–270
24. Xenakis A, Papadimitriou V, Sotiroudis TG (2010) Colloidal structures in natural oils. Curr Opin Colloid Int Sci 15:55–60
25. Galli C, Petroni A, Visioli F (1994) Nautral antioxidants, with special reference to those in olive oil, and cell protection. Eur J Pharm Sci 2:67–68
26. Cert A, Moreda W, Pérez-Camino MC (2000) Chromatographic analysis of minor constituents in vegetable oils. J Chromatogr A 881:131–148
27. Hidalgo FJ, Zamora R (2006) Peptides and proteins in edible oils: stability, allergenicity, and new processing trends. Trends Food Sci Technol 17:56–63
28. Tannahill R (1989) Food in history. Three Rivers Press, USA
29. Baiano A, Gambacorta G, La Notte E (2010) Aromatization of olive oil. Transworld Research Network
30. Li Y, Fabiano-Tixier AS, Chemat F (2017) Vegetable oils as alternative solvents for green extraction of natural products. In: Chemat S (ed) Edible oils: extraction, processing and applications. Taylor & Francis, pp 205–222
31. Li Y, Fabiano-Tixier AS, Tomao V, Cravotto G, Chemat F (2013) Green ultrasound-assisted extraction of carotenoids based on the bio-refinery concept using sunflower oil as an alternative solvent. Ultrason Sonochem 20:12–18
32. Li Y, Fabiano-Tixier AS, Ginies C, Chemat F (2014) Direct green extraction of volatile aroma compounds using vegetable oils as solvents: theoretical and experimental solubility study. LWT-Food Sci Technol 59:724–731
33. Li Y, Fabiano-Tixier AS, Ruiz K et al (2015) Comprehension of direct extraction of hydrophilic antioxidants using vegetable oils by polar paradox theory and small angle X-ray scattering analysis. Food Chem 173:873–880
34. Krichnavaruk S, Shotipruk A, Goto M, Pavasant P (2008) Supercritical carbon dioxide extraction of astaxanthin from *Haematococcus pluvialis* with vegetable oils as co-solvent. Bioresour Technol 99:5556–5560
35. Ma Q, Xu X, Gao Y, Wang Q, Zhao J (2008) Optimisation of supercritical carbon dioxide extraction of lutein esters from marigold (*Tagetes erect* L.) with soybean oil as a co-solvent. Int J Food Sci Tech 43:1763–1769
36. Gao Y, Liu X, Xu H et al (2010) Optimization of supercritical carbon dioxide extraction of lutein esters from marigold (*Tagetes erecta* L.) with vegetable oils as continuous co-solvents. Sep Purif Technol 71:214–219
37. Rossignol-Castera A (2010) Methods for extracting non-volatile compounds. International patent WO 2010112760 A1, 2010
38. Chevereau P (1998) Method for extracting and fixing aromas on non-aqueous substrate, machine for implementing the method, and product thereby. U.S. patent 5853726, 1998
39. Gu Y, Jérôme F (2013) Bio-based solvents: an emerging generation of fluids for the design of eco-efficient processes in catalysis and organic chemistry. Chem Soc Rev 42:9550–9570
40. Mamidipally PK, Liu SX (2004) First approach on rice bran oil extraction using limonene. Eur J Lipid Sci Technol 106:122–125
41. Liu SX, Mamidipally PK (2005) Quality comparison of rice bran oil extracted with d-limonene and hexane. Cereal Chem 82:209–215
42. Bertouche S, Tomao V, Hellal A, Boutekedjiret C, Chemat F (2013) First approach on edible oil determination in oilseeds products using alpha-pinene. J Essent Oil Res 25:439–443
43. Virot M, Tomao V, Ginies C, Chemat F (2008) Total lipid extraction of food using d-limonene as an alternative to n-hexane. Chromatographia 68:311–313
44. Virot M, Tomao V, Ginies C, Visinoni F, Chemat F (2008) Green procedure with a green solvent for fats and oils' determination: microwave-integrated Soxhlet using limonene followed by microwave Clevenger distillation. J Chromatogr A 1196–1197:147–152
45. Li Y, Fine F, Fabiano-Tixier AS et al (2014) Evaluation of alternative solvents for improvement of oil extraction from rapeseeds. C R Chim 17:242–251

46. Tanzi CD, Abert-Vian M, Ginies C, Elmaataoui M, Chemat F (2012) Terpenes as green solvents for extraction of oil from microalgae. Molecules 17:8196–8205
47. Tanzi CD, Abert-Vian M, Chemat F (2013) New procedure for extraction of algal lipids from wet biomass: a green clean and scalable process. Bioresour Technol 134:271–275
48. Golmakani MT, Mendiola JA, Rezaei K, Ibáñez E (2014) Pressurized limonene as an alternative bio-solvent for the extraction of lipids from marine microorganisms. J Supercrit Fluid 92:1–7
49. Sicaire AG, Abert Vian M, Fine F et al (2015) Experimental approach *versus* COSMO-RS assisted solvent screening for predicting the solubility of rapeseed oil. OCL 22:D404
50. Chemat-Djenni Z, Ferhat MA, Tomao V, Chemat F (2010) Carotenoid extraction from tomato using green solvent resulting from orange processing waste. J Essent Oil Bear Plant 13:139–147
51. Castro-Puyana M, Herrero M, Iratxe U et al (2013) Optimization of clean extraction methods to isolate carotenoids from the microalga *Neochloris oleoabundans* and subsequent chemical characterization using liquid chromatography tandem mass spectrometry. Anal Bioanal Chem 405:4607–4616
52. Filly A, Fabiano-Tixier AS, Lemasson Y, Roy C, Fernandez X, Chemat F (2014) Extraction of aroma compounds in blackcurrant buds by alternative solvents: theoretical and experimental solubility study. C R Chimie 17:1268–1275
53. Filly A, Fabiano-Tixier AS, Fernandez X, Chemat F (2015) Alternative solvents for extraction of food aromas. Experimental and COSMO-RS study. LWT-Food Sci Technol 61:33–40
54. Veillet S, Tomao V, Ruiz K, Chemat F (2010) Green procedure using limonene in the Dean-Stark apparatus for moisture determination in food products. Anal Chim Acta 674:49–52
55. Bertouche S, Tomao V, Ruiz K, Hellal A, Boutekedjiret C, Chemat F (2012) First approach on moisture determination in food products using alpha-pinene as an alternative solvent for Dean-Stark distillation. Food Chem 134:602–605
56. Aissou M, Chemat-Djenni Z, Yara-Varón E, Fabiano-Tixier AS, Chemat F (2017) Limonene as an agro-chemical building block for the synthesis and extraction of bioactive compounds. C R Chim 20:346–358
57. Yara-Varón E, Selka A, Fabiano-Tixier AS et al (2016) Solvent from forestry biomass. Pinane a stable terpene derived from pine tree byproducts to substitute n-hexane for the extraction of bioactive compounds. Green Chem. 18:6596–6608
58. vom Stein T, Grande PM, Kayser H, Sibilla F, Leitner W, Domínguez de María P (2011) From biomass to feedstock: one-step fractionation of lignocellulose components by the selective organic acid-catalyzed depolymerization of hemicellulose in a biphasic system. Green Chem 13:1772–1777
59. Tenne SJ, Kinzel J, Arlt M et al (2013) 2-Methyltetrahydrofuran and cyclopentylmethylether: two green solvents for efficient purification of membrane proteins like FhuA. J Chromatogr B 937:13–17
60. Sicaire AG, Abert-Vian M, Filly A, Li Y, Bily A, Chemat F (2014) 2-Methyl-tetrahydrofuran: main properties, production processes, and application in extraction of natural products. In: Abert-Vian M, Chemat F (eds) Alternative solvents for natural products extraction. Springer, Berlin, pp 253–268
61. Yara-Varón E, Fabiano-Tixier AS, Balcells M et al (2016) Is it possible to substitute hexane with green solvents for extraction of carotenoids? A theoretical versus experimental solubility study. RSC Adv 6:27750–27759
62. Sicaire AG, Vian M, Fine F et al (2015) Alternative bio-based solvents for extraction of fat and oils: solubility prediction, global yield, extraction kinetics, chemical composition and cost of manufacturing. Int J Mol Sci 16:8430–8453
63. Dai Y, van Spronsen J, Witkamp GJ, Verpoorte R, Choi YH (2013) Natural deep eutectic solvents as new potential media for green technology. Anal Chim Acta 766:61–68
64. Vanda H, Dai Y, Wilson EG, Verpoorte R, Choi YH (2018) Green solvents from ionic liquids and deep eutectic solvents to natural deep eutectic solvents. C R Chim 21:628–638

65. Dai Y, Rozema E, Verpoorte R, Choi YH (2016) Application of natural deep eutectic solvents to the extraction of anthocyanins from *Catharanthus roseus* with high extractability and stability replacing conventional organic solvents. J Chromatogr A 1434:50–56
66. Wei Z, Qi X, Li T et al (2015) Application of natural deep eutectic solvents for extraction and determination of pehnolics in *Cajanus cajan* leaves by ultra performance liquid chromatography. Sep Purif Technol 149:237–244
67. Dai Y, Verpoorte R, Choi YH (2014) Natural deep eutectic solvents providing enhanced stability of natural colorants from safflower (*Carthamus tinctorius*). Food Chem 159:116–121
68. Dai Y, Witkamp GJ, Verpoorte R, Choi YH (2013) Natural deep eutectic solvents as a new extraction media for phenolic metabolites in *Carthamus tinctorius* L. Anal Chem 85:6272–6278
69. Dai Y, Witkamp GJ, Verpoorte R, Choi YH (2015) Tailoring properties of natural deep eutectic solvents with water to facilitate their applications. Food Chem 187:14–19
70. Radošević K, Ćurko N, Srček G et al (2016) Natural deep eutectic solvents as beneficial extractants for enhancement of plant extracts bioactivity. LWT-Food Sci Technol 73:45–51
71. Espino M, Fernández MA, Gomez FJV, Silva MF (2016) Natural designer solvents for greening analytical chemistry. TrAC-Trends Anal Chem 76:126–136
72. Alvira P, Tomás-Pejó E, Ballesteros M, Negro MJ (2010) Pretreatment technologies for an efficient bioethanol production process based on enzymatic hydrolysis: a review. Bioresour Technol 101:4851–4861
73. Balat M (2011) Production of bioethanol from lignocellulosic materials via the biochemical pathway: A review. Energ. Convers. Manage. 52:858–875
74. Xia S, Yuan Z, Wang L, Chen P, Hou Z (2012) Catalytic production of 1,2-propanediol from glycerol in bio-ethanol solvent. Bioresour Technol 104:814–817
75. Hu J, Guo Z, Glasius M et al (2011) Pressurized liquid extraction of ginger (*Zingiber officinale* Roscoe) with bioethanol: an efficient and sustainable approach. J Chromatogr A 1218:5765–5773
76. Moity L, Durand M, Benazzouz A et al (2012) Panorama of sustainable solvents using the COSMO-RS approach. Green Chem 14:1132–1145
77. Moity L, Benazzouz A, Molinier V et al (2015) Glycerol acetals and ketals as bio-based solvents: positioning in Hansen and COSMO-RS spaces, volatility and stability towards hydrolysis and autoxidation. Green Chem 17:1779–1792
78. Moity L, Molinier V, Benazzouz A et al (2014) In silico design of bio-based commodity chemicals: application to itaconic acid based solvents. Green Chem 16:146–160
79. Moity L, Molinier V, Benazzouz A et al (2016) A "top-down" in silico approach for designing ad hoc bio-based solvents: application to glycerol-derived solvents of nitrocellulose. Green Chem 18:3239–3249
80. Hansen CM (1999) Hansen solubility parameters—a user's handbook. CRC Press, Florida, USA
81. Stefanis E, Panayiotou C (2008) Prediction of Hansen solubility parameters with a new group-contribution method. Int J Thermophys 29:568–585
82. Benazzouz A, Moity L, Pierlot C et al (2013) Selection of a greener set of solvents evenly spread in the Hansen space by space-filling design. Ind Eng Chem Res 52:16585–16597
83. Laboukhi-Khorsi S, Daoud K, Chemat S (2017) Efficient solvent selection approach for high solubility of active phytochemicals: application for the extraction of an antimalarial compound from medicinal plants. ACS Sustain Chem Eng 5:4332–4339
84. Petchey THM, Comerford JW, Farmer TJ et al (2018) Optimization of amidation reactions using predictive tools for the replacement of regulated solvents with safer biobased alternatives. ACS Sustain Chem Eng 6:1550–1554
85. Benazzouz A, Moity L, Pierlot C, Molinier V, Aubry JM (2014) Hansen approach versus COSMO-RS for predicting the solubility of an organic UV filter in cosmetic solvents. Colloids Surf A Physiochem Eng Aspects 458:101–109
86. Koudous I, Kunz W, Strube J (2015) Panorama of sustainable solvents for green extraction processes. In: Chemat F, Strube J (eds) Green extraction of natural products: theory and practice. Wiley-VCH, Germany, pp 180–187

87. Klamt A, Eckert F, Arlt W (2010) COSMO-RS: an alternative to simulation for calculating thermodynamic properties of liquid mixtures. Annu Rev Chem Biomol Eng 1:101–122
88. Moity L, Durand M, Benazzouz A, Molinier V, Aubery JM (2014) In silico search for alternative green solvents. In: Chemat F, Abert-Vian M (eds) Alternative solvents for natural products extraction. Springer, New York, pp 1–23
89. Cascant MM, Breil C, Garrigues S et al (2017) A green analytical chemistry approach for lipid extraction: computation methods in the selection of green solvents as alternative to hexane. Anal Bioanal Chem 409:3527–3539
90. Breil C, Abert Vian M, Zemb T, Kunz W, Chemat F (2017) "Bligh and Dyer" and Folch methods for solid-liquid-liquid extraction of lipids from microorganisms. Comprehension of solvatation mechanisms and towards substitution with alternative solvents. Int J Mol Sci 18:708–729
91. Rapinel V, Breil C, Makerri C et al (2017) Feasibility of using liquefied gas HFO-1234ze (trans-1,3,3,3-tetrafluoroprop-1-ene) as an alternative to conventional solvents for solid-liquid extraction of food ingredients and natural products. LWT-Food Sci Technol 83:225–234
92. Rapinel V, Rombaut N, Rakotomanomana N et al (2017) An original approach for lipophilic natural products extraction: Use of liquefied n-butane as alternative solvent to n-hexane. LWT-Food Sci Technol 85:524–533
93. Bundeesomchok K, Filly A, Rakotomanomana N, Panichayupakaranant P, Chemat F (2016) Extraction of α-mangostin from *Garcinia mangostana* L. using alternative solvents: computational predictive and experimental studies. LWT-Food Sci Technol 65:297–303
94. Grazhdannikov AE, Kornaukhova LM, Rodionov VI et al (2018) Selecting a green strategy on extraction of birch bark and isolation of pure betulin using monoterpenes. ACS Sustain Chem Eng 6:6281–6288
95. Everitt BS (1993) Cluster analysis. Wiley, New York
96. Ward JH (1963) Hierarchical grouping to optimize an objective function. J Am Stat Assoc 58:236–244
97. Fréville V, van Hecke E, Marinkovic S, Emenwein C, Pezron I (2011) Mapping and multivariate analysis for a rapid evaluation of new green solvents characteristics and efficiency. J Disper Sci Technol 32:1742–1752
98. Clark JH, Tavener SJ (2007) Alternative solvents: shades of green. Org Process Res Dev 11:149–155
99. Hackl K, Kunz W (2018) Some aspects of green solvents. C R Chim 21:572–580
100. Li Y, Fabiano-Tixier AS, Abert-Vian M, Chemat F (2013) Solvent-free microwave extraction of bioactive compounds provides a tool for green analytical chemistry. TrAC-Trend Anal Chem 47:1–12
101. Zhong Y, Shahidi F (2012) Antioxidant behavior in bulk oil: limitations of polar paradox theory. J Agric Food Chem 60:4–6
102. Shahidi F, Zhong Y (2011) Revisiting the polar paradox theory: a critical overview. J Agric Food Chem 59:3499–3504
103. Laguerreab M, Bayrasya C, Panyabd A et al (2015) What makes good antioxidants in lipid-based systems? The next theories beyond the polar paradox. Crit Rev Food Sci Nutr 55:183–201
104. Laguerreab M, Bily A, Roller M, Bitrić S (2017) Mass transport phenomena in lipid oxidation and antioxidation. Annu Rev Food Sci Technol 8:391–411
105. Plotka-Wasylka J, Rutkowska M, Owczarek K, Tobiszewski M, Namieśnik J (2017) Extraction with environmentally friendly solvents. TrAC-Trends Anal Chem 91:12–55

Chapter 4
Plant-Based Chemicals Extraction and Isolation

Hichem Ben Salah and Noureddine Allouche

Abstract Plants represent a huge reservoir of bioactive molecules that are still little explored. Modern science is now focused on identifying the beneficial compounds from these sources to be used in pharmaceutical drugs. The search for new bioactive natural molecules is based on the choice of extraction, separation and structural identification techniques. This chapter provides a brief review of the specificity and usefulness of the most popular extraction methods in the phytochemistry field. In addition, this study describes the development of chromatographic separation and structural elucidation techniques of secondary metabolites, including analytical and preparative chromatography, mass spectrometry (MS), one- (1D) and two-dimensional (2D) Nuclear magnetic resonance (NMR) experiences. Today, the immense progress of these techniques has enabled chemists, biologists and pharmacists to discover bioactive molecules that have found applications in drug and food industries.

Keywords Extraction · Identification · Phytochemistry · Separation · Spectroscopy

4.1 Introduction

The research of therapeutic powers in plants is an ancient idea. People from all continents have long applied infusions and poultices of thousands of indigenous plants, dating back to prehistory. The extraction of natural substances dates back to Mesopotamian and Egyptian times, where production of perfumes, waxes and pharmaceutically active oils was an important business. In addition, some Sumerian texts confirm the existence of a sophisticated chemical and pharmaceutical technology [1].

H. Ben Salah · N. Allouche (✉)
Laboratory of Organic Chemistry, Natural Substances Team, Faculty of Sciences of Sfax, University of Sfax, P.B.1171, P.C. 3000, Sfax, Tunisia
e-mail: noureddineallouche@yahoo.fr

Y. Li and F. Chemat (eds.), *Plant Based "Green Chemistry 2.0"*,
Green Chemistry and Sustainable Technology,
https://doi.org/10.1007/978-981-13-3810-6_4

Natural products are chemical substances or compounds produced by a living organism. They can be extracted from tissues of plants, marine organisms or microorganism fermentation [2]. Plants represent a source of bioactive compounds which are of potential use in medicine, nutraceuticals, food supplements, pharmaceutical intermediates and chemical entities for synthetic drugs. Bioactive natural products are secondary metabolites of plants generated through various biological pathways in secondary metabolism processes [3]. Plants contain a panel of chemical compounds, such as flavonoids alkaloids, phenols, steroids, tannins, glycosides, volatile oils and fixed oils, which are present in their different parts such as flowers, leaves, bark, root, fruits, seeds and fruits [4]. Nowadays, more than 30% drugs and about 80% food active additives are produced from bioactive natural products and the annual growth rate of natural products-derived drugs is up to 20% [5]. Furthermore, about 3.4 billion people in the developing countries (representing 88% of the world's inhabitants) depend on plant-based traditional medicines [2]. Consequently, the research on the extraction, separation and purification of natural products from plants has attracted much attention in academia and industry. In view of low contents of target compounds (0.01–10%) in a plant, separation is often a complex process, which may last from days to months, depending on the problem being tackled.

The investigation of natural products from plants has had a number of rewards. It has led to the discovery of various drugs useful for the treatment of different ailments. This fact induced the progress of extraction and separation technologies, spectroscopic methods useful for structure elucidation and synthetic methodologies that now constitute the basics of analytical organic chemistry.

The phytochemical investigation of plants may involve different steps. It starts with identification, collection and preparation of the plant material, usually by drying and grinding. After that, the prepared material is subjected to extraction which consists in the separation of target portions of plant tissues using selective solvents through standard methods. Such extraction procedures isolate the soluble plant metabolites and leave behind the insoluble residue. The extracts so recovered from plants are relatively complex mixtures of natural compounds, in liquid or semisolid state or in dry powder form after solvent evaporation [6]. Prior to isolation of pure compounds, several purification steps are necessary to remove most of the unwanted matrix. Finally, the purified compounds are identified and characterized using spectroscopic techniques [7].

This chapter aims to give readers an overview of the methodology adopted to determine what the appropriate methods for plant metabolites extraction as well as their separation and purification from crude extracts. In addition, we go on to explore how isolates are characterized and identified with a focus on spectroscopic techniques.

4.2 Extraction Techniques

Plants constitute a source of natural compounds which are of potential use in medicine and other applications. Plants contain many active compounds such as alkaloids, steroids, volatile oils, fixed oils, phenols and flavonoids which are present in their specific parts (leaves, flowers, bark, seeds, fruits and root). Generally, the first step in the isolation and identification of natural compounds is extracted to separate the products from the cellular matrix. Ideally, the extraction procedure should be exhaustive for the components to be recovered, simple, rapid and inexpensive. The increasing interest in secondary metabolites encourages researchers to improve and modify the traditional extraction methods. These latter include maceration, infusion, decoction, percolation and Soxhlet extraction. These conventional methods present major disadvantages mainly use of large quantities of organic solvents, long extraction times, unsatisfactory extraction efficiency and degradation of labile components. To remedy these inconvenient, innovative extraction techniques have been developed for the extraction of secondary metabolites from solid matrices with significant advantages over conventional methods: (i) reduction of extraction and cleanup times, (ii) avoidance of additional sample concentration and cleanup steps, (iii) decrease in sample degradation and organic solvent consumption, (iv) enhancement of extraction efficiency, selectivity and kinetics, (v) ease of automation. The modern extraction methods include mainly supercritical fluid extraction (SFE), ultrasonication-assisted extraction (UAE), pressurized liquid extraction (PLE), microwave-assisted extraction (MAE), solid-phase microextraction (SPME), extraction with supercritical or subcritical water and soxhwave. The latter constitutes a combination of Soxhlet with microwaves. This combines the simplicity of Soxhlet with the rapid heating capacity of microwaves.

4.2.1 Conventional Extraction Methods

4.2.1.1 Maceration, Infusion and Decoction

Maceration has been widely adopted and used in medicinal plants chemistry. Maceration consists in soaking plant materials in a stoppered container with an appropriated solvent and allowed to stand at room temperature with frequent agitation for a period of minimum 3 days until the soluble matter is dissolved [7]. In this step, the plant's cell wall is broken to release the soluble phytochemicals. The mixture then is strained and the marc (the damp solid material) is pressed. Finally, the combined liquids are clarified by filtration or decantation after standing. In this method, the type of compounds extracted from the plant material depends in the choice of solvent [8]. Decoction and infusion possess the same principle as maceration; both are soaked in cold or boiled water. However, for infusion, the maceration period is shorter and the plant material is boiled in specified volume of water (e.g. 1:4 or 1:16) for a defined time. In addition, decoction is used only for the extraction of heat-stable compounds from hard plant materials (roots and barks). Comparing with maceration and infusion, decoction usually resulted in more oil-soluble products [9].

4.2.1.2 Percolation

Percolation is used frequently for extracting active ingredients in the preparation of fluid extracts. This extraction procedure is performed in a percolator which is a narrow, cone-shaped vessel open at both ends. The plant material is moistened with an appropriate amount of menstruum and allowed to stand for about 4 h in a closed container after which the mass is packed and the top of the percolator is closed. After that, another amount of menstruum is added to form a thin layer above the mass. The obtained mixture is allowed to macerate for 24 h. The percolator outlet is then opened to allow the liquid contained therein to drip slowly. Additional menstruum is added until the percolate measures about three-quarters of the desired volume of the finished product. The marc is then pressed and the expressed liquid is added to the percolate. Sufficient quantity of menstruum is added to produce the required volume. Finally, the mixed liquid is clarified by standing or by filtration followed by decantation [10].

4.2.1.3 Soxhlet Extraction

Soxhlet extraction is a hot continuous extraction procedure. The finely ground plant material is put in a porous bag or "thimble" made of strong filter paper, which is placed in the Soxhlet apparatus. The extracting solvent is heated in the down flask. The solvent condensed vapor drips into the thimble containing the plant material and extracts it by contact. When the level of liquid rises to the top of siphon tube, it siphons again into the flask. This process is repeated continuously until a drop of solvent from the siphon tube does not leave residue when evaporated. Comparing with the previously described techniques, Soxhlet extraction is more advantageous and useful. In fact, a large quantity of plants can be extracted using a much smaller amount of solvent. This can provide economy in terms of time, energy and consequently financial inputs.

4.2.2 Supercritical Fluid Extraction (SFE)

Supercritical fluids are considered as alternatives to conventional solvents. Supercritical fluid is a substance having the physical properties of both gas and liquid at its critical point. Temperature and pressure are the crucial factors that can be modified to transform a substance into its supercritical state. The supercritical fluid extraction (SFE) is performed in high-pressure equipment in batch or continuous system. In

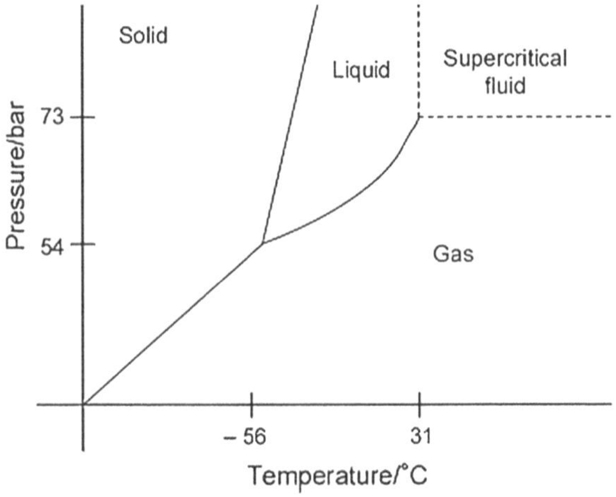

Fig. 4.1 Phase diagram of carbon dioxide. Reprinted with permission from [14]. © 2008 Royal Society of Chemistry

both cases, the supercritical solvent is put in contact with the plant material from which desirable products to be separated. The favorable transport properties of fluids near their critical points allow deeper penetration into solid plant matrix and more efficient and faster extraction than with conventional organic solvents. Generally, cylindrical extraction vessels are used for SFE [11]. In batch processing, plant material is placed into extraction vessel and the supercritical solvent is fed in until the target extraction conditions are reached. In semi-batch extraction process, the supercritical solvent is fed continuously through a high-pressure pump at a fixed flow rate. Supercritical fluid technology is now recognized as an effective analytical technique with efficiency comparable to existing chemical analysis methods. SFE is favorably applicable for the qualitative and quantitative identification of constituents of natural products, including heat-labile compounds [12]. Various factors are important during SFE mainly temperature, pressure, sample volume and flow and pressure control. Supercritical fluids exhibit properties intermediate between those of the gaseous and liquid phases [7]. Under appropriate conditions, any fluid can attain its supercritical state. The use of supercritical fluid as the extraction solvent is directly linked to its density. A supercritical fluid which is considered is as a dense gas is a fluid above its critical temperature (TC) and critical pressure (PC). To be supercritical, the reduced temperature Tr (i.e. T/TC) must not exceed 1.2 or 1.3, whereas the reduced pressure Pr (i.e., P/PC) may be as high as allowed by technological limits [11]. Among various supercritical fluids (i.e., pentane, butane, hexane nitrous oxide and fluorinated hydrocarbons), carbon dioxide (CO_2) stands out as the only convenient supercritical extraction solvent due to its comparatively low critical temperature (31.1 °C) and pressure (73.8 bar/7.38 MPa) (Fig. 4.1) [13].

CO_2 is generally the most convenient solvent in SFE, because its critical temperature is only 304 K, which makes it attractive for the extraction of heat-labile compounds. Moreover, CO_2 is safe (non-flammable, non-explosive), inert, inexpensive, odorless, colorless, non-corrosive, environment friendly and clean solvent that generates product without any solvent residue when used in SFE procedure [7]. In addition, carbon dioxide has a low viscosity and surface tension and a high diffusivity making it advantageous as a supercritical solvent. Furthermore, CO_2 is non-toxic, is removable from the extract by simple expansion and is accepted as a harmless ingredient in food and pharmaceuticals [15]. On the other hand, CO_2 is convenient only for extracting polar compounds and it presents, hence, some polarity limitations. Consequently, an organic polar solvent (also called modifier) may be added to the supercritical CO_2 to alleviate these polarity limitations and thus to enhance its solvating properties. The more commonly used modifiers are ethanol, methanol and acetonitrile. Organic solvents are frequently added to the carbon dioxide to alleviate these polarity limitations. Of late, instead of carbon dioxide, argon is being used because it is inexpensive and more inert. The component recovery rates generally increase with increasing pressure or temperature: the highest recovery rates in case of argon are obtained at 500 atm and 150 °C. Supercritical CO_2 is a good solvent for extracting natural compounds, mainly heat-sensitive substances because of its mild critical condition and molecules displaying some degree of lipophilicity, such as lactones, esters and ethers. SFE was used at laboratory scale since 1970. However, its commercial application happened slowly due to the expensive and sophisticated high-pressure equipment and technology required [16]. Initially, interest in the extraction of plant organs using SFE has been prominent. In the beginning, supercritical CO_2 was used in decaffeinating coffee and isolating compounds from hops and [17]. More recently, SFE was adopted as an analytical extraction method for sample preparation prior to chromatographic systems, such as SFC and GC [18]. The application of SFE in both analytical and processing scales is widespread in the food industry in extracting fats and oils from seeds, foodstuffs, and other materials [19, 20]. In this context, it was reported that the extraction of grape seeds oil using SFE with CO_2 presents some advantages concerning the oil quality and the efficiency of the technique [21]. Supercritical CO_2 extraction was also used for the extraction of oil from fennel seeds [13]. On the other hand, it was established that SFE conditions could be fine-tuned for selective extraction antioxidants without residual aroma from rosemary [22]. Moreover, SFE was applied for the extraction of biomolecules from medicinal plants, such as terpenes, steroids, alkaloids and phenolic compounds [23]. Recently, we investigated an innovative biorefinery cascade processing of tomato by-products starting with SFE-CO_2 extraction of carotenoids mainly lycopene within the oil fractions from tomato seeds and tomato peels [24]. The residues from the SFE-CO_2 extraction step were then submitted to a batch protein extraction. The amino acids composition of proteins from tomato industrial by-products was determined and their high-nutritional quality was confirmed. Furthermore, a subcritical liquid hot-water hydrolysis was applied for the pretreatment of the lignocellulosic matter recovered after SFE-CO_2 and protein extraction for the hydrolysis of hemicellulose and cellulose sugars [25].

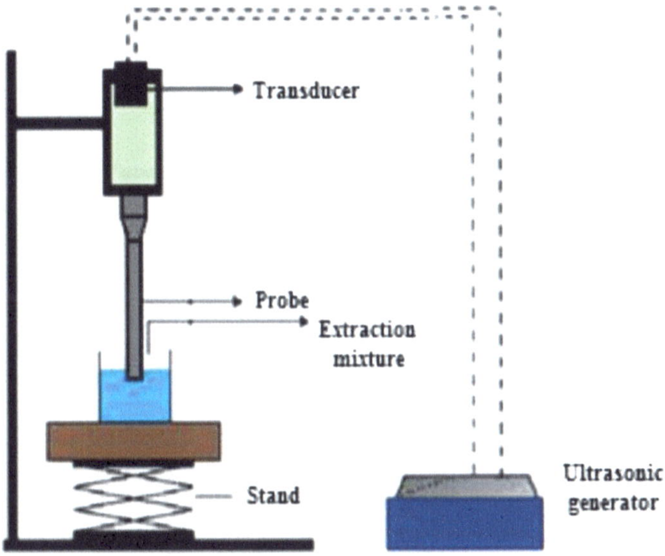

Fig. 4.2 Schematic representation of UAE system. Reprinted with permission from [26]. © 2009 Elsevier

4.2.3 Ultrasonication-Assisted Extraction (UAE)

Ultrasonication-assisted extraction (UAE) involves the application of ultrasound waves with high intensity and frequency (ranging from 20 to 2000 kHz) as well as their interaction with materials [11]. UAE is a potentially useful technology as it is relatively low-cost and does not require complex instruments as represented in Fig. 4.2. It can be used both on small and large scale [6]. In this technique, the plant material is covered by the extraction solvent in a glass container and subjected to ultrasound waves in an ultrasonic bath. These experimental conditions decrease extraction time and improve extraction yield due to mechanical tension which induces cavitations and cellular breakdown.

Under ultrasonic waves, liquid and solid particles are vibrated and consequently solute quickly diffuses out from solid phase to solvent [27]. The mechanical action of acoustic cavitation arising from ultrasound waves increases the contact surface between and sample and solvent and permeability of cell walls. Chemical and physical characteristics of the plant material subjected to ultrasound are altered and disrupt the plant cell wall. This fact facilitates compounds release and enhances mass transport of the solvents into the plant cells [28].

UAE technique was used for the extraction of several and various natural products including mainly anthocyanidins, flavonols, phenolic acids and essential oils from aromatic plants. It was established that UAE is the most effective method for extraction of biologically active compounds from propolis based on low extraction time (10–30 min), high yield and high selectivity [29]. In addition, UAE has been

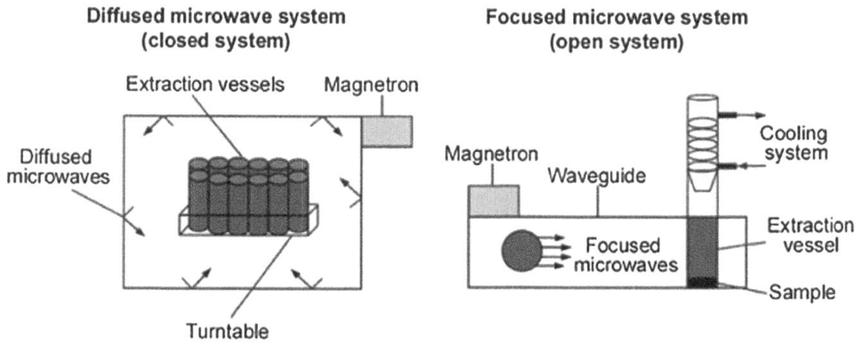

Fig. 4.3 Scheme of the two microwave systems, using diffused or focused microwaves. Reprinted with permission from [14]. © 2008 Royal Society of Chemistry

applied for the extraction of thermo-labile compounds mainly anthocyanin from plant flowers, to reduce extraction time and avoid exposure to high temperature [30]. Furthermore, a good efficacy of UAE was observed in phenolic compounds extraction from *Cratoxylum formosum* using ultrasound at 45 kHz, 50.33% ethanol v/v, at 65 °C for 15 min [31].

Although the benefits of UAE, mainly reduction in extraction time and solvent consumption, the use of ultrasound energy more than 20 kHz may alter the active natural compounds through the formation of free radicals [32].

4.2.4 Microwave-Assisted Extraction (MAE)

Microwave-assisted extraction (MAE) is a new technique for the extraction of soluble compounds into a fluid from plant materials using microwave energy. It allows the extraction of products more rapidly and more selectively (usually in less than 30 min) comparing with conventional extraction methods. Besides the advantage of high extraction speed, MAE permits a significant decrease in the organic solvent consumption. Microwaves directly heat the solvent or solvent mixture. The commonly used frequency for microwave instruments is 2450 MHz (2.45 GHz), which has energy of 600–700 W [33]. MAE can be performed in two different modes; the first one called *closed-vessel operation* is under controlled (elevated) pressure and temperature, the second is *open-vessel operation* carried out at atmospheric pressure. The two technologies are commonly named as pressurized microwave-assisted extraction (PMAE) with a multi-mode cavity or focused microwave-assisted extraction (FMAE) using the wave guide as a single-mode cavity, respectively. Both systems are shown in Fig. 4.3 [14].

Microwaves are composed of electric and magnetic fields which are perpendicular to each other. The electric field induces heating via two simultaneous mechanisms which are dipolar rotation and ionic conduction. The dipolar rotation produces collisions with surrounding molecules causing the liberation of thermal energy into the medium. Using a frequency of 2.45 GHz, this phenomenon is 4.9×10^9 times faster, and thus, the resulting heating is very fast. Consequently, unlike usual conductive heating methods, microwaves heat the whole sample simultaneously. In the case of extraction, the advantage of microwave heating is the disruption of weak hydrogen bonds promoted by the dipole rotation of the molecules [34]. When plant material is introduced into a microwave transparent solvent, the heat caused by microwave radiation directly attains the solid without being absorbed by the solvent and thus resulting in an immediate heating of the residual moisture in the plant. Heating provokes the evaporation moisture and creates a high vapor pressure that cracks the cell wall of plant material and releases the content into solvent [35]. Solvents used for most MAE experimentations are those with a high dielectric constant and capacity to absorb microwave energy. Polar solvents are usually believed to be better than nonpolar ones. Furthermore, extraction yields would be increased by the addition of water to the solvent [36]. MAE is an environment-friendly and economical technique for the recovery of bioactive natural products from various plant materials [15]. MAE is highly efficient for recovering extracts under mild conditions. MAE requires short extraction time and less solvent amount and improves purity of the extract and increases extraction yield in comparison with Soxhlet extraction. Microwaves provoke little or no quality deterioration when applied to samples of plant origin [37]. For the first time in 1975, microwave domestic ovens were used for the treatment of biological samples for metal analysis [38]. In 1986, MAE was used for the first time to produce natural products from plant materials [34]. Microwave radiations can be used for the extraction of essential oils from aromatic plants. In this case, the obtained essential oils consisted of relatively low-volatile fractions comparing to hydro-distillation [11]. MAE has been reported to be quicker than the reflux method for extracting phenolic compounds [39]. The effective study of solvent composition, solvent volume, matrix characteristics and extraction temperature on the MAE of leaves of peppermint and rosemary showed that for a sample matrix containing water, the use of pure microwave transparent solvents such as hexane accelerates extraction of essential oils. This would be due to the interaction of microwaves with the water molecules present in the cell, which induced cell rupture and release of essential oil into hexane [22]. Extraction of phenolic compounds from tomato with MAE has been reported to be more effective using higher microwave temperature and a short extraction time [40]. Both microwave- and sonication-assisted extraction proved to be effective in increasing the yield of phenolics and tannins [41]. Recently, we tested green extraction procedures of lipids from Tunisian date palm seeds. It was deduced that the extraction of oils from date pits with ultrasound and microwave was better in terms of extraction time (30 min versus 8 h) comparing to Soxhlet procedure and of extraction yield comparing to maceration [42]. MAE has been reported as an interesting alternative to conventional methods in extracting phenols such as chlorogenic acids from green coffee beans [43]. MAE allowed the recovery of highest amount of alkaloid fraction from leaves of *Mitragyna speciosa* plant and a high content of mitragynine was presented in the prepared extract [44].

4.3 Separation and Purification Methods

The separation and isolation of natural products from the complex extracts are arduous, and hence, need to combine effective and innovative separation techniques. The choice of the separation technique, which depends on the solubility, polarity, volatility and stability of the compound to be separated, is of great importance. In recent years, chromatography has made spectacular progress in terms of the varying needs of molecular separation. Today, several types of chromatography are used for different purposes in analytical laboratories. Among the commonly used chromatographic techniques, some important types are briefly discussed below.

4.3.1 Thin-Layer Chromatography (TLC)

TLC is a chromatographic technique widely applied in phytochemistry for separation of almost class of natural products. This technique is used to visualize the separation of the molecules of the extract during its fractionation on a column and to evaluate the complexity of the fractions. It is a sensitive and rapid adsorption chromatography based on the difference of the solute interactions to a thin layer of adsorbent fixed on a plate of plastic or aluminum or glass [45]. Products in the mixture are separated according to the differences in their affinity with the stationary phase and the mobile phase. Compounds analyzed by TLC chromatoplate can be detected by visual detection, under a UV lamp (254 nm), in iodine chamber and by using suitable spray reagents (vanillin, sulfuric acid, chromic acid, etc.). It is also used to identify a compound in a mixture by comparing its Rf (retention factor) with that of a known compound [46]. Preparative TLC (glass plates) can be used for preparative separation. Its sensitivity makes separations on very small quantities of extract (less than 1 mg) can be achieved.

4.3.2 Paper Chromatography (PC)

Paper chromatography (PC) has become standard and useful chromatographic system for the separation of complex mixtures of flavonoids, phenolic compounds, amino acids, carbohydrates and steroids. Among the advantages of PC is that separations are performed simply on filter paper sheets, which play both the role of support and separation medium. Another quality of the PC is the considerable reproducibility of the Rf values determined on paper [47]. In paper chromatography, a sheet of paper is used for the inert phase. A sample is placed near the bottom of the filter paper. Then, it is immersed in a solvent placed in a chromatographic chamber. The molecules transported by the solvent back the paper by capillary action. The low porosity paper will produce a slow moving speed of the solvent and the thick papers will have an increased sampling capacity.

In paper chromatography, a sheet of paper is used for the inert phase. A sample is placed near the bottom of the filter paper. Then, this latter is immersed in a solvent placed in chromatographic chamber. The solvent moves up the paper by capillary action carrying soluble molecules along with it. The low porosity paper will produce a slow rate of movement of the solvent and thick papers have increased sample capacity [48]. One of the applications of PC is cited by Ben Salah et al. during the purification and isolation of flavonol triglycosides from the leaves of *Hammada scoparia* (POMEL) ILJIN [49].

4.3.3 Column Chromatography (CC)

Column chromatography is the oldest form of chromatography widely used for the isolation of natural molecules with some analytical tools [50]. It is still mostly used due to its economy, convenience and availability in various stationary phases [51]. Silica, alumina, cellulose, RP-8, RP-18 and polyamide are frequently used for phytochemicals separation. The separation by repartition of compounds is achieved between two phases: stationary and mobile phases. The analytes that have the least interaction with the stationary phase or the most interaction with the mobile phase will move faster through the column.

4.3.4 Gel Permeation Chromatography

Gel permeation chromatography (size exclusion chromatography) is based on the ability of molecules to move through a column of gel according to their molecular weight and spatial structures [52]. The molecules with larger size cannot enter the pores, and therefore, they come out of the column more quickly and elute first. Slightly smaller molecules can enter some pores, and so take longer to elute, while small molecules can be delayed further. The advantage of the technique is simplicity, isocratic and large molecules rapidly elute. However, the columns are expensive and sensitive to contamination; consequently, they are mainly used in applications where alternative separation techniques are not available, and samples are fairly clean. The commonly used gel in natural products is Sephadex LH-20 to separate chlorophyll from compounds of interest, where usually chlorophyll elutes first [53].

4.3.5 Gas Chromatography (GC)

Gas chromatography is the analytical technique of choice for the separation and determination of volatile compounds. Several features such as ease of use, speed of analysis, high resolution, sensitivity, excellent quantitative results and moderate costs have helped the GC to become the most widespread in the field of analytical chemistry [54]. However, the success of the GC technique is largely associated with high sensitivity of multiple detectors, which can also be used in series with a mass spectrometer since MS is a non-destructive technique. The highly efficient separation with the capillary column permits the analysis of essential and vegetable oils, hydrocarbons, fatty extracts, etc.

(GC) is a separation technique using gas flow through a glass or metal column that separates compounds based on both volatility and interaction with the liquid stationary phase.

4.3.6 High-Pressure Liquid Chromatography (HPLC)

HPLC is one of the most useful techniques in chemical analysis laboratories. It provides both quantitative and qualitative data in the single analysis and hence allows the identification and purification of one or more compounds from a mixture [55]. Unlike gas chromatography, which is unsuitable for nonvolatile and thermally fragile molecules, HPLC can safely separate a very large range of natural compounds.

Today, preparative and semi-preparative HPLC are used for the separation of plant metabolites. The application of high pressure increases the sensitivity and speed of HPLC which increases its effectiveness in the isolation of the product compared to that of column chromatography. In this technique, the sample is introduced into the column at high pressures of nearly 400 atm, resulting a faster and more efficient separation. The solutes are dissolved first, and then they will be introduced into the mobile liquid phase. Thanks to the selective distribution of the analytes between the mobile and stationary phases, each solute is thus subjected to a mobility force due to the mobile phase and to a retention force exerted by the stationary phase. Depending on the nature of the molecules, they interact more or less with the stationary phase in the chromatographic column.

The ultraviolet (UV) detector is perfectly compatible with HPLC and offers adequate detection limits and calibration linearity over a wide range (1 ppm–2.5 ppb) [56]. HPLC is mainly used for different classes of natural compounds which are nonvolatile, such as terpenoids, all types of phenolics, alkaloids, lipids and sugars [49, 57].

4.3.7 High-Performance Thin Layer Chromatography (HPTLC)

HPLTC is a sophisticated means of purification where the separation of compounds is performed by high-performance layers with detection and data acquisition. These high-performance layers are pre-coated plates coated with a sorbent of particle size about 5–7 microns and a layer thickness of 150–200 microns. The efficiency of the plate and the nature of the separation are related to the reduction of the layer thickness and the particle size. The main difference between TLC and HPTLC are the particle and pore size of sorbents. Thanks to its automation, full optimization, selective detection principle, minimum sample preparation and hyphenation, HPTLC can be advantageous compared to high-performance liquid column chromatography in many analytical situations [58, 59].

4.3.8 Flash Chromatography

Flash chromatography is a rapid, simple and inexpensive approach to separate complex mixtures of compounds. It is a great alternative when other separation techniques cannot be used or are too difficult [60]. Flash chromatography system allows separations and purifications in an automated way, which saves user time, solvent saving and reproducibility. The operating principle of this technique is based on the separation of the compounds between a solid stationary phase and a mobile phase (the eluent). The separation will therefore be performed according to the greater or lesser affinity that the compounds have for the stationary phase, depending on their structure and the nature of the stationary phase.

Flash column chromatography uses compressed gas (such as nitrogen or air) or a pump to push solvent through the column. This technique is particularly advantageous because it allows for faster flow rates of the solvent, as opposed to simple gravity flow. Special considerations for flash chromatography include the usage of a finer particle size for the stationary phase (such as silica gel).

4.3.9 Centrifugal Partition Chromatography (CPC)

CPC, sometimes called countercurrent chromatography, is an analytical method used more and more for the separation and purification of individual products from plant extracts. It is a preparative liquid–liquid chromatography technique based on the difference of partition between two immiscible liquid phases of a biphasic system. One of the two liquid phases is used as a stationary phase and is maintained in the column (rotor) by centrifugal force. The second phase of the biphasic system is used as a mobile phase, which is then pumped through the stationary phase. Molecules, having different partition coefficients in the biphasic system, are thus separated. CPC has relatively a large volume of stationary phase which makes the separation of substances effective [51]. CPC as the analytical technique is much more advantageous than other types of chromatography such as high-performance liquid chromatography (HPLC). In contrast to the HPLC, CPC gives complete recovery of the sample introduced. There is no denaturation of the molecules, no irreversible absorption on the solid absorbent and the solvent consumption is very low. The main advantage of this technique is due to significant reduction of analysis costs, allowing the use of this method on a larger scale.

4.4 Identification and Structure Elucidation Methods

The structural identification of natural products is a key step that could reconcile the reliability of traditional medicine with the rigor of scientific evaluation and validation through the biological and clinical tests (Valorization of medicinal plants). The structural elucidation of biomolecules is largely established by spectroscopic evidence, mainly MS, IR, UV and NMR methods. These essential techniques are briefly described below.

4.4.1 Mass Spectrometry (MS)

The mass spectrometry is the analytical technique that has undergone the greatest growth in recent years. MS is an extremely sensitive detection technique that can be used to determine molecular structures. All mass spectrometry processes are based on the same principle: the mass spectrometer transforms molecules into ions (ionization technique) and sorts them according to the ratio of mass to charge (separation technique of ions) and the number of ions of each "unit" (m/z) is recorded as a spectrum [61–64].

Many ionization, analysis and detection techniques have been developed. The choice of the method to implement is, therefore, particularly important. Among the frequently used techniques for characterization and structural identification of natural products, we can basically quote.

4.4.1.1 Electronic Ionization (EI)

Electronic ionization is the most widely used ion production technique in mass spectrometry. Electrons, accelerated by an electric field (70 eV in general), bombard the sample, in a chamber maintained under a high vacuum (10^{-2} Pa = 10^{-7} atm) by continuous pumping. This system, applicable to gases and volatile liquids, is the most commonly used. However, at 70 eV, the molecular ion is not always observable [61, 64].

4.4.1.2 Chemical Ionization (CI)

Chemical ionization is among the techniques involving a gentle ionization that allows the observation and detection of the molecular peak on the mass spectrum. The primary ions are provided by a reactive gas (NH_3, CH_4, isobutane, etc.) introduced into the source and ionized by electronic impact. A series of reactions follows producing ions that can react with the molecules of the sample arriving in the source. In fact, the molecules of the sample collide with the ionized molecules of the reactive gas (CH_5^+, $C_4H_9^+$, etc.) in the chemical ionization source where the pressure is relatively elevated to lead to secondary ionization by proton transfer producing an ion $[M+H]^+$. This gentle ionization makes it possible to obtain a "pseudo-molecular" (M+1) ion in positive mode or (M−1) in negative mode, thus making it possible to determine the molecular weight of the sample [63, 64].

4.4.1.3 Desorption Chemical Ionization (DCI)

The process is identical to IC except that the sample is deposited in solution on a filament. After evaporation of the solvent, the sample is introduced into the source where there is a certain pressure of reactive gas. Rapid heating of the filament leads to the evaporation of the sample and its ionization by the reactive gas before the pyrolysis phenomena have taken place. The reactive gases used are the same as in conventional IC. This method provides, in positive mode, $[M+H]^+$ protonated molecules and $[M+G+H]^+$ adducts that are likely (or not) to fragment [63, 64].

4.4.1.4 Field Desorption (FD)

This method is well suited to nonvolatile and temperature-unstable molecules. The ionization is carried out in an intense electrostatic field (107–108 $V.cm^{-1}$). These conditions allow the molecules to lose an electron without heating and prevent their fragmentation [64].

4.4.1.5 Matrix-Assisted Laser Desorption–Ionization (MALDI)

This technique uses an aromatic solid matrix (Cyanocinnamic acid, sinapinic acid, 2,5-dihydroxy-benzoic acid, etc.) in which the sample is dispersed. Laser irradiation results in an electronic and thermal excitation of the matrix molecules that can give protons to the sample, leading to its ionization. The matrix thus serves as an energy vector between the laser beam and the sample molecule. Coupled with a time-of-flight analyzer, matrix-assisted laser desorption is a method of choice for the study of high-mass compounds [62].

4.4.1.6 Fast-Atom Bombardment FAB

The sample is dissolved in a liquid matrix with low vapor pressure, capable of giving off protons (glycerol, thioglycerol, meta-nitrobenzyl alcohol). In the FAB, atoms (Xe, Ar, Kr) accelerated to near 10 keV bombard the sample solution. In addition to the protonated molecules $[M+H]^+$, FAB also conducts, in positive mode, the formation of cationized molecules such as $[M+Na]^+$, which results in an increase of 22 units compared to the value of m/z expected for the ion $[M+H]^+$. This can be a problem when the compound is unknown and the $[M+Na]^+$ion is accompanied by a $[M+2Na-H]^+$ ion. If there is less ambiguity about the molecular weight, it may be useful to switch to negative mode to change the properties of the matrix. In this technique, polar molecules of high-molecular weight (peptides, proteins, saccharides, etc.) can be analyzed and structural determinations made, but unfortunately, we often get a significant background due to the liquid matrix [64].

4.4.1.7 Electrospray Ionization (ESI)

Electrospray ionization mass spectrometry (ESI-MS) is a method of choice for determining the mass of biomolecules. The electrospray is based on the introduction of an aqueous solution of the compound to be analyzed by a very fine capillary carried to high potential. This tension creates charges in the solution and, at the end of the capillary; a "spray" of fine droplets (1 μ) is formed, favored by a pneumatic nebulization. In the source at atmospheric pressure, these droplets encounter a flow of hot nitrogen against the current which leads to the gradual evaporation of the solvent. The size of the droplets decreases and the concentration of charged ions increases. When the electrostatic repulsion between ions reaches a critical point, the droplets explode, releasing ions into the gas phase. The desolvated ions are focused by several samplers to the mass analyzer [63, 66].

4.4.1.8 Coupling of a Chromatographic Technique with a Spectroscopic Technique

The identification and quantification of the constituents of a natural mixture always remain delicate operations that often require the use of multiple complementary techniques. The technological development of ionization and chromatography techniques has given rise to LC-MS and GC-MS couplings which have played a very important role in spectrometric identification of natural compounds and especially in the separation of optically active drugs [67].

GC-MS

This method is very sensitive and it is able to identify most of the compounds present, despite their range of concentration which can be very extensive. It is compatible with EI and CI ionization sources. After chromatographic separation, it is necessary to analyze compound by compound. Thanks to the carrier gas and according to the retention time of the different compounds, they will pass one by one in a mass spectrometer where they will undergo an electronic (70 eV) and chemical ionization. The mass spectrometer unit of the GC-MS provides structural information for a particular peak on the chromatogram [68]. When the compound eluted from the column is unknown, the mass spectral data can provide detailed information about its structure, empirical formula, family and the presence of important functional groups. The GC-MS is frequently used in identification of plant metabolites like essential oils, fatty acids, triglycerides and polysaccharides [69].

LC-MS

Currently, the coupling of liquid chromatography (HPLC) with mass spectrometry uses in the vast majority of cases atmospheric pressure sources (electrospray, APCI or APPI), which offers a powerful analytical tool for chemists and biochemists [70]. The use of LC-MS in the detection, identification and purification of components in a crude mixture of natural products such as flavonoids, alkaloids and complex lipids has been successfully demonstrated [72–76].

4.4.2 UV–Visible Spectroscopy (UV–VIS)

UV–visible spectroscopy is based on electronic transitions of the bonding electrons (valence layer) that characterize a molecule. It allows assays to be carried out thanks to the Beer–Lambert law which shows a relation of proportionality between the absorbance and the concentration, as well as a structural study of the complexes by the study of absorption spectra. The UV–VIS absorption is due to the presence of certain chromophore groups namely C=C, C=O, C≡N and OH. This technique is frequently used to identify phenolic compounds such as flavonoids and to establish the sitting of unsubstituted phenolic hydroxyl groups by adding shift reagents [49].

4.4.3 Infrared Spectroscopy (IR)

Infrared spectroscopy is considered among the most widely used spectroscopic techniques in the structural elucidation of new molecules. IR is related to the vibrational changes that occur within a molecule when excited by infrared radiation. It provides the chemical bonds (elongation and deformation vibrations) indicating the presence of characteristic functional groups in natural products and even the determination of the structure of simple molecules [62]. Otherwise, the infrared sometimes remains the only way to remove the ambiguities which can remain as for the structure of a molecule.

Today, Fourier-transformed infrared spectroscopy (FTIR) is a high-resolution analytical technique that offers a fast and non-destructive investigation for the fingerprint of plant extracts and to elucidate the structures of their chemical constituents [63].

4.4.4 Nuclear Magnetic Resonance (NMR)

In recent decades, NMR has considerable growth and has become an indispensable tool for chemists, biochemists, physicists and more recently in medical research. Indeed, NMR is one of the most valuable analytical tools available today. It enables chemists to identify the structure of organic compounds and biomolecules in order to develop new drugs. The nuclei most studied by NMR in organic chemistry are hydrogen ^{1}H (99.8% abundance), fluorine ^{19}F (100%), phosphorus ^{31}P (100%) and carbon ^{13}C (1.1%) and ^{15}N nitrogen (0.37%). ^{1}H and ^{13}C NMR provide, through the variety of its one- and two-dimensional experimental techniques, essential information for the structure identification of natural molecules [77, 78].

4.4.4.1 One-Dimensional NMR (1D-NMR)

The structural study of a natural product requires at first the recording of the 1D spectra (NMR, ^{1}H, ^{13}C) under the best possible conditions. This provides important information on all type of molecular protons and carbon through the determination of chemical shift and coupling constants.

Proton NMR (^{1}H-NMR)

^{1}H-NMR provides important information on chemical shift and coupling constants of hydrogen nuclei, thus giving a clear idea of their position in the molecule. The integral curve or integration provides indications on the number of protons present in the molecule, and the chemical shift of the signals reveals information about the chemical and electron environment of the protons [79, 80]. In addition, selective irradiation and decoupling experiments can be performed to identify the complex signals of a rotation system and will show which atoms are present in neighboring groups [81, 82].

Carbon NMR (^{13}C-NMR)

As for proton NMR, carbon NMR is a set of signals that appear as singlets due to decoupling with attached protons. Different 1D carbon NMR recording techniques have been developed to differentiate different carbon types such as primary, secondary, tertiary and quaternary. The range of chemical shift values differs between the ^{1}H (normally 0–14) and ^{13}C-NMR (normally 0–230) that results from the two nuclei having different numbers of electrons around their corresponding nuclei as well as different electronic configurations [81, 82]. Currently, the widely used carbon NMR techniques are DEPT (improvement without distortion by polarization transfer) or APT (proton test attached) spectra. In these experiments, resulting signals are observed that can be either positive or negative, depending on the number of protons attached.

4.4.4.2 Two-Dimensional NMR (2D-NMR)

Unfortunately, in complex molecules as is often the case for natural products, information on chemical shifts and coupling constants is sometimes inaccessible in one-dimensional NMR spectra because of signal overlaps. To overcome these problems, 2D-NMR experiments offer considerable improvements in the resolution of the spectra and allow the characterization of complex structures [83].

High-resolution multi-pulse and two-dimensional experiments provide ^{1}H-^{1}H and ^{1}H-^{13}C scalar correlations (by binding) and ^{1}H-^{1}H (in space) dipolar molecular connectivity data that essentially map out the molecular structure. In the field of sensitivity, the stronger magnetic fields produced by superconducting magnets, microprobe technologies and cryogenic electronics have significantly decreased the amount of material needed for NMR analysis to less than one milligram. The 2D NMR spectra allow a correlation of the resonance frequencies of different nuclei. Homonuclear or heteronuclear correlations can be distinguished according to whether the two frequency axes represent frequencies of nuclei having the same or different nature.

^{1}H-^{1}H COSY (Correlated SpectroscopY)

Correlation spectroscopy ^{1}H-^{1}H COSY is one of the most useful 2D experiments. This experiment makes it possible to visualize correlation spots indicating the coupling among neighboring protons [84]. In fact, the ^{1}H-^{1}H COSY spectrum reveals homonuclear correlations (spin couplings) between vicinal hydrogen separated by three links ($^{3}J_{H\text{-}H}$). This allows to identify neighboring carbon atoms connected by a chemical bond [85].

TOCSY (TOtal Correlation SpectroscopY)

The TOCSY experiment is an indispensable tool for interpreting the spectra of complicated molecules, such as polysaccharides and amino acids. It provides molecular connectivity information based on proton–proton correlations, not only between the geminal or vicinal protons as in H-H COSY, but also between all the protons in a given spin system. Correlations are observed between far protons via ^3J and ^4J couplings [86].

DQF-COSY (Double Quantum Filter-Correlated SpectroscopY)

Compared to the ^1H-^1H COSY and TOCSY experiments, the DQF-COSY affords a very good resolution for measuring proton–proton couplings. Also, the DQF version has the advantage of making the spectrum "cleaner" by reducing the diagonally dispersive peaks of the COSY experiment, which makes the interpretation significantly easier [86]. But, in view of the diminution of its sensibility, this experiment will only be useful for large molecules. This can be best supported from the DQF- COSY and TOCSY spectra of the sugar moiety of isorhamnetin 3-O-β-D-xylopyranosyl-(1'''' → 3''')-α-L-rhamnopyranosyl-(1''' → 6'')-β-D galactopyranoside given in Fig. 4.4 [49].

Noesy/Roesy

Nuclear Overhauser Enhancement Spectroscopy (NOESY) provides a key method for identifying spatially close pairs of protons (through dipolar interactions) even if they are separated by a large number of bonds and therefore provides information about conformations and relative configurations of the molecules [78]. Figure 4.5 shows a nicely NOESY spectrum of α-D-Galactosyl-(1'' → 6')-α-D-glucosyl-(1' → 2)-β-D-frutosylundecaacetate isolated from *Pinus halepensis* [77].

Otherwise, rotating-frame nuclear overhauser enhancement spectroscopy (ROESY) has positive peaks for molecules of all molecular weights. For high-molecular weight molecules, the NOE effect may be zero. For this, NOESY spectra are useful for small molecules [88, 89] and ROESY for molecules of intermediate molecular weight (about 700–1500 Da) [90] such as flavonoids glucosides. The sequence ROESY is therefore complementary to the sequence NOESY.

HSQC/HMQC

Heteronuclear multiple-quantum coherence spectroscopy (HMQC) is a sequence that gives correlation spots due to direct couplings (^1JC-H) between specific carbon atoms and the protons bound to them.

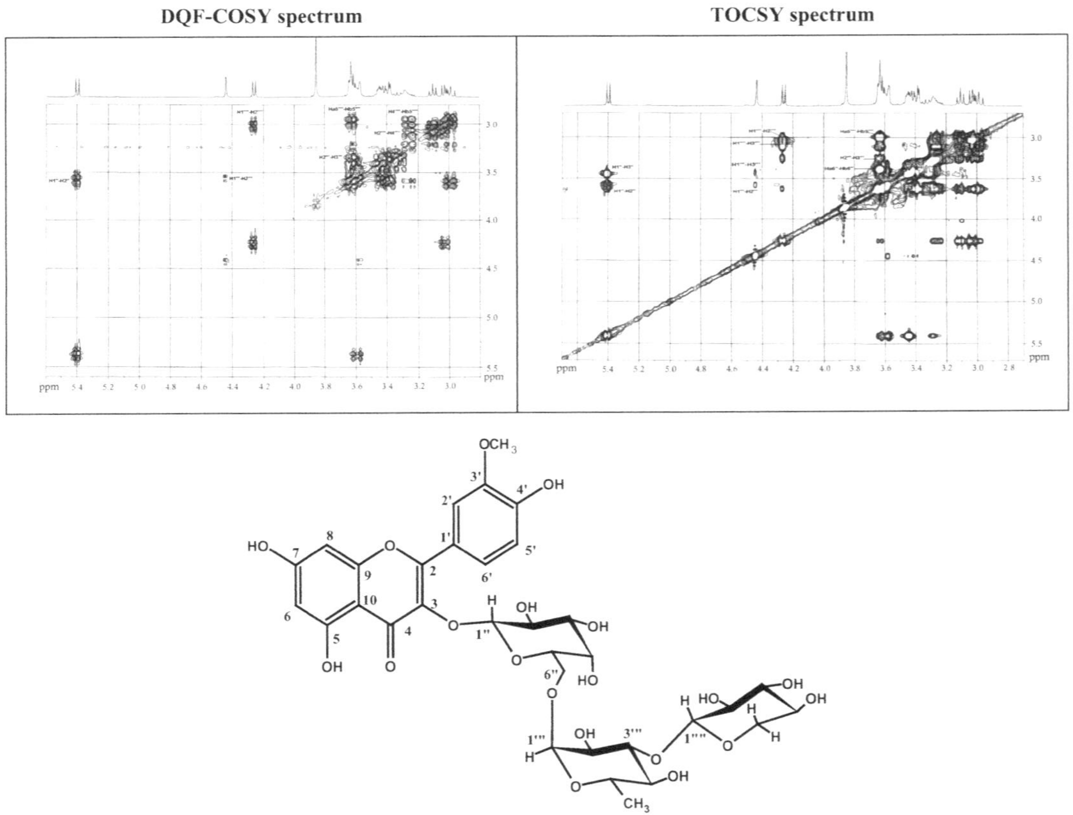

Fig. 4.4 DQF-COSY and TOCSY spectra of the sugar moiety of isorhamnetin 3-O-β -D-xylopyranosyl-(1'''' → 3''')-α-L-rhamnopyranosyl-(1''' → 6'')-β-D-galactopyranoside in DMSO-d_6

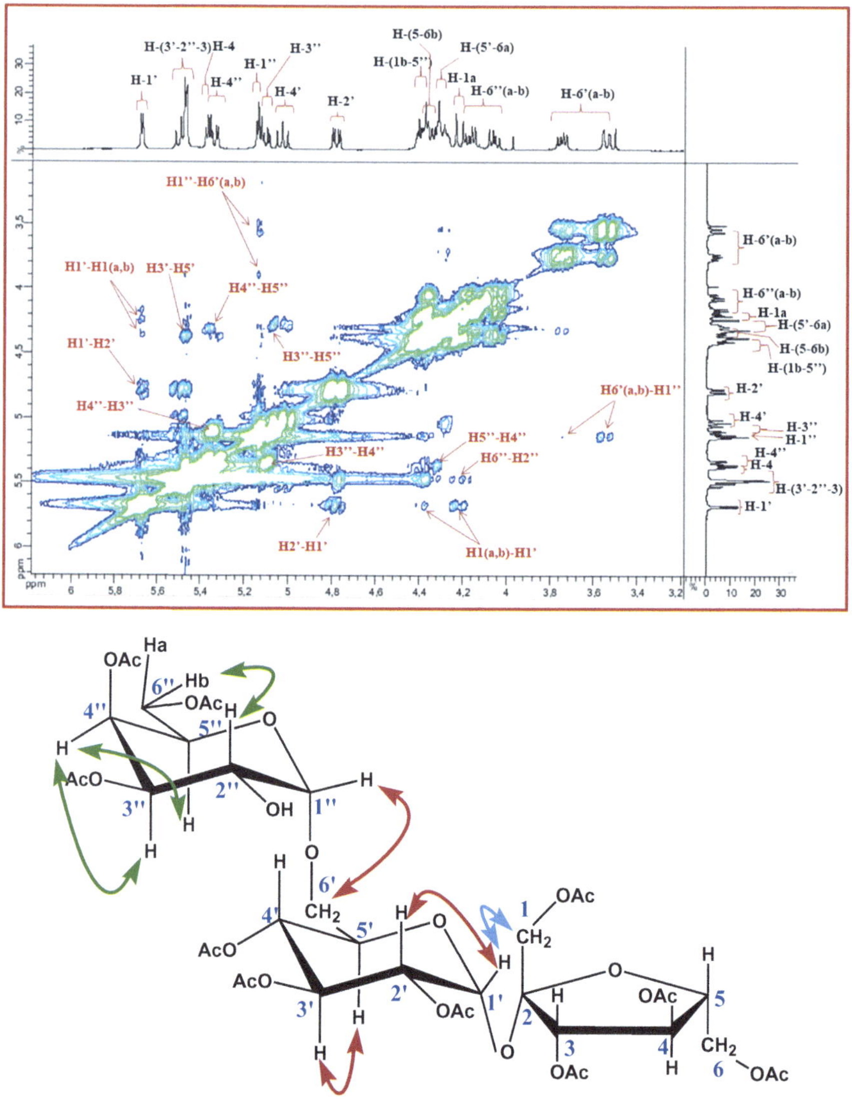

Fig. 4.5 Significant spin–spin correlations of the NOESY spectrum of α-D-Galactosyl-(1'' → 6')-α-D-glucosyl-(1' → 2)-β-D-frutosylundecaacetate in $CDCl_3$

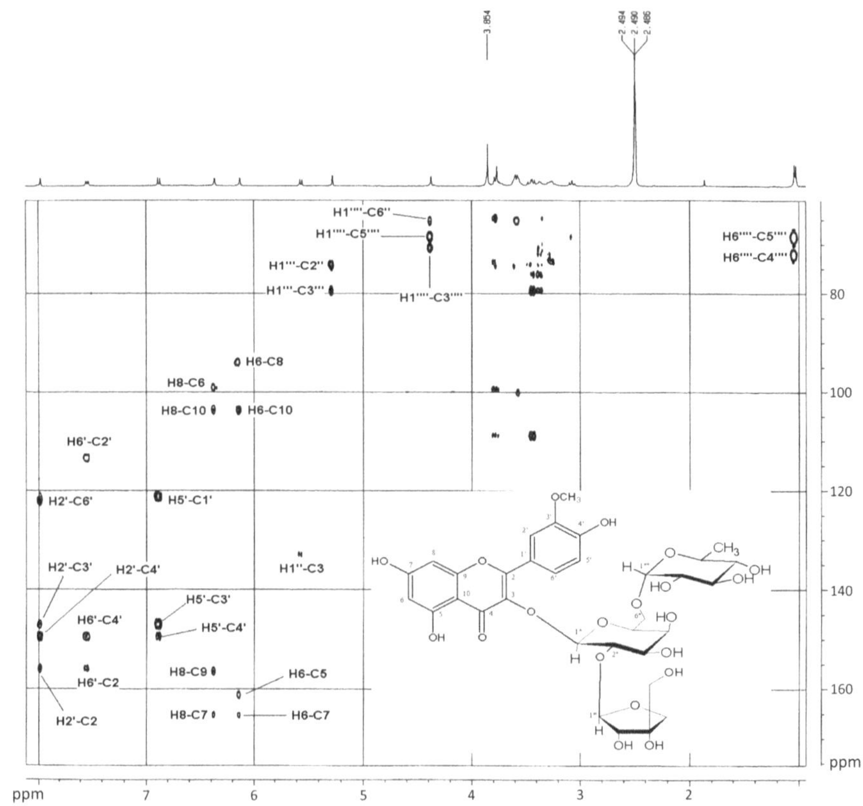

Fig. 4.6 Some $^1J_{H\text{-}C}$ couplings observed in HSQC spectrum of isorhamnetin 3-*O*-β-D-xylopyranosyl-(1'''' → 3''')-α-L-rhamnopyranosyl-(1''' → 6'')-β-D-galactopyranoside in DMSO-d_6

Heteronuclear single-quantum coherence spectroscopy (HSQC) provides the same connectivity information as HMQC but gives narrower resonances for ^1H-^{13}C correlations. This technique, better than HMQC, has the advantage of decreasing or even removing the artifacts [86]. This is shown in HSQC spectrum of isorhamnetin 3-O-β-D-xylopyranosyl-(1'''' → 3''')-α-L-rhamnopyranosyl-(1''' → 6'')-β-D-galactopyranoside (Fig. 4.6) [49].

HMBC

Heteronuclear multiple-bond correlation spectroscopy (HMBC) provides long-range connectivity between protons and carbon atoms separated by 2–4 covalent bonds. Direct one-bond cross-peaks are suppressed. Hence, this tool can detect quaternary carbon atoms as well as determine the junctions between the different parts of the molecule as evidenced by inter-glycosidic bonds in sugar moieties. Figure 4.7 is a

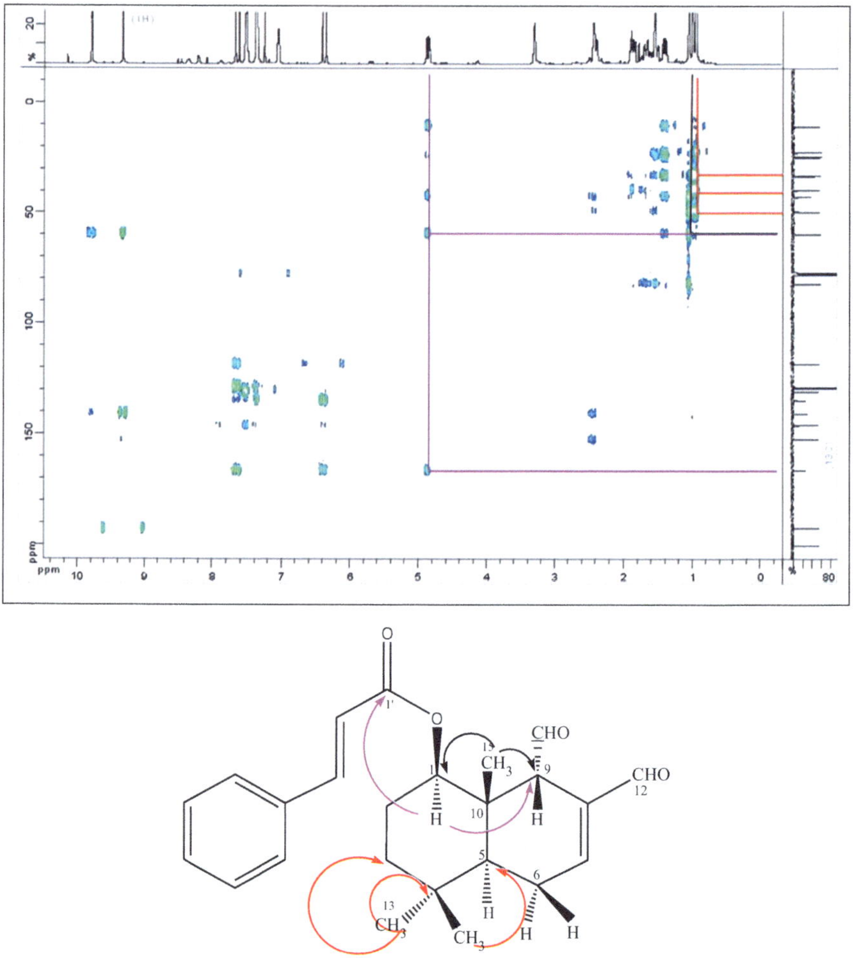

Fig. 4.7 Main correlations observed in HMBC spectrum of 1β-*O*-*E* cinnamoyl polygodial

HMBC spectrum of 1β-o-Ecinnamoyl polygodial [91] with color projections that show the major correlations indicated by the crossed peaks.

4.5 Conclusion

The isolation of natural products has undergone several transitions over the years. All stages of extractions are equally important in the investigation of medicinal plants. The plant material preparation such as drying and grinding affects the extraction efficiency and the phytochemical constituents of the final extract. The evaluation and selection of extraction methods are depending on the study objectives, samples and target compounds. An excellent extraction can be achieved in shorter time, and the obtained extract can have improved quality and yield than that recovered by a conventional method. Methods such as SFE, MAE and UAE are better suitable for the extraction of volatile and heat-labile compounds, which is not the case with the conventional methods. For alternative solvents, SCF is considered to replace conventional solvents since they are regarded as environmental benign solvents and are better in the diminution of energy consumption, solvent consumption and wastes emission in the separation process. Furthermore, some physical methods, such as microwave-assisted extraction, ultrasonic extraction and supercritical chromatography, have been performed to improve the separation efficiency of natural products on laboratory scale. Nevertheless, the application of such methods in industry scale still remains to be fully explored. Mutual combination of different separation methods is important for various bioactive compounds. Hence, it is important to investigate the combination ways of diverse separation methods in the future. On the other hand, the improvement of separation technologies is highly dependent on the amelioration of separation media. Consequently, researchers have to focus their researches on the development of novel adsorbents. Finally, great improvements in structure determination were made over the past decades. While NMR spectrometry has noticeably improved through the use of 2D techniques, especially proton-detected heteronuclear correlations, the application of mass spectrometry has extended through the use of ESI and APCI interfaces that led to a wide application of LC-MS methods.

References

1. Levey M (1959) Chemistry and chemical technology in ancient Mesopotamia. Elsevier, Amsterdam, The Netherlands
2. Barnes J, Anderson LA et al (2007) Herbal Medicines. A guide for Healthcare Professionals, 3rd edn. Pharmaceutical Press, London, pp 1–23
3. Rocha LG, Almeida J et al (2005) A review of natural products with antileishmanial activity. Phytomed 12:514–535
4. Balandrian MFJ, Kjoke A et al (1985) Natural plant chemicals: source of industrial and medicinal materials. Sci J 228:1154–1160
5. Sticher O (2008) Natural product isolation. Nat Prod Rep 25:517–554
6. Dai J, Mumper RJ (2010) Plant phenolics: extraction, analysis and their antioxidant and anticancer properties. Mol 15:7313–7352
7. Evans WC (2002) General methods associated with the phytochemical investigation of herbal products. In Trease and Evans Pharmacognosy, 15 edn. New Delhi: Saunders (Elsevier), pp 137–148
8. Borhan MZ, Ahmad R et al (2013) Impact of nanopowders on extraction yield of Centellaasiatica. Adv Mater Res 667:246–250
9. Arya V, Thakur NM et al (2012) Preliminary phytochemical analysis of the extracts of Psidium leaves. J Pharmacogn Phytochem 1:1–5
10. Rathi BS, Bodhankar SL et al (2006) Evaluation of aqueous leaves extract of *Moringa oleifera* Linn for wound healing in albino rats. Indian J Exp Biol 44:898–901

11. Handa SS, Khanuja SPS et al (2008) Extraction technologies for medicinal and aromatic plants, 1st edn, no. 66. United Nations Industrial Development Organization and the International Centre for Science and High Technology, Italy
12. Mohameda RS, Mansoor GA (2002) The use of supercritical fluid extraction technology in food processing. Food Technol Magazine. The World Markets Research Centre, London, UK
13. Reverchon E, Marco I (2006) Supercritical fluid extraction and fractionation of natural matter. J Supercrit Fluids 38:146–166
14. Sticher O (2008) Natural product isolation. Nat Prod Rep 2008(25):517–554
15. Hemwimon S, Pavasant P et al (2007) Microwave assisted extraction of antioxidative anthraquinone from roots of Morindacitrifolia. Sep Purif Technol 54:44–50
16. Tonthubthimthong P, Chuaprasert S et al (2001) Supercritical CO_2 extraction of Nimbin from neem seeds an experimental study. J Food Eng 47:289–293
17. Pereira CG, Meireles MAA (2010) Supercritical fluid extraction of bioactive compounds: fundamentals, applications and economic perspectives. Food Bioprocess Tech 3:340–372
18. Liza MS, Rahman RA et al (2010) Supercritical carbon dioxide extraction of bioactive flavonoid from *Strobilanthescrispus*(*PecahKaca*). Food Bioprod Process 88:319–326
19. Ganan N, Brignole EA (2011) Fractionation of essential oils with biocidal activity using supercritical CO_2-experiments and modeling. J Supercrit Fluids 55:58–67
20. Vidovic S, Mujic I et al (2011) Extraction of fatty acids from boletus edulis by subcritical and supercritical carbon dioxide. J Am Oil Chem Soc 88:1189–1196
21. Aleksovski S, Sovova H et al (1998) Supercritical CO_2 extraction and Soxhlet extraction of grape seeds oil. Bull Chem Technol Maced 17:129–134
22. Huie CW (2002) A review of modern sample-preparation techniques for the extraction and analysis of medicinal plants. Anal Bioanal Chem 373:23–30
23. Serra AT, Seabra IJ et al (2010) Processing cherries (*Prunus avium*) using supercritical fluid technology. Part 1: recovery of extract fractions rich in bioactive compounds. J Supercrit Fluids 55:184–191
24. Kehili M, Kammlott M et al (2017) Supercritical CO_2 extraction and antioxidant activity of lycopene and β-carotene-enriched oleoresin from tomato (*Lycopersicum esculentum* L.) peels by-product of a Tunisian industry. Food Bioprod Process 102:340–349
25. Kehili M, Schmidt LM et al (2016) Biorefinery cascade processing for creating added value on tomato industrial by-products from Tunisia. Biotechnol Biofuels 261:1–12
26. Jadhav D, Rekha BN et al (2009) Extraction of vanillin from vanilla pods: a comparison study of conventional Soxhlet and ultrasound assisted extraction. J Food Eng 93:421–426
27. Cares MG, Vargas Y et al (2009) Ultrasonically assisted extraction of bioactive principles from *Quillaja Saponaria* Molina. Phys Procedia 3:169–178
28. Huaneng X, Yingxin Z et al (2007) Ultrasonically assisted extraction of isoflavones from stem of *Pueraria Lobata* (Willd.) Ohwi and its mathematical model. Chin J Chem Eng 15:861–867
29. Trusheva B, Trunkova D et al (2007) Different extraction methods of biologically active components from propolis: a preliminary study. Chem Cent J 1:1–4
30. Ebrahim N, Kershi M et al (2014) antioxidant activity and anthocyanin content in flower of *Mirabilis jalab* L. collected from Yemen. World Appl Sci J 29:247–251
31. Yingngam B, Monschein M et al (2014) Ultrasound-assisted extraction of phenolic compounds from *Cratoxylumformosum* ssp. Formosum leaves using central composite design and evaluation of its protective ability against H_2O_2-induced cell death. Asian Pac J Trop Med 7:497–505
32. Cho YJ, Hong JY et al (2006) Ultrasonication assisted extraction of resveratrol from grapes. J Food Eng 77:725–730
33. Jain T, Jain V et al (2009) Microwave assisted extraction for phytoconstituents—An overview. Asian J Res Chem 2:19–25
34. Kaufmann B, Christen P (2002) Recent extraction techniques for natural products: microwave-assisted extraction and pressurized solvent extraction. Phytochem Anal 13:105–113
35. Ahuja S, Diehl D (2006) Sampling and Sample preparation. In: Ahuja S, Jespersen N (eds) Comprehensive analytical chemistry, vol. 47. Elsevier (Wilson & Wilson), Oxford, UK pp 15–40

36. Kothari V, Gupta A et al (2012) Comparative study of various methods for extraction of antioxidant and antibacterial compounds from plant seeds. J Nat Remedies 12:162–173
37. Sasaki K, Honda W et al (1998) A study of microwave sterilizer for injection ampoules (no. 4): application to sterilization of thermally labile drug solutions. J Pharm Sci Technol 58:125–135
38. Letellier M, Budzinski H (1999) Microwave assisted extraction of organic compounds. Analusis 27:259–270
39. Mandal V, Mohan Y et al (2007) Microwave assisted extraction: an innovative and promising extraction tool for medicinal plant research. Pharmacog Rev 1:7–18
40. Li H, Deng Z et al (2012) Microwave-assisted extraction of phenolics with maximal antioxidant activities in tomatoes. J Food Chem 130:928–936
41. Thomas R, Tripathi R et al (2012) Comparative study of phenolics and antioxidant activity of phytochemicals of T. chebula extracted using microwave and ultrasonication. Int J Pharm Sci Res 3(1):194–197
42. Ben-Youssef S, Fakhfakh J et al (2017) Green extraction procedures of lipids from Tunisian date palm seeds. Ind Crops Prod 108:520–525
43. Upadhyay R, Ramalakshmi K et al (2012) Microwave-assisted extraction of chlorogenic acids from green coffee beans. Food Chem 130:184–188
44. Orio L, Alexanderu L et al (2012) UAE, MAE, SFE-CO_2 and classical methods for the extraction of Mitragynaspeciosa leaves. Ultrason Sono chem 19:591–595
45. Hahn-Deinstrop E (2000) Applied thin layer chromatography: best practice and avoidance of mistakes. Wiley-VCH, Weinheim, Germany
46. Sasidharan S, Chen Y et al (2011) Extraction, isolation and characterization of bioactive compounds from plants' extracts. Afr J Tradit Complement Altern Med 8:1–10
47. Altemimi A, Lakhssassi N et al (2017) Phytochemicals: extraction, isolation, and identification of bioactive compounds from plant extracts. Plants 6:42
48. Harborne JB (1998) Phytochemical methods: a guide to modern techniques of plant analysis. 2nd edn. Chapman and Hall publishers, Springer, Germany
49. Krishnananda PI, Amit GD et al (2017) Phytochemicals: extraction methods, identification and detection of bioactive compounds from plant extracts. J Pharmacogn Phytochem 6:32–36
50. Salah HB, Jarraya R et al (2002) Flavonoltriglycosides from the leaves of Hammada scoparia (POMEL) ILJIN. Chem Pharm Bull 50:1268–1270
51. Zhang Z, Pang X et al (2005) Role of peroxidase in anthocyanin degradation in litchi fruit pericarp. Food Chem 90:47–52
52. Hostettmann K, Wolfender JL et al (1997) Rapid detection and subsequent isolation of bioactive constituents of crude plant extracts. Planta Med 63:2–10
53. Guiochon G (2001) Basic principles of chromatography. In: Günzler H, Williams A (eds) Handbook of analytical techniques. Wiley-VCH Verlag GmbH, Weinheim, Germany, pp 189–194
54. Yang F-Q, Zuo H-L et al (2013) Preparative gas chromatography and its applications. J Chromatogr Sci 51:704–715
55. Hancock WS (1990) High performance liquid chromatography in biotechnology. Wiley-Interscience, New Jersey, USA
56. Babaee S, Beiraghi A (2010) Micellar extraction and high performance liquid chromatography-ultra violet determination of some explosives in water samples. Anal Chim Acta 662:9–13
57. Michel T, Khlif I et al (2015) UHPLC-DAD-FLD and UHPLC-HRMS/MS based metabolic profiling and characterization of different Olea Europaea organs of Koroneiki and Chetoui varieties. Phytochem Lett 11:424–439
58. Nyiredy S (2001) Planar chromatography: a retrospective view for the third millennium. Springer Scientific Publisher, Debrecen
59. Rajopadhye AA, Namjoshi TP et al (2012) Rapid validated HPTLC method for estimation of piperine and piperlongumine in root of piper longum extract and its commercial formulation. Rev Bras de Farmacogn 22:1355–1361
60. Chaitanya D (2014) Phani.r.s.ch. flash chromatography and preparative HPLC. Res Desk 3:434–439

61. Rouessac F, Rouessac A et al (2004) Analyse Chimique, Méthodes et techniques instrumentales modernes, 6ème édn. Dunod, Paris
62. Silverstein RM, Webster FX et al (2007) Identification spectrométrique de composés organiques, 2ème édn. De Boeck
63. Bernard AS, Clède S et al (2014) Techniques expérimentales en chimie, 2ème édn. Dunod, Paris
64. Richard BVB, LeRoy BM et al (1988) Comparison of electron impact, desorption chemical ionization, field desorption, and fast atom bombardment mass spectra of nine monosubstituted Group VI metal carbonyls. Anal Chem 60:1314–1318
65. Huang EC, Wachs H, Conboy JJ et al (1990) Atmospheric Pressure Ionization Mass Spectrometry. Anal Chem 62:713–725
66. Cech NB, Enke CG (2001) Practical implications of some recent studies in electrospray ionization fundamentals. Mass Spectrom Rev 20:362–387
67. Ching J, Soh WL et al (2012) Identification of active compounds from medicinal plant extracts using gas chromatography-mass spectrometry and multivariate data analysis. J Sep Sci 35:53–59
68. Bouchonnet S, Libong D (2004) Le couplage chromatographie en phase gazeuse-spectrométrie de masse. Actual Chim 7–14
69. Kolsi RBA, Salah HB et al (2017) Sulphated polysaccharide isolated from Sargassumvulgare: characterization and hypolipidemic effects. Carbohydr Polym 170:148–159
70. De Hoffman E, Stroobant V (2001) Mass spectrometry: principles and applications, 2ème edn. Wiley
71. Mroczek T, Ndjoko K et al (2004) On-line structure characterization of pyrrolizidine alkaloids in Onosmastellulatum and *Emilia coccinea* by liquid chromatography–ion-trap mass spectrometry. J Chromatogr A 1056:91–97
72. Bouaziz-Ketata H, Zouari N et al (2015) Flavonoid profile and antioxidant activities of methanolic extract of *Hyparrheniahirta* (L.) Stapf. Indian J Exp Biol 53:208–215
73. Dugo P, Mondello L et al (2000) LC-MS for the identification of oxygen heterocyclic compounds in citrus essential oils. J Pharm Biomed Anal 24:147–154
74. La Torre GL, Saitta M, Vilasi F, Pellicanò T, Dugo G (2006) Direct determination of phenolic compounds in Sicilian wines by liquid chromatography with PDA and MS detection. Food Chem 94:640–650
75. Andries PB, Thomas RC et al (1987) Ion spray interface for combined liquid chromatography/atmospheric pressure ionization mass spectrometry. Anal Chem 59:2642–2646
76. Kolsi RBA, Salah HB et al (2017) Effects of Cymodocea nodosa extract on metabolic disorders and oxidative stress in alloxan-diabetic rats. Biomed Pharmacother 89:257–267
77. Gunther H (1996) La Spectroscopie de RMN: principes de base, concepts et applications de la Spectroscopie de Resonance Magnétique Nucléaire du Proton et du Carbone 13 en Chimie. Elsevier Masson, Paris
78. Bross-Walch N, Kühn T et al (2005) Strategies and tools for structure determination of natural products using modern methods of NMR spectroscopy. Chem Biodivers 2:147–177
79. Giraudeau P, Tea I et al (2014) Reference and normalization methods: essential tools for the intercomparison of NMR spectra. J Pharm Biomed Anal 93:3–16
80. Abraham RJ, Fisher J et al (1988) Introduction to NMR spectroscopy. Wiley, Chichester
81. Derome AE (1987) Modern NMR techniques for chemistry research. Pergamon Press, Oxford
82. Sanders JKM, HunterBK (1993) Modern NMR spectroscopy, 2nd edn. Oxford University Press, Oxford
83. Mahrous EA, Farag MA (2015) Two-dimensional NMR spectroscopic approaches for exploring plant metabolome: a review. J Adv Res 6:3–15
84. Friebolin H (2010) Basic one- and -two-dimensional NMR spectroscopy. 5th edn. Wiley VCH
85. Breton RC, Reynolds WF (2013) Using NMR to identify and characterize natural products. Nat Prod Rep 30:501–524
86. Fuloria NK, Fuloria S (2013) Structural elucidation of small organic molecules by 1D, 2D and multi-dimensional-solution NMR spectroscopy. Anal Bioanal Tech 11:1–8

87. Ben Youssef S, Fakhfakh J et al (2016) Efficient purification and complete NMR characterization of galactinol, sucrose, raffinose, and stachyose isolated from *Pinus* halepensis (Aleppo pine) seeds using acetylation procedure. J Carbohydr Chem 35:224–237
88. Aganova O, Galiullina L et al (2014) The study of the conformation and dynamics of the new quaternary phosphonium salts by NMR spectroscopy. Appl Magn Reson 45:653–665
89. Khodov I, Efimov S et al (2014) Determination of preferred conformations of ibuprofen in chloroform by 2D NOE spectroscopy. Eur J Pharm Sci 65:65–73
90. Canet D, Boubel J-C et al (2002) La RMN: concepts, méthodes et applications, 2ème édn. Dunod
91. Keskes H, Litaudon M et al(2014)Antioxidant and α-amylase inhibitory activities of extract and isolates from Zygogynumpancheri subsp. Arrhantum. J Asian Nat Prod Res 16:1132–1138

Chapter 5
Phytochemistry: A Multifunctional Ecological Heritage and a Rewarding Natural Wealth

Nourreddine Bouarourou, Kamal Aberkani and Adyl Oussaid

Abstract Since long times, human has been treated with plants. In nature, there are various species used for different purposes such as therapeutic, food, cosmic, dietary, chemical, pharmaceutical, agri-food, and industrial. Currently, worldwide population uses traditional medicines to meet primary healthcare needs and reduce the use of chemical drugs. Our ecosystem is a very original natural setting offering a complete range of climates favoring a rich and varied medicinal flora. In recent decades, sector of medicinal plants became very popular and focus on traditional medicine, pharmaceutical industry, food industry, manufacture of natural perfume, soap, esthetics products, insecticides, and disinfectants, etc. This chapter presents a bibliographic overview on the biodiversity of medicinal plants and theirs chemicals compounds.

Keywords Alkaloids · Biodiversity · Carbohydrates · Essential oil · Flavonoids · Medicinal plants · Phytochemistry · Phytotherapy · Polyphenols · Rosmarinic acid

5.1 Introduction

For millennia, herbal medicines were of great use to human health because of their great medical virtues. Moreover, plants were also valued by their aromatic and cosmetic interest. The therapeutic activity of aromatic and medicinal plants (AMP) is required thanks to their active compounds [1].

N. Bouarourou (✉)
Taourirt, Morocco
e-mail: pharmacie7000@gmail.com

K. Aberkani
Department of Biology and Geology, Faculty Multidisciplinary of Nador, University Mohamed First, Postal Box 300, 62700 Selouane, Morocco

A. Oussaid
Department of Chemistry, Faculty Multidisciplinary of Nador, University Mohamed First, Postal Box 300, 62700 Selouane, Morocco

Y. Li and F. Chemat (eds.), *Plant Based "Green Chemistry 2.0"*,
Green Chemistry and Sustainable Technology,
https://doi.org/10.1007/978-981-13-3810-6_5

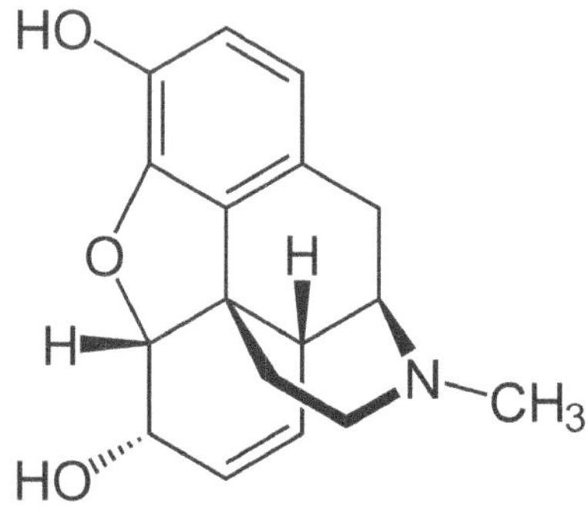

Fig. 5.1 Structure of Morphine

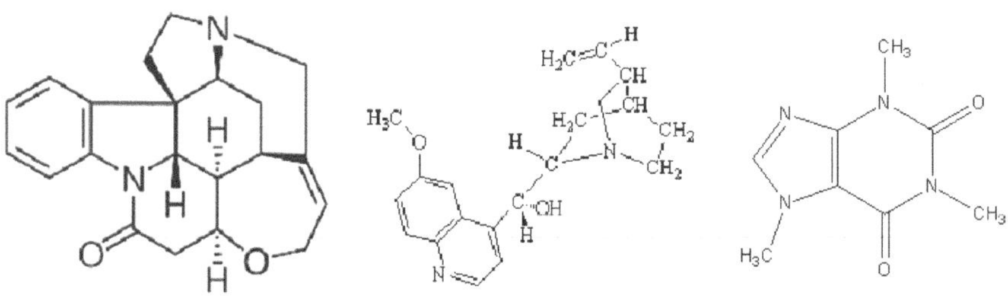

(a) Strychnine **(b)** Quinine **(c)** Caffeine

Fig. 5.2 Structures of three naturals substances: **a** strychnine; **b** quinine; and **c** caffeine

5.1.1 History

- 400 years BC, the doctor of medicine Hippocrates has listed 200 plants of medicinal interest.
- 50 After J.-C.: Dr. Dioscoride discovered 600 plants each of which is intended for the cure of a specific pathology.
- After J.-C., the Doctor of Pharmacy Galen revolutionized the world of medicinal plants by giving them a galenic form adapted to the use.
- A few decades later, homeopathy has just put its place in field of AMP.
- In 1800 AD, Dr. Baunie Antoine, pharmacist and French chemist, implemented the equation Baumé $= 1443 * (1 - 1/\text{scientific weight of the plant})$.
- In 1800 AD, Dr. Derson Charles, French pharmacist and chemist, isolated morphine from the plant (Fig. 5.1).
- In 1820 after J.-C., Doctors Pelletier and Carventon, French pharmacists and chemists, discover the strychnine, the quinine and the caffeine, origin of AMP (Fig. 5.2).

In 1870 AD, Dr. Claud Bernard, a physiologist physician, demonstrated the correlation between chemical structures of a substance derived from the plant.

5.1.2 What Is a Medicinal Plant?

The decree of January 6, 1989, appeared in the French Official Journal of February 18, 1989, defined in a legal way the term Medicinal plant:

Medicinal Plants are plant drugs within the meaning of the European Pharmacopoeia of which "at least some of them have medicinal properties."

"Plants with medicinal properties may also have food or condiment uses, or be used for the preparation of hygienic beverages" decree 2008 of the French Pharmacopoeia 10th edition.

Definition published in the note to the manufacturers of the notebooks of the agency number 3 (Official Journal of February 7, 1998) [2]:

- Herbal medicines are medicines whose active ingredients and are exclusively vegetable drugs and/or preparations based on herbal drugs.
- Plant drugs are parts of fresh or dried plants used for therapeutic purposes. They are sometimes whole plants, most often plant parts (root, bark, flowery top, leaf, flower, fruit, seed) whole or fragmented. They are also vegetable drugs, the juices extracted from plant (oleoresins, gum, latex, etc.) having not undergone any galenic operation.
- The vegetable drug preparations are in extract, tincture, fatty or essential oil, fragment, powder or juice expressed by pressure. Their production involves operations of fractionation, purification, or concentration.
- Substances such as solvents, diluents, preservatives may be included in the composition of vegetable drug preparations. The presence of these substances must be indicated in the composition.
- Active constituents are substances or groups of substances chemically defined in the contribution to the therapeutic effect of an herbal drug or preparation which is well known.
- Tracers are chemically defined constituents of a plant drug that is of interest for performing controls. A tracer can be used to calculate the quantity of herbal drugs or preparations present in the finished products, since the quantity of tracer in the herbal drug or the preparation has been determined during the tests on the raw materials. Wherever possible, the tracer should not be a degradation product of a constituent of the drug.
- The active ingredients are vegetable drugs or preparations based on vegetable drugs, whether the constituents with therapeutic effect are known or not.
- The raw materials are products (active ingredients, excipients, solvents, gases, etc.) used for the manufacture of drugs, and their quality is defined by a monograph (pharmacopoeia or internal monograph).

A plant after harvesting must be stabilized; this stabilization operation consists of removing the water from the plant and destroying the enzymes with boiling alcohol (Bourquelot process). Some active substances pass into the alcohol called extractive stabilization to obtain a stabilized extract [3].

Moreover, in order to stabilize robust drugs (seeds and nuts), we use the process of wet heat (steam under pressure: Goris and Arnaud process) or alcohol vapor

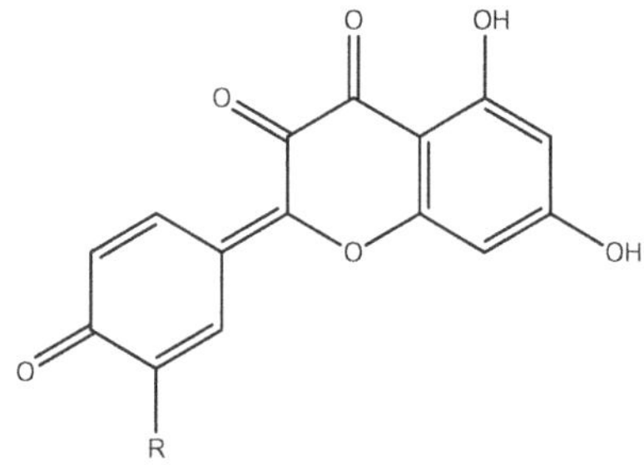

Fig. 5.3 Structure of polyphenol

(Goris and Perrot process) and in some cases, the dry heat process is used (ultra-high temperature: UHT) [4].

Packaging and storage is usually prepared in cardboard or double envelope kraft paper containers protected from light, air and insects.

To ensure precise control of the quality of a medicinal plant, several parameters must be taken into consideration:

- Identity control: to control the species, genus, and family.
- Purity control: Consists of looking for foreign elements such as pesticide residues, bacterial contaminations (mycotoxins).
- Activity control: must be within the limits described in the standard of the drug as described in the pharmacopoeia:
  - Botanical trials: Consists of a microscopic and macroscopic examination of the drug.
  - Physicochemical tests: To calculate the water content and the search for ashes and to estimate a qualitative research which consists in detecting the characteristics of the active principles (coloring, precipitation, fluorescence, chromatographic methods, spectroscopic methods, etc.). Quantitative methods allow the determination of total active ingredients by analytical methods (colorimetric, spectrometric, gravimetric, etc.).
  - Biological tests: Allow the determination of the acute and chronic toxicity of active ingredients (e.g., lethal dose fifty LD50).

5.2 Compounds of Medicinal Plants

5.2.1 Polyphenols

Polyphenols (Fig. 5.3) are considered the largest metabolic pathway of aromagenese. They are highly valued for their chelating character of free radicals.

The biogenesis of phenols comprises three main routes:

1. Aromagenese pathway which makes it possible to obtain shikimates and polyacetates.
2. Pathway of shikimic acid that includes:

   2.a. The biogenesis of shikimic acid.
   2.b. The biogenesis of cinnamic acids.

Polyphenols are characterized by many physicochemical properties.

- They are the supports of yellow colors, red to blue;
- They are fat-soluble hydro;
- They are chelators of heavy metals;
- They increase the absorbance of the anthocyanin pigments;
- They have a notable acidity due to the oxidability of the phenate ion;
- They are highly valued by the antioxidant character, thus free radicals chelator to block the chain reaction cascade of free radicals and prevent the formation of cytosines by interrupting the inflammatory process;
- They inhibit the oxidation of low-density lipoprotein (LDL) and therefore prevent the formation of atheroma plaques;
- They naturally absorb in the visible for pigments;
- They absorb at least in the UV after exposure or not to ammonia vapor;
- They react by staining in the presence of FeCl3, phosphomolybdates, etc.

With regard to the determination of polyphenols, these derivatives can be determined by two methods:

- In the presence of beef serum albumin (BSA), polyphenols precipitate proteins;
- The polyphenols reduce the ciocalteu folin reagent (phosphotungstic acid and phosphomolybdic acid complex) to a blue-violet oxide mixture assayed spectrophotometric at 700 nm.

These compounds have long been used for the treatment of various diseases. They have expanded medicinal properties:

- Diuretic and antiseptic urinary tract;
- Employed for fossil and influenza conditions;
- Analgesics;
- Anti-inflammatory especially in case of joint pain;
- Salicylated acetyl acid at a dose of 50–150 mg per day inhibits platelet cyclooxygenase and consequently a decrease in the production of thromboxanes (anti-platelet effect);
- Antipyretic;
- Epidemiological studies on aspirin and dose between 75 and 300 mg show that use between 2 and 10 years can reduce the frequency of colorectal cancer, and other studies also show the impact on reducing breast cancer rates and lungs;
- Keratolytic;
- Aspirin is contraindicated in children and adolescents in case of viral infection (chickenpox);

Fig. 5.4 Rosmarinic acid

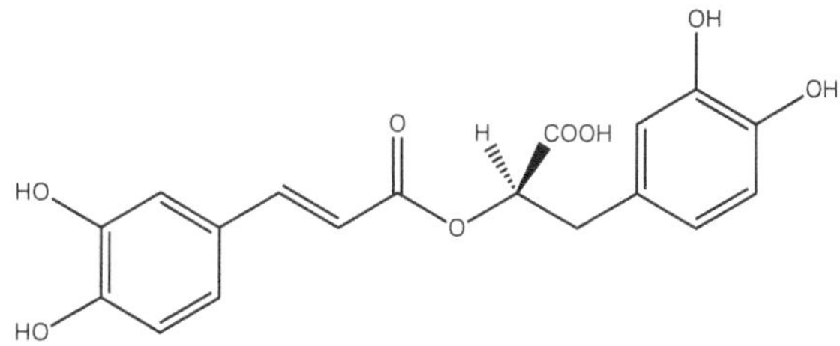

- In case of overdose for infants, it induces neurological disorders.

➢ **Cinnamic acids**: rosmarinic acid (Fig. 5.4) as an antioxidant and antiviral.

➢ **Balms**

- **Tolu balm**
  - Contains 0.25–0.5 free or combined acids;
  - Insoluble in water;
  - Very soluble in alcohol;
  - Insoluble in petroleum ether.

- **Pearl balm**
  - Consisting of benzyl benzoate and benzyl cinnamate;
  - Soluble in alcohol to give a green colored solution in the presence of $FeCl_3$;
  - Contains 5–6% free benzoic acid and cinnamic acid and 50–60% of their benzoic esters.

Use

- It has the character of an antiseptic and healing;
- Indicated in burns and erythema;
- Indicated for pruritus;
- Used in the case of ulcer crevices and frostbite.

- **Benzoin of Laos**
  - Contains 45–55% total acids calculated as benzoic acid;
  - The benzoic styrax is evidenced by the addition of $FeCl_3$ to a solution of benzoin of Laos in the presence of ethanol which gives a greenish coloration.

Use

- Decongestant respiratory tract;
- Against canker sores;
- Against gingivitis;
- Against periodontal diseases;

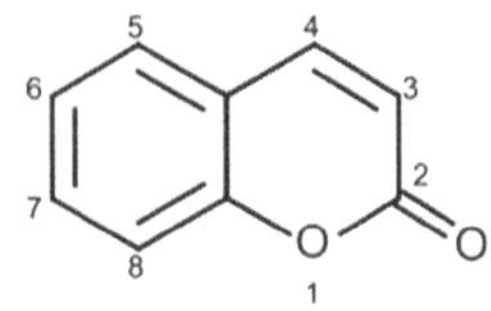

Fig. 5.5 Basic structure of coumarin

- Healing gum injury;
- Indicated in irritate dermatitis and insect bites.

➢ **Active ingredient of coumarins**

In the free state, coumarins (Fig. 5.5) are soluble in alcohols. They are extractable by organic solvents which are not very polar and give in ultraviolet–visible spectrum which has a pace depending on the pH, the position, and the nature of the substituents.

These derivatives have demonstrated flavoring activity and considered as a source of vitamin P (increases the resistance and decreases the permeability of the capillaries). They can be used in puvatherapy thanks to their photo-sensitizing power (treatment of psoriasis at a dose of 20–60 mg per day followed by a session of UV at long wavelengths) and with nucleic bases, form covalent adducts (risk of carcinogen). They have the characteristics of anticoagulant and can be served in the case of digestive disorders and sleep.

➢ **Active ingredient of lignans**

They are characterized by the dimerization of two shikimate units by radical coupling under their alcohol or cinnamic acid forms. These natural species possess essential pharmacological particularities:

- Anti-helminthic;
- Hepatoprotective;
- Slow down the course of rheumatoid arteritis;
- Anti-inflammatory;
- Antihistamine;
- Inhibit the phosphodiesterase of cyclic AMP and cyclic GMP;
- Inhibit the formation of leukotriene;
- Inhibit the production of cytokines including TNF-α;
- Attenuate the symptoms of menopause by interacting with estrogen receptors;
- Cytostatic properties of true lignans (antineoplastic activity);
- The heteroside derivatives of the podophyllotoxin used for anticancer of the "S" phase and their inhibitory activity of topoisomerase II and consequently the cleavage of DNA chains for malignant cells.

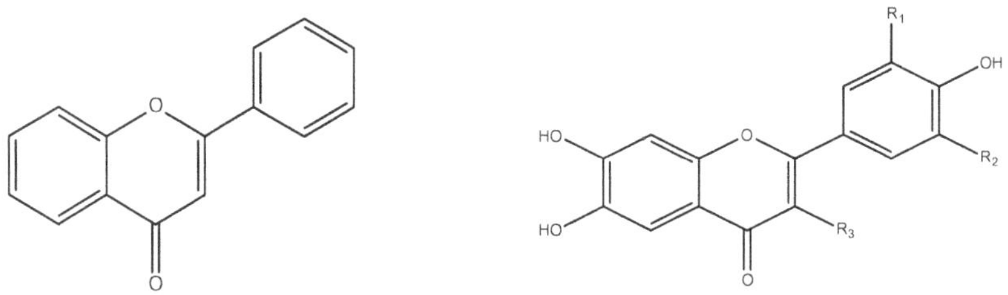

(a) Flavone (or 2-phenylchromone) **(b)** Flavonoid

Fig. 5.6 Structure of: **a** flavone; **b** flavonoid

Fig. 5.7 Structure of anthocyanin

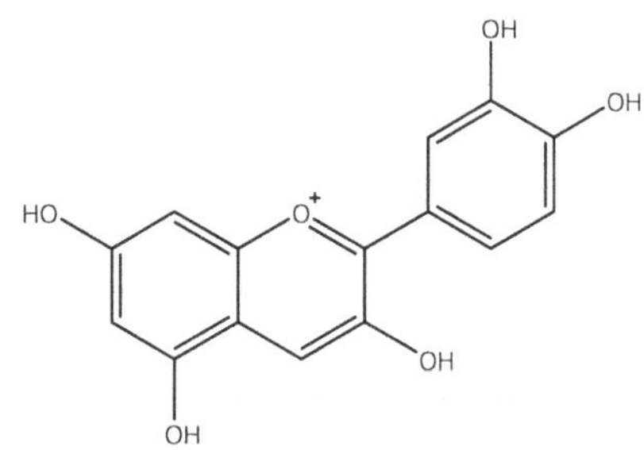

5.2.2 Active Ingredients of Mixed Polyphenol Compounds

➢ **Flavonoids**

They are derivatives of flavone nuclei or 2-phenyl chromone (Fig. 5.6). They are in the form of glycosides stored in vacuolar juices and characterized by their considerable power of chelation of heavy metals [5].

Generally, the flavonoids are soluble in water or in hydroalcoholic media and their determination is carried out by color reaction in the presence of cyanidin, spectrophotometry after passage to high-performance liquid chromatography (HPLC) or by biological methods.

These substances of natural origin are catalysts of electronic transport during photosynthesis and vitamin factors “C2” or “P” (antioxidant effect). They can also stabilize collagen by inhibiting histidine decarboxylase and stimulate proline hydroxylase.

From a pharmaceutical point of view, they have the character of veinotonics (indicated in cases of venous insufficiency, varicose veins, etc.) and are indicated in the metrorrhagia induced by taking micro-progestins and wearing the IUD.

➢ **Active ingredients of anthocyanin**

They are colored heterosides which can be assayed by spectrophotometric (Fig. 5.7).

Their extraction is done by alcoholic solvents in acidulated medium, and their coloration is according to the value of the pH. They are used as hair protectors thanks to vitamin P and allow the increase of the regeneration of retinal purple as they can play the role of food additives (E163).

➢ **Tannins**

This family of natural compounds comes in two forms: hydrolysable and condensed tannins.

The first form which relates to hydrolysable tannins contains polymers having in the center a poly-esterified glucose with gallic and/or ellagic acids. While the second is formed flavanoic polymers linked by C–C bonds.

They form the two colored chelates in the presence of heavy metal salts. Their demonstration is confirmed by staining or formation of a black precipitate after addition of iron perchloride, and their quantitative determination is carried out by the method of Folin–Ciocalteu.

These derivatives are indicated in digestive disorders (epigastralgia, meteorism, bloating, etc.), in the case of venolymphatic insufficiency and capillary fragility and for sleep disorders.

5.2.3 Active Ingredients of Alkaloids

They are substances with a basic character and their solubility is a function of the pH. In basic medium, they are soluble in low polar organic solvents, while in acidic medium, they are very soluble in water.

➢ **Alkaloids with tryptophan derivate**

Indole alkaloids and quinoline nuclei are distinguished. For alkaloids with indole nucleus, we can meet:

- Indoloisoprenic containing rye ergot.
- Indolomonterpenics containing periwinkle, reserpine, antileukemic dimers, strychnine, curares, elliptin, etc.
- For alkaloids containing a quinoline nucleus, we could find: countothecin and quinine.

➢ **Alkaloids with lysine derivate**

This group of active ingredients contains nicotinic acid-derived alkaloids, piperidine alkaloids, quinolizidine alkaloids, and piperidine amides.

➢ **Alkaloids with phenylalanine derivate**

Alkaloids with a tropolone nucleus (colchicine) and alkaloids with an isoquinoleic nucleus (opium and emetic) are distinguished.

➢ **Alkaloids with ornithine derivate**

They are tropane-core alkaloids containing local anesthetics and parasympatholytics.

➢ **Quinolizidine alkaloids**

Contains flavone glycosides, aromatic amines, and sparteine. Cystine remains a neuro- and cardiotoxic alkaloid without neglecting the liver toxicity which can be fatal.

➢ **Alkaloids and piperidine amide**

Piperidine alkaloids are respiratory, bronchodilator, and β-adrenergic analgesics. Moreover, they are well recommended in cases of smoking, antispasmodic. Piperidine amides have anticonvulsant and depressant properties of the central nervous system.

➢ **Pyridinic alkaloids**

They are derivatives of nicotinic acid which is present in tobacco having carcinogenic properties due to nitrosamines.

Nicotine has the following properties:

- Ganglioplegic acting on neuromuscular plaques by ganglionic excitation;
- Increases peristalsis and tone by acting on the smooth intestinal musculature;
- Exciting central nervous system resulting in seizures and tremors.

This active ingredient is a muscarinic receptor agonist then nicotinic (parasympathomimetic) and having a cytotoxic action.

➢ **Tropic alkaloids**

There are supinidine, retronecine and pyrrolizidine. They possess local anesthetic and parasympatholytic properties. They also possess other pharmacological cleanliness:

- Accelerated heart rate by suppressing vagal tone;
- Slowdown of peristalsis;
- Decrease sees inhibition of salivary secretion inducing dry mouth and acting by the same effect on sweat and pancreatic secretion;
- In the case of passive mydriasis;
- Indicated for protection against vagal manifestations in pre-anesthesia;
- Indicated for the prevention and treatment of atrioventricular blocks;
- Sinus bradycardias following myocardial infarction;
- In the case of atrioventricular or atrioventricular block;
- For the treatment of functional disorders of the digestive, biliary, and urinary tracts.

➢ **Alkaloids derived from pseudotropanol**

This set of naturally occurring derivatives contains cocaine which has the capacity to inhibit sodium exchange at the level of the neuronal membrane, thus blocking the sensory message and inducing a local anesthetic effect. It also has parasympathomimetic effects on the autonomic nervous system by blocking the reuptake of dopamine and noradrenaline manifesting as hyperthermia, mydriasis, and vasoconstriction.

In the central nervous system, this derivative acts by stimulating the motor and psychic sensory functions, thus resulting in intellectual stimulation, euphoria, and a rising of inhibition that can lead to suicide or aggression. These effects are followed by a depressive phase, resulting in physical and mental asthenia, respiratory depression, and vasomotility.

➢ **Isoquinoleic alkaloids**

These substances are isoquinoleic and tetrahydroisoquinoleic.

- The precursor amino acids, by decarboxylation, make it possible to obtain dopamine and by deamination make it possible to obtain dihydroxyphenylpyretic acid.
- Benzyletetrahydroisoquinoline, by oxidative coupling, makes it possible to obtain the skeleton aporphines and morphinans; by cyclization of Mannich type make it possible to obtain the skeletons of protoberberine.

Their use is strictly hospital or under restricted prescription in case of rebellious pain. Apomorphine is often indicated as anti-parkinsonian by stimulation of dopamine D1 and D2 receptors.

Isoquinoleics of the aporphine type are well indicated in case of digestive and hepatobiliary insufficiencies.

By intermolecular oxidative coupling of phenylalanine or tyrosine and benzyltetrahydroxyisoquinoline, the resulting curare is obtained.

Parenteral curare alone acts on skeletal striated muscles by inhibiting nerve transmission at the neuromuscular junction. They have a depressant effect on the central nervous system and a ganglioplegic effect on the autonomic nervous system. Their toxicity causes a myo-relaxation inducing a stop of neuromuscular conductivity. In addition, the curares are used in gastric and thoracic surgery as adjuvant of general anesthesia to facilitate endotracheal intubation and assisted ventilation.

Isoquinoleic protoberberic type: are obtained by Mannich-type coupling followed by rearrangement.

Phthalyl isoquinoline alkaloids are low-dose bacteriostatic, and high-dose bactericidal dye and extract are indicated in case of venolymphatic insufficiency and gingival ulceration.

➢ **Alkaloids of protopine type (protopine)**

They are regulators of the bile flow and antispasmodic and facilitate the digestive and urinary elimination.

➢ **Benzophenanthridine Alkaloids (Chelidoine)**

They are digestive antispasmodics; their extract is used for its anti-bacterial and antiviral properties.

➢ **Isoquinoleic alkaloids of the tropolonic type (colchicine)**

They are obtained by phenolic oxidative coupling. They are soluble in water and chloroform. Colchicine is a very toxic substance, in high doses causes cardiovascular collapse and respiratory arrest. It is widely used as an antimitotic. In pharmacies, it is indicated in case of gout attack.

➢ **Isoquinoleic alkaloids of amaryllidaceae (galanthamine)**

It is an inhibitor of acetyl-cholinesterase and crosses the hematomeningeal barrier allowing it to have central effects by increasing cholinergic transmission at cortical synapses (well indicated for Alzheimer's disease).

➢ **Isoquinoline-monoterpene alkaloids (Ipeca)**

They are characterized by different important pharmacological properties. They are considered expectorants, emetics, amoebicides, and virulicides.

➢ **Phenethylamine alkaloids (ephedrine)**

Ephedrine is an indirect sympathomimetic by release of catecholamines from post-ganglionic fibers, thus having an intro-positive and stimulating effect on biliary respiratory centers. Ephedrine crosses the hematoencephalic barrier and induces a psych stimulant action similar to that of amphetamine.

➢ **Indolic Alkaloids**

They are substances with indole nuclei, tryptophan, or tryptamine. We define four groups:

1. Isoprene: they contain derivatives of lysergic acid. (Ergo of segle);
2. Monoaldehydes containing β-carbolines and tetrahydro-β-carbolines;
3. Ester alkaloids: obtained by intramolecular cyclization;
4. Terpenics: They contain indolomonterpenic derivatives.

For isoprene alkaloids, ergoline is the basic skeleton to form three groups:

a. Alkaloids of the clavine type and chanoclavine;
b. Simple amides: They are lysergic amides that are more or less soluble in water (ergine, ergometrine);
c. Peptide amides: These compounds are ergopeptins which are insoluble in water.

Indoloisoprenics are extracted with ether. In acidic medium by adding Na_2CO_3, a violet coloration is obtained characteristic of the presence of ergo.

The assay is performed on the degreased ergo powder after extraction with ether in an alkaline medium ($NaHCO_3$). The ether solution is added to a 1% tartaric acid solution; the van Urk reagent is added to the aqueous phase to obtain a staining evaluated by spectrophotometer which is expressed in ergotamine tartrate.

These same derivatives have demonstrated important pharmacological properties:

- Ergo and its galenic preparations are vasoconstrictors and cytocics;
- Peptide derivatives have a high affinity for α-adrenergic receptors; on the other hand, simple amides have an affinity for serotonergic and dopaminergic receptors;
- Ergometrine is elective oxytocic. It increases the intensity and frequency of uterine contractions; their metabolism is hepatic and their elimination is biliary;
- Ergotamine and ergotoxin are partial agonists of α-adrenergic receptors. They are vasoconstrictors at low doses; this effect is more expressed at the cephalic than peripheral level and is high-dose vasodilators (α-adrenolytic action). In these same substances at the central level, they inhibit the secretion of prolactin and stimulate the central sympathetic structures, thus inducing mydriasis, hyperglycemia, hyperthermia, and vomiting;
- Clavines are indicated for the treatment of intermittent claudication of chronic obliterate peripheral arterial disease and for the symptomatic treatment of neurosensory and cognitive pathological deficit;
- Hemisythetics from ergine and bromocriptine have anti-parkinsonian properties;
- Indolomonterpenic alkaloids (strychnine and brucine) stimulate the respiratory and cardiovascular bulbar centers. In case of overdose, they induce seizures with hyperflexibility;
- Reserpine is an antihypertensive drug that inhibits reuptake of norepinephrine at the post-ganglion level. At the central level, it induces a sedative neuroleptic effect;
- Yohimbine and Ajmalicine are α-blocking sympatholytics inducing an elective peripheral vasodilator effect of arterioles by increasing cerebral flow; at a high dose, Yohimbine is a hypotensive agent;
- Vincamine is indicated in the case of cognitive and neurosensory deficit; and
- Vinblastine and Vincristine have antimitotic properties.

5.2.4 Active Ingredients Based on Carbohydrates

These are compounds based on saccharides or carbohydrates which have the formula $Cn(H_2O)\,n$ with $3 < n < 9$. It is a set of simple sugars called monosaccharides that can bind together to form saccharides or glycosides.

a. **Monosaccharaides**

They are poly-hydroxylated aldehydes or ketones. Their physicochemical properties are due to several parameters, due to:

- Asymmetric carbons: The monosaccharaides are dextrogyrous or levogyrous series, but most of them are dextrogyre series;
- Hydroxyls: The monosaccharaides are soluble in water, oxidizable, and reducible; and
- Carbonyls: The monosaccharaides possess a reducing power on the liquor.

Conformation, isomerism, anomerism: Aqueous sugar solutions are in permanent equilibrium between the two forms α and β.

This family of compounds is composed of a variety of monosaccharaides:

- Pentose: D-xylose, L-arabinose;
- Hexose: D-glucose (dextrose), D-fructose (levulose);
- Deoxy-sugars: L-Rhamnose, D-digitoxose;
- Amino sugars: D-glucosamine.

The extraction of monosaccharaides is due to their solubility in water thanks to the polar hydroxyl group and their precipitation in the presence of alcohol.

They are characterized by their rotatory power, their refractive index, and their characteristic melting point.

These derivatives can be assayed in the presence of the Fehling liquor to form a precipitate which will be weighed. The enzymatic assay consists of adding Penicillium glucose oxidase (notatin), which is a glucose-specific enzyme that will oxidize it in the presence of oxygen to form hydrogen peroxide (H_2O_2).

The pharmaceutical use of these substances is diversified, we can distinguish:

Monosaccharaides

- **Hexoses**: are obtained from the fruits. It is widely used in hospitals in the form of injectable solutions: isotonic (5–10%) or hypertonic (15–30%), for rehydration and as a source of caloric energy [6]. In industrial (agri-food), they are used to synthesize sorbitol and caramel.
- **Galactose**: obtained from algae, gums, and mucilages and used in radio-diagnostics as a contrast medium in two-dimensional echocardiography to increase signal intensity. This product is for old people over fifteen.
- **D-fructose**: exists in fruits and honey. Its interest is due to its use in case of loss of water greater than the loss of sodium chlorides and to prevent intra- and extracellular dehydration. It is also indicated for the prevention and treatment of ketoses.

Itoles

- **D-Mannitol:** used in hospital as a solute (10–20%) for infusion or injectable intravenously. In addition, this use is also recommended as diuretic indicated in case of cerebral edema and intraocular hypertension. In the food industry, this derivative can be used as a sweetener.

- **D-Sorbitol**: obtained by hydroxylation of glucose or by hydrolysis of starch. It is indicated:
  - In case of digestive and transit disorders;
  - As a diuretic.

In the pharmaceutical industry, it is used as raw material for the synthesis of ascorbic acid.

Derivatives of the Itoles

In particular, vitamin C which is indicated in case:

- From iron deficiency anemia;
- Influenza;
- Allergic rhinitis;
- Rhinopharyngitis;
- Transient asthenia; and
- Free radicals.

Amino Derivates (D-Glucosamine)

This compound is indicated as analgesic in case of mild to moderate gon-osteoarthritic pain.

b. **Osides**: (Dextrans, Starches, Celluloses, Fructoses, Alginates, Agar, Carrageenans, Pectins, Erasers, Mucilages).

They are in the form of holosides or glycosides obtained from sugar cane (*Saccharum officinarum*) or sugar beet (*Beta vulgaris*). They are used in the pharmaceutical industry as sweeteners, excipients for syrups in the form of tablets and capsules, and in the food industry as food additives. They can also be used for the production of bio-ethanol.

Thanks to their acid function, they acquire the ability to emulsify, in the presence of water, and form gels which gives them anticoagulant properties (e.g., heparins).

In medicine, they are indicated as:

- Gastrointestinal dressing (alginates);
- Source of energy (source of carbohydrates);
- Culture center;
- Excipient;
- Anticoagulant;
- Softening of the respiratory mucous membranes;
- Anti-martial anemia in case of failure of treatment with iron (Dextran 7000);
- Replenishing and stocking iron in case of emergency;
- Platelet anti-aggregate (Dextran 4000);
- Anticoagulant for external use (Dextran 4000); and
- Local analgesic in case of tendinitis (local treatment).

Alginates owe their name to the presence of alginic acid (mixture of guluronic acid and mannuronic acid) which is the main constituent of brown algae walls. They have the following properties:

- Insoluble in water and absorb more than 100 times their volume of water;
- Calcium alginates are insoluble in water;
- The monovalent cation salts of alginates are water soluble forming a colloidal pseudoplastic solution which is transformed into a non-thermo-reversible elastic gel by progressive addition of divalent cations;
- The agar (agar-agar E406) extracted from the red algae in boiling water as an extract which will be filtered hot. Its structure is a mixture of three heterogeneous polymers (agaro-pectin + agarose + pyruvyl agarose). It is totally soluble in hot water and swells in cold water;
- Carrageenans (E407) are polysaccharides obtained from different rhodophycea by extraction with boiling water or with the aid of alkaline solutions. They are separated by precipitation in alcohol (2-propanol) or potassium chloride by the cylinder process or by freezing.

Use of polysaccharides from algae

This family of naturally occurring chemical compounds is of great importance. They are used to modify the rheological behavior of the media with which they are in contact (stabilize, gel, thicken). In agri-food, they can be used for the production of creams, ice creams, desserts, etc. While in the pharmaceutical field, these substances are used as excipients in the manufacture of ointments, creams, protectors, suspensions, emulsions, and dental impressions or as active ingredients in the manufacture of gastric bandages, neutralizers, or mechanical laxatives.

5.2.5 Lipid Substances

This group of derivatives is a set of so-called fixed oils due to their nonvolatile nature. They are composed of esters, fatty acids, and alcohols and are classified into three categories depending on the nature of the alcoholic function:

- Triglycerides: are fixed oils whose nature of the alcohol is a glycerol;
- Cerides: are waxes whose alcoholic function is a fatty alcohol;
- Steroids: whose alcohol function is a sterol or phytosterol.

These lipid substances are also distinguished by their consistency (fat, oil, or butter), and they are mainly located in seeds as cytoplasmic inclusion. Their content is according to the state of maturation of this seed.

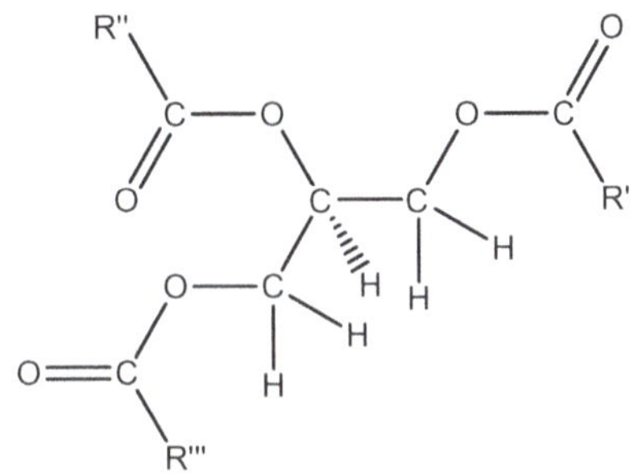

Fig. 5.8 Structure of triglyceride

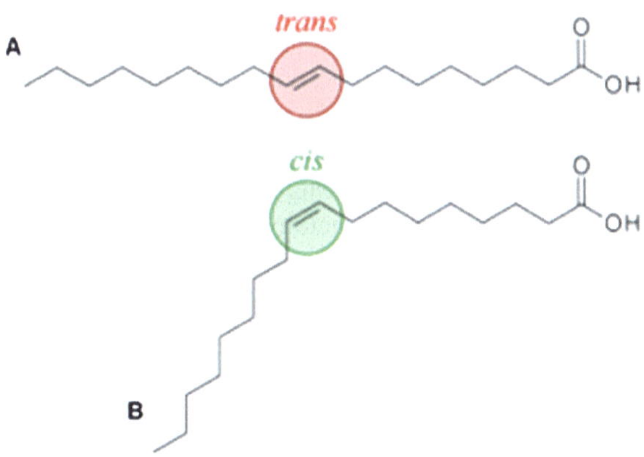

Fig. 5.9 Structure of a fatty acid

Physicochemical proprieties

a. **Triglycerides** (Fig. 5.8)

- In butter and fat, the fatty acids are mostly saturated at high molecular point and are solid at ordinary temperature;
- In oils, the fatty acids are especially unsaturated which facilitates their peroxidation and consequently the formation of aldehyde;
- They are hydrophobic in nature;
- They are soluble in low polar organic solvents (benzene, petroleum ether, etc.); and
- They are saponifiable.

b. **Fatty acids**

A fatty acid is a carboxylic acid with an open hydrocarbon chain (with or without a carbon–carbon double bond) (Fig. 5.9). In general, the hydrocarbon backbones of natural fatty acids contain from 4 to 36 carbon atoms with most often an even number of carbon atoms.

Fatty acids are generally known by different physicochemical properties:

- Liquids at ordinary temperature (carbon chain below C10);

- Insoluble in neutral or acidic water;
- Form soaps in an alkaline medium;
- Soluble in organic solvents;
- Form volatile methyl esters;
- Double bonds of unsaturated fatty acids absorb in the UV; and
- Fix the halogens: Br_2 and I_2. This characteristic is exploited in radiology as a contrast medium.

Their extraction takes place under cold or hot pressure by organic solvents.

These lipid compounds have several applications, in particular:

- For the basic treatment of certain dermatoses;
- As a contrast medium in diagnostic radio;
- Vascular embolization in interventional radiology;
- Prevents disorders related to iodine deficiency;
- As a source of vitamin E (antioxidant, lipid-lowering, supplanting lack of fertility);
- Used in the pharmaceutical industry as an excipient;
- Also used in dermocosmetology; and
- Used in food industry as food as rapeseed oil, olive oil, etc.

Fatty acids are ubiquitous in different classes of agri-food products, especially oils. On this subject, we can mention:

✓ Food oil

- Olive oil.
- Peanut oil.
- Sunflower oil.

In the Maghreb culture, olive oil is often eaten fresh (without cooking).

✓ Concrete food oil

- Copra oil.
- Palm oil.
- Sesame oil.

✓ Oil for industrial use

Linseed oil and castor oil are used as sources of raw materials in an industrial environment.

5.2.6 Active Ingredients with Poly-acetate Derivatives

The polyacetates are obtained by polycondensation of the acetates. The main polyacetates are:

- Isocoumarins and chromons;

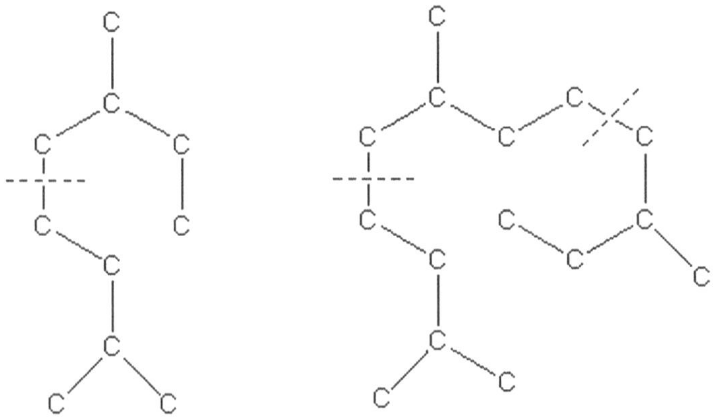

Fig. 5.10 Structure of: left—monoterpene; right—sesquiterpene

- Orsellinic acids (obtained by aldolization);
- Fluoroacetophenones;
- Poly-β-ketothioacids;
- Quinones: possibly anthraquinones, naphthoquinones, and benzoquinones.

For quinones, the double bond conjugated with a carbonyl allows them to fix nucleophilic substances and thus form covalent bonds, and thus offer them a strong electrophilic character.

5.2.7 Terpenoids Active Ingredients

This group of chemical substances (Fig. 5.10) consists of:

- The regular monoterpenes and sesquiterpenes.
- Irregular monoterpenes.

Terpenoids-based essential oils are mainly obtained either by distillation or by steam or cold expression. They can have the antiseptic power of the respiratory and urinary tracts or analgesic by local application. Some essential oils are used as diuretics, others as gastrointestinal antispasmodics, and a third category as minor anxiolytics.

5.2.8 Oleoresin Active Ingredients

They are substances composed of a mixture of essential oils and resins. Their consistency is either semiliquid or soft.

5.2.9 Active Ingredients of Iridoids

They are glycosides which are soluble in polar solvents and hydrolysable by hydrochloric acid and β-glucosidases. In the presence of ammonia, they give false alkaloids, and their assay takes place by HPLC.

These active ingredients have several important medicinal properties:

- Antirheumatic (analgesic effect);
- Laxatives;
- Tonic;
- Antioxidants;
- Sedatives acting at the level of the central nervous system;
- Antispasmodic; and
- Antihypertensive acting by inhibiting the conversion enzyme, also having an antiarrhythmic and coron-dilating effect.

5.2.10 Sesquiterpene Lactones Active Ingredients

These are active ingredients that are soluble in organic solvents (ether/petroleum ether/methanol or dichloromethane). They are demonstrated by thin-layer chromatography (TLC). Different biological activities have been identified for this family of natural substances:

- Pest and anti-helminthic effect;
- Antagonistic effect of GABA (aminobutyric acid);
- Anti-malarial effect;
- Diuretic.

5.2.11 Active Ingredients of Diterpenes

They are obtained by cyclization, heterolysis, or by protonation. The most used diterpenic compounds are the taxanes, known by their anticancer properties acting by stabilization of cell microtubules. They are well indicated for breast, ovarian, and branch cancer.

5.2.12 Active Ingredients of Saponosides

These are glycosides with a tri-terpene or steroid origin. They are characterized by:

- A high membrane permeability;

- Aphrogenic power;
- Hemolytic power; and
- An ichthyotoxic character.

From a physicochemical point of view, the glycoside saponosides are insoluble in apolar organic solvents and soluble in water and diluted alcohol, whereas the saprogenic saponosides have inverse properties of being soluble in organic solvents and insoluble in organic solvents. Their use is very wide:

- Antifungal;
- Pest control;
- Antimicrobial;
- Anti-inflammatory;
- Anti-hemorrhoids;
- Expectorant; and
- Diuretic.

5.3 Botany of Medicinal Plants

Medicinal plants have long been known and used for therapeutic and cosmetic purposes. There are several species of these plants in nature. This plant biodiversity varies from one region to another, and several factors affect this biodiversity such as microclimate, soil and environment as well as socioeconomic activities. Table 5.1 summarizes 88 species of these medicinal plants and indicates the family of each type of plant.

*5.3.1 Case of Rosemary (***R. Officinalis** *L.)*

In this section, we have chosen to present, briefly, a portrait of rosemary, which is of great medicinal value. Over the last decades, the scientists in the field have been interested in the ecological, physiological, and biochemical aspects of this species in order to fully understand the mechanisms of the valorization of this plant by being a source of therapeutic molecules used in alternative medicine or modern medicine.

R. officinalis belongs to the botanical family of Lamiaceae. Its genus name *Rosmarinus* means in Latin "rose of the sea." Rosemary is native to the Mediterranean region and it grows south in Europe and also in other regions of the globe such as Africa, America, Australia, or Asia. There are more than 10 varieties of rosemary. The main varieties of Rosemary are SLT, VAU 3, CNPMAI, Benenden Blue, Flora Rosa, Tuscan Blue, Majorca pink, Arp, Albiflorus, Huntington carpet, Mcconnell's Blue, Irene, Holly Hyde, and Hill Hardy. Rosemary is harvested preferably in spring, but can be harvested year-round depending on the region. It is found in the wild and also in gardens and it is a decorative plant. The plant can grow in altitude. Recently,

Table 5.1 List of the main medicinal plants found in nature

Plant	Latin name	Family
Absinthe	*Artemisia absinthium* L.	Asteraceae
Yarrow	*Achillea millefolium* L.	Asteraceae
Aigremoine	*Agrimonia eupatoria* L.	Rosacea
Bear's garlic	*Allium ursinum* L.	Liliaceous
Garlic mustard	*Alliaria officinalis* Andrz.	Brassicaceae
Fragrant dill	*Anethum graveolens* L.	Apiaceae
Dioecious antennae	*Antennaria dioica* L.	Asteraceae
Sea buckthorn	*Hippophaë rhamnoides* L.	Elaeagnaceae
Wormwood	*Artemisia vulgaris* L.	Asteraceae
Sweet scented leaf	*Galium odoratum* L.	Rubiaceae
Hawthorn	*Crataegus oxyacantha L.*	Rosacea
Burdock	*Arctium lappa L.*	Asteraceae
Basil	*Ocimum basilicum* L.	Lamiaceae
Cornflower	*Centaurea cyanus* L.	Asteraceae
White broth	*Verbascum thapsus* L.	Scrophulariaceae
Borage	*Borago officinalis* L.	Boraginaceae
Shepherd's purse	*Capsella bursa-pastoris* (L.) Medik.	Brassicaceae
Bearberry	*Arctostaphylos uva-ursi* (L.) Spreng	Ericaceae
Brunelle	*Prunella vulgaris* L.	Lamiaceae
Common heather	*Calluna vulgaris* L.	Ericaceae
Nasturtium	*Tropaeolum majus* L.	Tropaeolaceae
Cassis	*Ribes nigrum* L.	Grossulariaceae
Carline acaule	*Carlina acaulis* L.	Asteraceae
Chervil	*Anthriscus cerefolium var. cerefolium*	Apiaceae
Musky chervil	*Myrrhis odorata* L.	Apiaceae
Circée	*Circaea lutetiana* L.	Onagraceae
Celandine	*Chelidonium majus* L.	Papaveraceae
Wild chicory	*Cichorium intybus* L.	Asteraceae
Colchicum	*Colchicum autumnale* L.	Liliaceous
Poppy	*Papaver rhoeas* L.	Papaveraceae
Comfrey	*Symphytum officinale* L.	Boraginaceae
Coriander	*Coriandrum sativum* L.	Apiaceae
Couco	*Primula officinalis* *Primula veris* L.	Primulaceae
Purple Foxglove	*Digitalis purpurea* L.	Scrophulariaceae
Echinacea	*Echinacea purpurea* L.	Asteraceae
Eglantier	*Rosa canina* L.	Rosacea
Barberry	*Berberis vulgaris* L.	Berberidaceae

(continued)

Table 5.1 (continued)

Plant	Latin name	Family
Tarragon	*Artemisia dracunculus* L.	Asteraceae
Fennel	*Foeniculum vulgare* (Miller)	Apiaceae
Ficaire	*Ranunculus ficaria* L.	Renonculaceae
False holly	*Ruscus aculeatus* L.	Asparagaceae
Génépi	*Artemisia umbelliformis* Lam.	Asteraceae
Broom	*Sarothamnus scoparius* L. *Cytisus scoparius* L.	Fabaceae
Juniper	*Juniperus communis* L.	Cupressaceae
Juniper cade	*Juniperus oxycedrus* L.	Cupressaceae
Yellow gentian	*Gentiana lutea* L.	Gentianaceae
Mistletoe	*Viscum album* L.	Loranthaceae
Marshmallow	*Althaea officinalis* L.	Malvaceae
Holly	*Ilex aquifolium* L.	Aquifoliaceae
Hyscope	*Hyssopus officinalis* L.	Lamiaceae
Mountain Houseleek	*Sempervivum montanum* L.	Crassulaceae
White Lamier	*Lamium album* L.	Lamiaceae
Laurel	*Laurus nobilis* L.	Lauraceae
Climbing ivy	*Hedera helix* L.	Araliaceae
Ground ivy	*Glechoma hederacea* L.	Lamiaceae
Linnaeus	*Sideritis hyssopifolia* L.	Lamiaceae
Sweet Clover	*Melilotus officinalis* Lam.	Fabaceae
Melissa	*Melissa officinalis* L.	Lamiaceae
Green mint Mint on the cob	*Mentha spicata* Hudson *Mentha viridis* L.	Lamiaceae
St. John's wort perforated	*Hypericum perforatum* L.	Hypericaceae
bergamot	*Monarda didyma* L.	Lamiaceae
Swamp Myosotis	*Myosotis scorpioides* L. *Myosotis palustris* Hill.	Borraginaceae
Yellow water lily	*Nuphar lutea* L.	Nymphéaceae
Nigella	*Nigella sativa* L.	Renonculaceae
Oregano	*Origanum vulgare* L.	Lamiaceae
Stonecrop of England	*Sedum anglicum* Huds.	Crassulaceae
Nettle	*Urtica dioica* L.	Urticaceae
Daisy	*Bellis perennis L.*	Astéraceae
Periwinkle	*Vinca minor* L.	Ményanthacées
Wild thought	*Viola tricolor* L.	Violaceae

(continued)

Table 5.1 (continued)

Plant	Latin name	Family
Parsley	*Petroselinum crispum* Mill. *Petroselinum sativum*	Apiaceae
Dandelion	*Taraxacum officinale* *Taraxacum dens leonis* L.	Asteraceae
Peony	*Paeonia officinalis* L.	Paéoniaceae
Coucou	*Primula officinalis* *Primula veris* L.	Primulaceae
High primrose	*Primula elatior* L.	Primulaceae
Pulmonary	*Pulmonaria officinalis* L.	Borraginaceae
Rosemary	*Rosmarinus officinalis* L.	Lamiaceae
blackberry	*Rubus fruticosus* L.	Rosacea
Saffron	*Crocus sativus* L.	Iridaceae
Savory mountains	*Satureia montana* L.	Lamiaceae
Sage	*Salvia officinalis* L.	Lamiaceae
Souci	*Calendula officinalis* L. *Calendula arvensis* L.	Asteraceae
Black elderberry	*Sambucus nigra* L.	Caprifoliaceae
Tansy	*Tanacetum vulgare* L.	Asteraceae
Thyme	*Thymus vulgaris* L.	Lamiaceae
Coltsfoot	*Tussilago farfara* L.	Asteraceae
Fragrant verbena	*Lippia citriodora* L.	Verbenaceae
Violet	*Viola odorata* L.	Violaceae

Source Au Jardin de Flore

some growers started domestication of rosemary for commercial purposes and to maximize dry yields and quality of extractions. Rosemary is known for its particular smell, its blue flowers and especially its particular leaves with a green color on the top, and white on the underside. The leaves look like needles, and they are long and very tight. It develops in semiarid and sub-humid climates with hot to cool variants. It is a drought-resistant plant with apparent characters of xerophytes. Rosemary generally does not support a very wet winter and is a resistant and temperate plant that can tolerate frost and heat. Optimum growth temperatures are in the range of 20–25 °C. Areas with a rainfall of more than 500 mm/year are favorable for rosemary growth. Rosemary prefers sand soil, fine clay, and clay (max 30%) with a pH between 5.5 and 8.0. Rosemary does not grow well in watered or high clay soils, active limestone, and soil rich in copper.

Rosemary is a tree plant that reaches a height of 1–2 m. The upright stems are divided into many long, thin, scaly branches. The branches face opposite, thick, and leathery leaves that are lustrous, linear, dark green. The purplish blue flowers, visible from January to May, are clustered at the end of the twigs. The fruit has an ovoid shape, surrounded by a brown and persistent calyx. The inflorescence and the calyx have a very short hairiness. Reproduction can be in the form of seed and in the form of cuttings and bursts of tuft. Rosemary is harvested between early May and late November. It is harvested in full bloom if the collected material is destined for distillation to produce the essential oil. To ensure a good harvest, it is necessary to make sure of the authorities concerned in order to know the laws and the norms of the management plans of the rosemary resources.

Regarding its use, rosemary is a medicinal plant and is often in the form of herbal tea, or infusion. The constituents are essential oils (such as cineol, alpha-pinol, and camphor), rosmarinic acid, carnosic acid, caffeic acid, tannins, bitter diterpenes, and flavonoids. The parts used are the leaves and the flowers. Extracts of rosemary usually have antirheumatic [7], depurative, diuretic, astringent, anti-inflammatory, antimicrobial [8, 9], and healing effects. Rosemary is used for cough, to prevent Alzheimer's, headache, migraine, concentration problem, and stomach ulcers [10]. Essential oil of rosemary has many roles and is used as anti-proliferative, antioxidant and anti-bacterial [11]. Moreover, rosemary is used against rheumatism, arthritis, osteoarthritis, and torticollis and muscle pain. Rosemary is a plant widely used in cosmetics in the form of cream or shampoo. Among the side effects of rosemary is hypertension. At a high dose, rosemary can cause deep sleep, spasms, and gastritis. It is inadvisable to ingest the essential oil of rosemary, because of toxic side effects. In addition, it is recommended to avoid the use of rosemary in pregnant women. Rosemary can be prepared in the form of herbal tea, infusion, ointment or cream, essential oil, capsule or capsule, poultice, tincture, mother tincture, wine, and as a bath.

5.4 Conclusion

Phytochemistry and medicinal plants have always attracted the attention of researchers and therapists. Always between benefits, risks and the discovery of new concepts and new impacts on human health, the scientists are engaged in hard and endless efforts to better identify these substances and to further engage their deliberate will towards the permanent discovery of new substances and new uses.

Medicinal plants are considered as a diverse set of many plants rich in particular compounds called active ingredients. The results of scientific research found that these derivatives contained in these plants have therapeutic capabilities to prevent, relieve, and cure various ailments. They can also be used in the field of medicinal plants, aromatherapy, cosmetology, and modern pharmacology. By respecting the dose, this variety of plants can lead to remarkable results for the treatment of various physical and even psychological pathologies.

References

1. Au jardin de Flore (2016) Index des plantes médicinales. Available via http://aujardindeflore.net/herbier/herbier.php?page=medicinales. Accessed 10 Apr 2018
2. Chafai Elalaoui A, Boukil A, Bachar M, Lkhoumsi D, Guermal A Aafi A et al (2014) Manuel des bonnes pratiques de collecte du romarin « *Rosmarinus officinalis* » . Centre de Recherche Forestière, HCEFLCD, 12 p. Available via https://www.fellah-trade.com/ressources/pdf/MBPC_Romarin_Francais.pdf. Accessed 12 Feb 2018
3. Gurib-Fakim A (2006) Medicinal plants: traditions of yesterday and drugs of tomorrow. Mol Aspects Med 27(1):1–93
4. Padulosi S, Leaman D, Quek P et al (2002) Challenges and opportunities in enhancing the conservation and use of medicinal and aromatic plants. J Herbs Spices Med Plants 9(4):243–267
5. Darmady E (1961) Sterilization by dry heat. J Clin Pathol 14(1):38–44
6. Nguyenhiep P, Delaveau RR et al (1964) La dessiccation et la stabilisation des plantes medicinales: plantes à heterosides flavoniques. Ann Pharm Fr 22:573–580
7. Wanamaker BP, Massey K (eds) (2014) Applied pharmacology for veterinary technicians. Elsevier Health Sciences, USA
8. Miller NJ (1996) The antioxidant properties of theaflavins and their gallate esters-radical scavengers or metal chelators? FEBS Lett 392(1):40–44
9. Calvo M, Akerreta S, Cavero R et al (2011) Pharmaceutical ethnobotany in the Riverside of Navarra (*Iberian Peninsula*) J Ethnopharmacol 135(1):22–33
10. Pintore G (2002) Chemical composition and antimicrobial activity of *Rosmarinus officinalis* L. oils from Sardinia and Corsica. Flavour Fragr J 17(1):15–19
11. Hussain AI (2010) *Rosmarinus officinalis* essential oil: antiproliferative, antioxidant and antibacterial activities. Braz J Microbiol 41(4):1070–1078

Chapter 6
Bioactives from Plant Food Processing Wastes: Ultrasonic Approaches to Valuable Chemicals

Yaning Shi, Yue Wu, Yang Tao and Yongbin Han

Abstract The utilization of plant food processing wastes is a global issue. The recovery of bioactives has been widely recognized as an important way for the utilization and valorization of by-products generated after plant food processing. For this purpose, ultrasound can be a green approach to take part in the processes transforming bioactives in food wastes to valuable chemicals. In this chapter, the bioactivities of three major bioactives in plant foods (including polyphenols, peptides, and polysaccharides), as well as their stabilities in ultrasonic fields, were first discussed. Then, the applications of ultrasound in two procedures, namely extraction and purification referring to adsorption and desorption, were reviewed. For ultrasound-assisted extraction, more attention were paid to the combinations of ultrasound and other green methods, such as utilization of green solvents, enzymolysis, microwave irradiation, and supercritical fluid extraction. Due to the limited applications of ultrasound to adsorb bioactives from plant foods, the effects of different operating parameters were reviewed. Besides, a methodology to simulate the mass transfer process during extraction and adsorption by means of a diffusion model with numerical methods was introduced. It is expected that this knowledge could provide some new ideas about extraction and purification of bioactives from plant food wastes in a greener way, as well as control of the ultrasound processes intelligently.

Keywords Ultrasound · Bioactives · Plant food wastes · Extraction · Adsorption · Diffusion

---

Y. Shi · Y. Wu · Y. Tao (✉) · Y. Han
College of Food Science and Technology, Nanjing Agricultural University, Nanjing 210095, China
e-mail: yang.tao@njau.edu.cn

Y. Li and F. Chemat (eds.), *Plant Based "Green Chemistry 2.0"*,
Green Chemistry and Sustainable Technology,
https://doi.org/10.1007/978-981-13-3810-6_6

6.1 Introduction

Plant foods play an important role in our daily life. However, the processing of plant foods generates a huge quantity of wastes, which are disposed simply by landfilling, animal feeding, composting, and incineration [1]. To reduce the environmental impact and accomplish the requirement of sustainable development, plant food processing wastes must be treated properly. It has been extensively reported that wastes derived from plant food processing are rich in various bioactive compounds, such as phenolics, peptides, polysaccharides, and others [2–4]. Thus, the recovery of bioactives is a reasonable method for the valorization of plant food wastes.

To transform bioactives from plant food wastes to chemicals with commercial value, a series of processes are needed, including extraction, purification, concentration, drying, and sometimes encapsulation [5]. Ultrasonic technology can be well-incorporated into the whole processing chain, thus enhancing processing efficiency and improving product quality [6]. Meanwhile, ultrasound itself is a green, economical, feasible, and versatile technology, which can fit the "green" topic of this book ideally. The frequency of ultrasound suiting food processing mainly distributes the range between 20 to 800 kHz, which is usually named as low-frequency and high-energy power ultrasound [7]. The ultrasonic effect on food processing is mostly attributed to ultrasonic cavitation and related physical and chemical phenomena, such as locally extreme high temperature and pressure, free radicals, turbulence, and micro-streaming [6].

Ultrasonic approach to process bioactives in plant food wastes to valuable chemicals is a vast topic. Only the advances in applications of ultrasound in the extraction and purification (referring to adsorption and desorption) processes were reviewed in this chapter. Since this book is focused more on the revolutionary theories and applications in green chemistry area, the basic knowledge about ultrasound-assisted extraction of bioactives was shortened. Herein, five parts about ultrasound-assisted extraction, adsorption, and desorption were included, being bioactivities of plant origin biomolecules (polyphenols, peptides, and polysaccharides), stabilities of these bioactives in ultrasound fields, combinations of ultrasound and other green extraction methods to extract bioactives from plant food processing wastes, ultrasound-assisted adsorption and desorption for purification of bioactives, and diffusional mass transfer modeling of ultrasound-assisted extraction and adsorption processes. These parts can provide some new insights into applications of ultrasound for the recovery of bioactives from plant food wastes from both chemistry and engineering perspectives of view.

6.2 Bioactives from Plant Food Processing Wastes and Their Bioactivities

Plant foods contain many types of bioactive natural products. Among them, polyphenols, peptides, and polysaccharides are the three most widely studied bioactives. Thus, the bioactivities of these three bioactive families are discussed in the following parts.

6.2.1 Bioactivity of Plant Polyphenols

Plant polyphenols, as one of the important bioactive substances in all phytochemicals, usually consist of three main subgroups: phenolic acids, flavonoids, and phenolic diterpenes [8, 9]. Among phenolic acids, two important naturally occurring types of phenolic acids are hydroxybenzoic acids and hydroxycinnamic acids, which are derived from non-phenolic molecules of benzoic and cinnamic acid, respectively [10]. Also, there are five subgroups of flavonoids, namely anthoxanthins, flavanones, flavanonols, flavans, and anthocyanidins.

Polyphenolic compounds play an important role in cell protection ascribed to their capacity of scavenging different species of reactive oxygen, transferring electrons to free radicals, activating antioxidant enzymes, and inhibiting oxidase enzymes [11]. Furthermore, recent studies show that the antimicrobial potential of natural phenolic compounds is higher than that shown by synthetic antibiotics [12], mainly owing to the composition of extracts [13]. In this respect, numerous studies have added plant polyphenols to foods along with suitable packaging films as antioxidants and preservatives. For instance, durian leaf extract can be added in gelatin-based film as active packaging [14]; apple fruit bioactive substances can be preserved by olive wastes extract-chitosan film coating [15]. Besides, plant phenols have protective actions against the oxidative stress, which has been suggested as the root cause of aging and human disorders, such as arteriosclerosis, diabetes, cancer, and neurodegenerative diseases [16]. Thus, they are well behaved in various cosmetic products and pharmaceuticals [17]. The book of Holland et al. [18] also pointed out that phenolic compounds such as gallic acid, flavonols, ellagic tannins, anthocyanin, procyanidins, and ellagic acid present in fruit peels exhibit various pharmacological activities. In general, the bioactivities of phenolic compounds depend on the number and arrangement of hydroxyl groups [19].

As aforementioned, food by-products are rich in polyphenols, whereas they vary from different processed wastes. For example, aronia berries contain high levels of procyanidins, anthocyanins, and phenolic acids. Procyanidins have been identified as the major class of polyphenolic compounds in this plant species [20]. In the study of Haas et al. [21], malvidin glucosides were found to be the major phenolics in the grape non-pomace residues. Catechin, epicatechin, ferulic acid, caffeic acid, and trans-resveratrol with potent antioxidant ability were also found at high concentrations in the extractives. Health advantages of apple peels are correlated with the presence of flavonols, anthocyanins, flavan-3-ols, phenolic acids, and dihydrochalcones [22]. Onion skin wastes are rich in flavonoids, while quercetin and kaempferol were identified as major flavonoids in samples of subcritical water extraction [23]. In addition, the total phenolic content of the same fruits varies greatly with their varieties. Data from the study of Haas et al. [21] showed that the concentrations of total polyphenols in the extractives of Bordo and Isabel grape samples were 14910.34 and 6381.60 mg/kg, respectively, wherein Bordo sample had a 57.2% higher concentration of total polyphenols than Isabel counterpart. Actually, a series of factors impact polyphenolic composition and their quantity, such as habitat/location, harvest date, cultivar, fertilization, and maturation [24, 25]. As far as common concerned, pomegranate, aronia berry, grape, and olive are several oldest known edible fruits that contain a higher concentration of polyphenols in comparison with other fruits studied [26]. However, a large amount of polyphenol losses has switched researchers' focus to new agro-food wastes, such as carrot pulp, red beets, and tomato seeds [27–29].

6.2.2 *Bioactivity of Plant Peptides*

Bioactive peptides, which contain 2–20 amino acids, are generated from protein and become active once cleaved. In recent years, various bioactive peptides from food proteins are known to possess functional and nutritional properties, such as antioxidant, antitumor, hypotensive, and immunomodulatory peptides [30–32]. Hence, the interest in health-promoting bioactive peptides becomes hot topics. Initially, bioactive peptides are prepared from animal proteins. In recent years, many studies have also been performed on bioactive peptides from plants [33].

Antioxidants applied in food industry are common and necessary for inhibiting oxidative processes. They play important roles in the prevention mechanisms of the oxidative stresses related to various degenerative aging diseases, such as cancer and atherosclerosis [34, 35]. Due to consumers' request for healthy foods, natural antioxidants have been investigated as alternative to the synthetic ones. Commonly, antioxidant peptides can act as scavengers of free radicals, inhibitors of lipid peroxidation, and chelator of metal ions. For comprehensive utilization of plant materials, antioxidant peptides from plant food processing wastes can be isolated mainly from peanut kernels, rice bran, sunflower protein, alfalfa leaf protein, corn gluten meal, curry leaves, and others [36].

Cancer is a multistage process derived from a combination of multiple heritable and environmental factors. Proteins, peptides, and amino acids could act at different levels to prevent different types of cancer [37]. It is noteworthy that the populations consuming high levels of soybean products have both lower cancer incidence and mortality rates in western countries. One of the well-known anticancer peptides is lunasin, a 43-amino acid peptide. Lunasin is a novel and promising peptide initially discovered in soy and now also found in wheat, barley, and other seeds [38]. It has been proved that lunasin exhibits an inhibitory effect against the core histone acetylation of mammalian cells, which means lunasin might adjust cell-cycle control and suppress carcinogenesis through its involvement in the chromatin modification [39].

Hypertension can lead to various serious diseases, such as heart disease and arteriosclerosis. Angiotensin-converting enzyme (ACE) promotes blood vessels to constrict and increase blood pressure. Hence, ACE inhibitors play an important role to control hypertension. However, synthetic ACE inhibitor drugs have side effects, such as coughs, increased blood potassium levels, low blood pressure, skin rashes, headaches, fatigue, fetal disorder, and an abnormal taste [40]. The ACE inhibitors with minimal side effects are needed in urgent. So far, a number of studies show that peptides derived from plants possess ACE-inhibitory activity, including the peptides derived from rice, cherry, wheat, soybean, sweet potato, rapeseed, walnuts, and other plants [40–42]. Most ACE-inhibitory peptides including plant origin ones are composed of hydrophobic amino acids, such as Ile and Leu, at the last position of the N-terminus and proline residue [40].

The immune system includes innate and adaptive immunity, which acts vital for human survival and provides defense by adjusting the immune responses [43]. Since most immunomodulatory drugs have toxic side effects, immunomodulatory substances in food play a significant role in protecting the body from invading pathogens. Food-derived immunomodulatory peptides could increase mucosal immunity at gut, augment natural killer cell cytotoxicity, activate macrophage, and its phagocytosis, regulate secretion of cytokines, immunoglobulins, and nitric oxide, and exhibit other immunomodulatory activities [44]. At present, immunomodulatory peptides have been produced from various plant foods and their processing wastes, including rice, wheat, soybean, yellow pea, lupin, flaxseed, and amaranth [44].

6.2.3 Bioactivity of Plant Polysaccharides

Polysaccharides are important natural biomacromolecules, which are usually composed of more than ten monosaccharides through glycosidic linkages in linear or branched chains. These bioactive polymers widely exist in plants and are beneficial to many biological processes [45]. The bioactivities of polysaccharides are significantly influenced by their chemical structures and chain conformations. The macromolecular structures of plant polysaccharides are extremely complex due to the presence of different monosaccharides as building blocks, which contain variations in sequence, linkage, branching, and distribution of side chains [46]. Moreover, bioactive polysaccharides are usually combined together with proteins, polynucleotides, lipids, lignins, inorganic mineral substances, and other biomolecules [47]. Recently, the scientific community started to focus on polysaccharide–polyphenolic conjugates. Comparing with polysaccharides and polyphenols separately, polysaccharide–polyphenolic conjugates show improved physicochemical properties and biological activities [48]. The bioactivities of polysaccharide–polyphenolic conjugates are shown in Fig. 6.1.

Edible plant polysaccharides can be found from dietary fibers. The dietary fibers contain indigestible plant polysaccharides, such as cellulose, hemicelluloses, pectins,

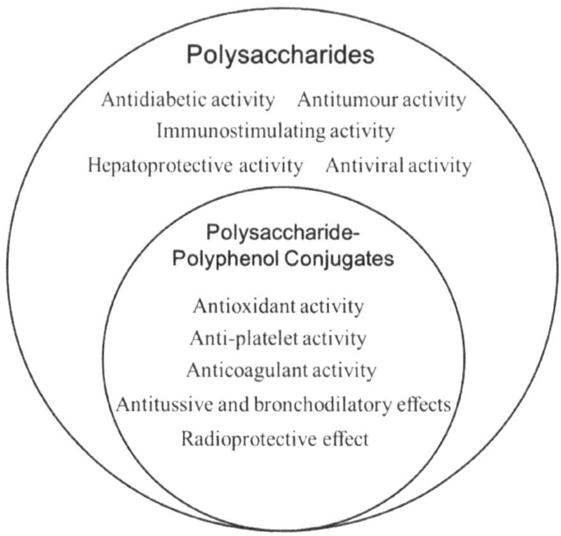

Fig. 6.1 The biological properties of polysaccharide–polyphenolic conjugates

oligosaccharides, gums, and other lignified compounds. These polysaccharides can slow the movement of food in the digestive tract, decrease the level of blood sugar and cholesterol, stimulate bowel movement, and protect gut from various gastrointestinal disorders [49, 50]. Hence, regular intakes of dietary fiber can significantly reduce risks for chronic diseases like diabetes, cardiovascular diseases, hypercholesterolemia, hyperlipidemia, obesity, and even cancer. The dietary fibers can be obtained from almost all categories of agro-food, including grains, fruits, vegetables, nuts, and their by-products [45].

6.3 Stability of Bioactives from Plant Food Processing Wastes Under Sonication

For ultrasound-incorporated food processes, special attention should be paid to the stability of food components including bioactives. Although ultrasonic energy can benefit the food processes, the extremely high temperature and pressure and highly active free radicals generated after the collapse of ultrasound cavitation bubbles occasionally lead to degradation problems [51]. The cavitation effect and energy dissipation as heat are directly related to the degradation of thermolabile compounds. Another possible explanation is that ultrasound could provoke severe structural modifications and degrade cell wall components [52], thus making the released enzymes to easily oxidize bioactives, such as phenolics [53]. Experiment data from the study of Cruz et al. [54] showed phenolic compounds in grape skin were sensitive to the application of ultrasound. The reduction of total phenolic content observed ranged between 18.4 and 31.0%. The highest correlations were found for total phenolic content, anthocyanin and petunidin 3-*O*-glucoside and malvidin 3-*O*-b-D-galactoside. These compounds can be considered as the key polyphenols with the greatest influence on antioxidant capacity. Thus, the optimal operating parameters for ultrasonication need to be studied.

The use of ultrasound to increase protein digestion is common and has been widely reported. On one hand, ultrasonic bubble cavitation causes rapid changes in temperature, pressure, shear forces, jets, and shock waves, which result in mixing of reaction components instantly and proteolysis [55]. On the other hand, ultrasound enhances the surface hydrophobicity and loosening of protein for easy release of peptides [56]. It has been reported that ultrasound treatment significantly improved antioxidant activities of peanut protein and wheat gluten hydrolysates [55]. Meanwhile, different ultrasonic frequencies could strongly alter the molecular weight distribution of peptides. Usually, molecular weight range of peptides is lower when a lower ultrasound frequency is applied. The study of Moure et al. [57] showed that the industrial waste liquors from the manufacture of soy protein concentrates treated by ultrasound contained lots of antioxidant peptides smaller than 1 kDa [57]. However, ultrasound treatment at a higher intensity for a longer time could cause high-temperature and pressure conditions, which could induce protein denaturation or excessive digestion of protein or peptide. Wave attenuation with distance in reaction systems is a unique problem for ultrasound food processing, especially with samples containing small gas bubbles [4].

For ultrasound processing of food matrices containing bioactive polysaccharides, such as *Zizyphus jujube* jujube, corn cobs, corn hulls, buckwheat hulls, almond shells, wheat bran, wheat straw, sugar cane bagasse, apple pomace, and many other by-products, ultrasound could induce cleavage of glycosidic bonds in the main polymeric chain and branching units [58]. The side effects of ultrasound depend on the applied power and operating parameters. However, depolymerization effect of ultrasound on polysaccharides has also been utilized in targeted reduction of their molar mass [59].

Overall, the ultrasonic modification of structures of plant food bioactives can be either positive or negative. The stability of target bioactives should be strictly controlled, thus maximizing the beneficial effect of ultrasound on the studied processes.

6.4 Ultrasonic Approach for the Extraction of Bioactives from Plant Food Processing Wastes: Combination with Other Green Technologies

Ultrasound-assisted extraction is an old topic, and a large number of papers in this field have been published. Compared with traditional solvent exaction, its combination with ultrasound can markedly accelerate extraction kinetics and enhance extraction yield [7]. The main principle behind it is related to ultrasonic cavitation, which can generate high shear forces and turbulences in liquid medium, corrode, or disrupt raw materials, thus making the release of target components easier [60].

Ultrasound-assisted extraction has been widely recognized as an eco-friendly technology, since the introduction of sonication to solvent extraction can enhance the extraction efficiency, reduce the solvent consumption, and lower the pollution to the environment [61, 62]. Nevertheless, the common ultrasound-assisted solvent extraction method is not a completely green technology, since the application of organic solvent is hard to be avoided, especially for the extraction of plant origin bioactives. To accomplish the requirement of sustainable development and make this technology "greener" in a new epoch, the way that ultrasound is applied in the extraction process needs to be improved.

Apart from ultrasound-assisted extraction technology, there are also other methods can be labeled as green extraction technologies, such as the application of green solvents for extraction, microwave-assisted extraction, enzyme-assisted extraction, supercritical fluent extraction, and subcritical water extraction. In both theory and practice, ultrasound has already been coupled with the aforementioned technologies for the extraction of bioactives from plant food processing wastes, bringing the extraction process to a higher level of greenness. Several applications about combinations of ultrasound with these green technologies for the extraction of plant origin bioactives are summarized in Table 6.1.

In recent years, deep eutectic solvent (DES) has been extensively regarded as a green alternative to organic solvent in the extraction field. The conception of DES was first proposed by Abbott et al. [78] in 2003, which refers to four main types of mixed solutions, including combination of organic salts and metal hydrates, combination of organic salts and metal salts, combination of metal chlorides and compounds being hydrogen bond donors, and combination of organic salts and compounds being hydrogen bond donors [79]. The green label of DES can be attributed to its physicochemical properties, such as low-volatile, non-flammable, chemically and thermally stable, cheap, low-toxic, biodegradable, water neural, and easy-preparing [80]. Meanwhile, DES has a high dissolution ability, making it an efficient solvent for the extractions of various compounds. Ultrasound can be well incorporated into DES-based extraction of bioactives from plant food wastes, making this technology more competitive. For example, Chanioti and Tzia [64] extracted phenolics from olive pomace using combined natural deep eutectic solvents and ultrasound. The results revealed that once the proper DES solvent was used, being choline chloride with citric acid, the total phenolic yield under sonication was higher than that obtained using aqueous ethanol as solvent. Similarly, the study of Bosiljkov et al. [63] also indicated that choline–chloride-based DES with malic acid as hydrogen bind donor coupled with ultrasound extraction was a green and efficient method to recovery anthocyanins from wine lees. In the case of extraction of polysaccharides from *Dioscorea opposita* Thunb, the extraction yield from combined DES (choline chloride and 1,4-butanediol) and ultrasound extraction were higher than those of hot water extraction and water-based ultrasound-assisted extraction [27]. Overall, the selection of DES type is a key step when it comes to combine ultrasound and DES for the extraction of plant origin bioactives.

Enzyme is sometimes added during extraction from plant materials, so as to hydrolyze polysaccharides in cell walls including hemicelluloses, starch, and pectins

Table 6.1 Examples about combinations of ultrasound and other green methods for the extraction of plant origin bioactives

Plant material	Extracted bioactive compounds	Extraction techniques	Extraction conditions			Enhancement of extraction yield after sonication	References
			Temperature (°C)	Time (min)	Ultrasound power (W)		
Wine lees	Anthocyanins	Ultrasound-assisted deep eutectic solvent extraction	35	30.6	341.5	From 2.89 to 6.42 mg/g	Bosiljkov et al. [63]
Olive pomace	Phenolics	Ultrasound-assisted natural deep eutectic solvent extraction	60	30	280	From 13 to 34 mg/g	Chanioti and Tzia [64]
Dioscorea opposita Thunb	Polysaccharides	Ultrasound-assisted deep eutectic solvent extraction	94	44.7	500	From 10.51 to 15.98%	Zhang and Wang [27]
Blueberry	Anthocyanins	Ultrasound-assisted water extraction	25	40	1500	From 2.12 to 5.01 mg/g	Jiang et al. [65]
Mango peel	Pectins	Ultrasound-assisted citric acid extraction	80	15	1200	From 16.70 to 17.15%	Wang et al. [66]
Olive pomace	Polyphenols	Ultrasound-assisted cyclodextrin extraction	70	20	1200	From 577 to 645 mg/kg	Albahari et al. [67]
Mulberry (*Morus nigra*) must	Phenolics and flavonoids	Ultrasound-assisted enzymatic extraction	20	11.5	60	From 230.58 mg/100 mL to 298.06 mg/100 mL (phenolics) From 347.79 mg/100 mL to 379.24 mg/100 mL (flavonoids)	Tchabo et al. [68]
Lentinus edodes	Polysaccharides	Enzymes-microwave-ultrasound-assisted extraction	48	10	440	From 5.88 to 9.38%	Yin et al. [69]

(continued)

Table 6.1 (continued)

Plant material	Extracted bioactive compounds	Extraction techniques	Extraction conditions			Enhancement of extraction yield after sonication	References
			Temperature (°C)	Time (min)	Ultrasound power (W)		
Adlay seed	Coixenolide	Ultrasound-assisted supercritical fluid extraction	40	210	110.5	From 76 to 90%	Hu et al. [70]
Oregano	Polyphenols	Ultrasound-assisted supercritical fluid extraction	40	60	151.6	From 2.1 to 4.2 mg/g	Santoszea et al. [71]
Clinacanthus nutans	Polyphenols, triterpenoids, and vitamin C	Combined ultrasound and microwave extraction	–	1.3	90	From 39.73 to 49.93 mg/g (total yield)	Yu et al. [72]
Scutellaria baicalensis	Flavonoids	Combined ultrasound and microwave extraction	62	15	300	From 5.5 to 8.71%	Xiang and Wu [73]
Sugar beet pulp	Pectins	Ultrasonic-assisted subcritical water extraction	120.72	30.49	1 W/cm^2	From 21.58 to 24.63%	Chen et al. [74]
Orange peels (*Citrus sinensis*)	Essential oils and antioxidants	Ultrasound-assisted instant controlled pressure drop technology	40	60	150	From 1.8 to 8.25 mg/g	Allaf et al. [75]
Sunflower seeds	Functional oils	Combined pulsed electric field and ultrasound extraction	20	2	100	From 574.3 to 577.1 ppm	Moradi and Rahimi [76]
Rice bran	Polysaccharides	Combined extrusion and ultrasound extraction	72	21	117	From 5.16 to 7.18%	Li et al. [77]

and release intracellular components [81]. Different enzymes, such as cellulase, hemicellulase, hemicellulase, xylanase, and β-glucanase, and their combinations can be considered. Ultrasound can be easily coupled with enzyme-assisted extraction to extract various bioactives from plant foods with high yield and quality. Tchabo et al. [68] optimized the ultrasound-assisted enzymatic extraction (UAEE) parameters for anthocyanins from mulberry must and found that UAEE could reduce extraction time and enhance must quality. Similarly, Zhang et al. [82] extracted *Cistanche tubulosa* polysaccharide by ultrasound-cellulase-assisted extraction. Compared to hot water extraction, the yield of ultrasound-cellulase-assisted extraction increased by 44%. In enzyme-assisted extraction process under ultrasonic fields, the enhanced mass transfer can be attributed to two aspects, one is the modification of morphological features of raw materials by ultrasound cavitation and the other one is the enhancement of enzyme activity due to favorable conformational changes and structural integrity under ultrasonic irradiation [81]. Extra attention should be paid to the interaction effects among ultrasound parameters, enzyme-related parameters (enzyme loading, type, etc.), and environmental parameters (pH, temperature, solvent type, etc.).

Microwave-assisted extraction, instantaneous controlled pressure drop process (abbreviation: DIC), pulsed electric field, extrusion extraction, supercritical fluid extraction, and subcritical water extraction are innovative physical extraction technologies. The principles of these technologies for extraction are diverse. Specifically, microwave irradiation can provide non-contact heat source due to molecular motions of ions and rotation of dipoles when samples are exposed to microwave fields [83]. DIC technique is based on a rapid transition from high stream pressure to vacuum, which leads a fast moisture vaporization and textural changes in raw materials [60]. The principle of pulsed electric field-based extraction is the enhanced permeability of raw materials under short duration pulses of moderate or high electric field [84]. Extrusion extraction is a compression step to extrude liquids containing target components from raw materials. Supercritical fluid extraction is using supercritical CO_2 as solvent for extraction, since CO_2 in supercritical phase possesses both liquid-like solvating features with high diffusivity and gas-like mass transfer properties [85]. Besides, subcritical water extraction replies on heated and pressurized water at temperatures between its boiling point and critical point and considerably high pressures, since subcritical water has low viscosity, high penetration ability into solid matrix, and strong capacity to dissociate target compounds [86]. Ultrasound apparatus with proper design can be well incorporated into these aforementioned physical techniques. The benefit of combinations of ultrasound with these innovative techniques for extraction of bioactives from plant food wastes can be found in Table 6.1. There is no doubt that the hybrid techniques are green and promising extraction techniques. More concerns are needed on the stability of bioactive compounds under the multiple physical fields.

One of the biggest challenges in the development of green extraction methods based on ultrasonic approach is the scale-up of equipment suiting commercial uses. Ultrasonic technique has been successfully utilized for the extraction of natural bioactives in the industry. However, difficulties emerge when ultrasound needs to be coupled with other physical techniques. Currently, the devices of pulsed electric field, supercritical fluid extraction, and subcritical water extraction are still very expensive, and the processing conditions need to be strictly controlled. In these environments, the built-up of industrial-scale hybrid equipment including ultrasonic element might be difficult. The distribution of ultrasound field and cavitation intensity should be different from that in laboratory-scale hybrid equipment. To construct solid hybrid equipment, the knowledge about engineering, chemistry, physics, and mathematics should be combined well. More theoretical studies about scaling-up hybrid extractors used for the recovery of bioactives from plant food processing wastes are expected in the future.

6.5 Ultrasonic Approach for the Purification (Adsorption and Desorption) of Bioactives from Plant Food Processing Wastes

In the literature, a large number of studies have reported the extraction of bioactive compounds from plant materials by means of various green and rapid methods. However, in most cases, the purity of target bioactives after a primary extraction process is still very low. Taking anthocyanin extracts from berry-related pomace for example, a considerably high portion of sugars, organic acids, and amino acids can be found in the raw anthocyanin extracts [87, 88]. The low purity can not only affect the stability of bioactives involved, but also lower the commercial value of final products. Therefore, a purification process is needed to transfer bioactive compounds from plant food processing wastes to value chemicals or products.

Compared to extraction of bioactive compounds from plant materials, fewer studies about purification have been reported in the literature. Currently, the main purification technologies for natural bioactives include graded precipitation method, salting-out method, metal coordination method, chromatography method, ultrafiltration method, solid-phase extraction, and others [89, 90]. Different technologies are usually combined together to get a satisfactory purity of target compounds in practical applications. However, the aforementioned purification technologies somewhat have several disadvantages, such as time-consuming, expensive with poor recovery, laborious, unsuitable for industrial scale-up, and unsafe due to the utilization of toxic solvents [91]. Therefore, natural bioactive compounds with high purity in the market are expensive.

On the other hand, adsorption is an attractive method to separate certain compounds from liquid phase. It may also be defined as the tendency of chemical species existing in a phase to adhere onto a solid. Adsorption is generally characterized by obvious advantages, including low cost, simplicity of operation and scale-up, favorable rate, and less applications of toxic solvent [92]. Various cheap and facile adsorbents can be used to separate target adsorbates, such as activated carbons, minerals, resins, industrial and agricultural wastes, and biosorbents. Following adsorption, the adsorbates can also be desorbed using proper elution procedures. In the literature, successful purification of phenolics by the aforementioned adsorbents has been reported [92–94].

In recent years, ultrasonic technology has been occasionally incorporated into the adsorption process, so as to accelerate adsorption uptake and kinetics [95, 96]. The principles for ultrasonic enhancement of adsorption process are still the cavitational properties of ultrasound, which can promote the movement of adsorbates across liquid film surrounding adsorbents, strengthen their binding capacity with adsorbents, and penetration into interior structure of adsorbents [97]. According to the literature, the application of ultrasound in the adsorption and desorption process is mainly focused on the separations of metals and organic compounds, such as *p*-chlorophenol [98], chromium [99], copper [100], Pb (II) [101], and nitrate [102] using resin and activated carbon as adsorbent.

To date, the reports about the application of ultrasound on adsorption and desorption of plant origin bioactives for purification are quite limited. A successful application has been reported by Wu et al. [103] when ultrasound was utilized to promote the absorption and desorption of phenolics from blueberry anthocyanins using macroporous resins. The reported studies revealed that the introduction of ultrasound during both adsorption and desorption processes accelerated the kinetics. The adsorption capacity of anthocyanins on macroporous resins and desorption capacity under sonication were significantly ($p < 0.05$) higher than that under shaking. Together with the enhanced adsorption and desorption kinetics, the recovery of anthocyanins obtained from ultrasound-assisted adsorption and desorption at 279 W/L was 1.8 times higher than that obtained under shaking. In general, the results of Wu et al. [103] clearly demonstrated that ultrasound was not only an efficient tool to shorten adsorption and desorption time for blueberry anthocyanins on macroporous resins, it can also benefit the recovery of anthocyanins through adsorption–desorption treatment. More studies are needed to clarify the influence of ultrasound on the adsorption and desorption of other bioactives from plant origin using various adsorbents.

To fulfill the potential of ultrasound to promote the adsorption and desorption of bioactives from plant food processing wastes for purification purpose, a series of factors should be taken into consideration in practical applications. The operating factors include ultrasonic and environmental factors.

The ultrasonic factors mainly refer to ultrasonic energy and ultrasonic duration. Ultrasonic energy can be expressed as ultrasonic power, ultrasound intensity, and acoustic energy density. In most ultrasound-included food processes including adsorption and desorption, the processing efficiency increases with input ultrasound energy [7]. An example of ultrasound energy on the adsorption and desorption of anthocyanins using macroporous resins is illustrated in Fig. 6.2 [103]. For ultrasound-assisted adsorption process, the increase of ultrasound energy can generate stronger ultrasound cavitation, thus making the external diffusion of target compounds across liquid film surrounding adsorbents easier and exposing more binding sites on adsorbents [102, 103]. For desorption process, the increase of ultrasound energy can promote the mass transfer among adsorbate/adsorbent/solution, thus enhancing the diffusion coefficient of adsorbate desorbed from adsorbent [103, 104]. Regarding the influence of sonication time, many studies demonstrate that both adsorption and des-

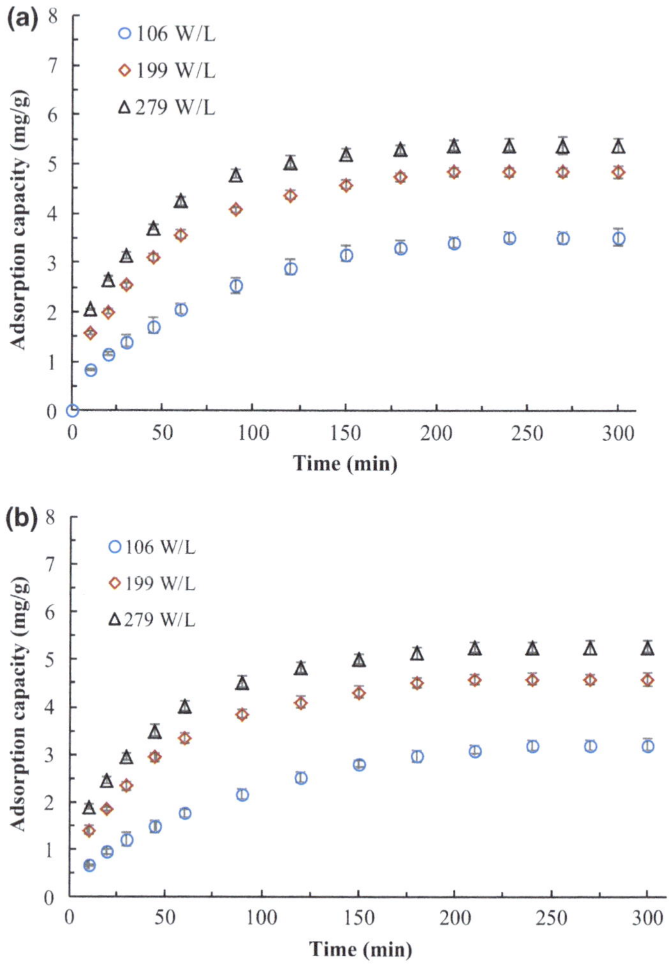

Fig. 6.2 Influence of ultrasound energy on adsorption (**a**) and desorption (**b**) kinetic curves of total anthocyanins from blueberry using macroporous resins [103]

orption capacities can increase with prolonging sonication time until reaching the equilibrium [96, 103].

The environmental factors include temperature, pH, initial concentrations of adsorbent and adsorbate, and existence of other components that compete with target adsorbent. Contradictory effects of temperature on adsorption process have been reported. On one hand, many adsorption processes could be favored by temperature diminishing due to non-spontaneous and exothermic features of adsorption processes themselves and the low solubility of adsorbents in solution at low temperatures [103, 105, 106]. On the other hand, temperature rising could strengthen the affinity between binding sites on adsorbents and adsorbates [107]. Meanwhile, temperature rising can facilitate desorption process attributing to the non-spontaneous and endothermic nature of desorption [105]. The pH value of solution can also affect the adsorption process profoundly, since the site dissociation of adsorbent surface and solution chemistry of adsorbate change with pH value [108]. In many cases, the increase of initial concentration of adsorbate can accelerate their uptake rate on adsorbent [108]. Moreover, the initial concentration of adsorbent can also influence the adsorption behavior. The increase of adsorbent concentration can increase sorption site number, while the decrease of adsorbent concentration can facilitate the penetration of adsorbate into intracellular parts [108, 109]. Besides, a competition phenomenon may take place if there are different adsorbates existing in the solution [108]. Taking the study of Wu et al. [103], for example, malvidin 3-galactoside had the highest absorption capacity on macroporous resins, whereas cyaniding 3-glucoside had the lowest adsorption capacity. The competitive effect can be attributed to the difference in affinity and selectivity of adsorbents to different adsorbates.

Generally speaking, current knowledge about the influences of ultrasonic and environmental factors, as well as their interaction effects on ultrasound-assisted adsorption and desorption of plant origin bioactives, is scarce. However, many applications have been reported in the environmental research field, and several preliminary studies have also proven the potential of ultrasonic technology to promote the purification of plant origin bioactives by means of adsorption and desorption. Thus, more fundamental studies are needed to study the adsorption and desorption of bioactives from plant foods with the assistance of ultrasound when different adsorbents are utilized, as well as scale up of the corresponding sono-equipment.

6.6 Diffusional Modeling of Mass Transfer Process for Ultrasound-Assisted Extraction and Adsorption of Bioactives from Plant Foods

Mathematical modeling is an important tool to understand the mass transfer mechanism behind ultrasound-related processes and achieve intelligent control of the processes. According to the authors' knowledge, there are three different methodologies for process simulation, being statistical models, empirical and semi-empirical models, and phenomenological models. The statistical tools include the extensively utilized artificial neural network and response surface methodology [110]. Although statistical models and empirical and semi-empirical models are capable to correlate the operating parameters with target response, these models have limited or even no information to explain the physical phenomena under investigation [7]. Instead, phenomenological models are usually based on complicated partial differential equations, which can describe heat and mass transfer and other phenomena occurred during food processing well. Thus, phenomenological simulation is a potent tool to investigate the mechanism about ultrasonic enhancement of mass transfer during extraction and adsorption.

The difficulties of phenomenological simulation lie in finding suitable models to represent the real conditions and solving the partial differential equations. In practical applications, mathematical language cannot depict the physical and chemical changes completely and a series of assumptions should be made to simplify the models. Meanwhile, numerical method is usually utilized to solve the partial differential equations, which needs a well-equipped computer to accomplish the calculation.

There have already been large quantities of studies reporting the mass transfer modeling of ultrasound-assisted extraction and adsorption processes using artificial neural network, response surface methodology, and different empirical and semi-empirical models [111–113]. However, limited studies can be found about phenomenological simulation of extraction and adsorption of bioactives from plant foods with the assistance of ultrasound. Theoretically, different physical models can be considered to model the mass transfer phenomenon occurred in extraction and adsorption processes, such as Newton's law about mass balance, momentum balance and energy balance, and acoustic theories [114, 115]. If the only purpose is to study the mass transfer phenomenon, the diffusion model based on Fick's second law can be a suitable and easy model [96, 112]. It should be addressed that several assumptions are usually made before the application of diffusion model to simulate extraction and adsorption processes, involving perfect mixing of liquid medium and consequently no external mass transfer resistance, a specific geometry of materials releasing or adsorbing bioactives (spherical, slab or column geometries), and constant effective diffusion coefficient of bioactives within solid matrix [112, 116, 117].

For the diffusional mass transfer modeling of ultrasound-assisted extraction of bioactives from plant foods, the governing equation is written as [116]:

$$\frac{\partial C_s(x,t)}{\partial t} = D_e \frac{1}{x^{\nu-1}} \frac{\partial}{x} \left[x^{\nu-1} \frac{\partial C_s(x,t)}{\partial x} \right] \tag{6.1}$$

where C_s is the concentration of bioactives in solid matrix, D_e is the effective diffusivity of bioactives within solid matrix, x is the radial distance in diffusion direction, ν is shape index, and t is time. The shape index can be 0, 1, and 2, corresponding to slab, cylindrical, or spherical symmetry, respectively.

The initial condition of Eq. 6.1 can be $C_s = C_S^0$ for dispersed phase and $C_L = 0$ for continuous phase. Here, C_S^0 refers to the initial concentration of bioactives in solid matrix, and C_L is the concentration of bioactives in liquid phase. The boundary condition at the central axis of solid matrix for Eq. 6.1 can be:

$$\left(\frac{\partial C_s}{\partial x} \right)_{x=0} = 0 \, \forall t \tag{6.2}$$

At the interface, the outcoming flux of bioactives from solid matrix is equal to the incoming flux in liquid. Thus, the equality relationship about flux is written as:

$$-D_e A \left[\frac{\partial C_s}{\partial x} \right]_{x=r} = V_L \frac{dC_L(t)}{dt} \tag{6.3}$$

where A is the surface area of solid matrix, V_L is volume of solvent, and r is the radius of spherical and cylindrical solid matrices or half thickness of slab solid matrix.

Using the aforementioned equations, the ultrasound-assisted extractions of phenolics from grape pomaces and oak chips were successfully simulated [112, 116]. According to the variations of obtained effective diffusivity of phenolics as a function of ultrasonic operating parameters, the mass transfer mechanism under sonication can be explained.

The mass transfer phenomenon occurred during adsorption is more complicated than that occurred during extraction. The adsorption process usually contains three steps, namely external mass transport, intraparticle diffusion, and adsorption on active sites distributed on the pore surface of adsorbents [118]. Meanwhile, intraparticle diffusion of bioactives within adsorbents includes surface diffusion, pore diffusion, or both [116]. Since most adsorbents can be regarded as spherical geometry, the diffusion model for ultrasound-assisted adsorption of bioactives from plant foods can be expressed as [119]:

$$\varepsilon \frac{\partial C_{A,r}}{\partial t} + \rho_p \frac{\partial q}{\partial t} = \frac{1}{r^2} \frac{\partial}{\partial r}\left[r^2\left(D_{\text{ep}} \frac{\partial C_{A,r}}{\partial r} + D_{\text{s}} \rho_p \frac{\partial q}{\partial r}\right)\right] \tag{6.4}$$

where ε is void fraction of adsorbent, ρ_p is adsorbent density, r is radial distance, $C_{A,r}$ is concentration of bioactives within adsorbent at distance r, D_{ep} is effective pore volume diffusion coefficient of bioactives, D_{s} is surface diffusion coefficient, and q is mass of bioactives adsorbed by adsorbent.

The left side of Eq. 6.4 includes the accumulation of bioactives in the pore volume of adsorbent particles, while the right side of Eq. 6.4 refers to both pore volume diffusion and surface diffusion.

The initial conditions for Eq. 6.4 can be expressed as:

$$t = 0, \quad C_A = C_{A,0} \tag{6.5}$$

$$t = 0, \quad C_{A,r} = 0 \tag{6.6}$$

$$t = 0 \quad q = 0 \tag{6.7}$$

where $C_{A,0}$ is the initial concentration of bioactives in liquid phase, and $C_{A,r}$ is the concentration of bioactives within adsorbent particle at distance r.

The boundary conditions for Eq. 6.4 are shown as follows:

$$\left.\frac{\partial C_{A,r}}{\partial r}\right|_{r=0} = 0 \tag{6.8}$$

$$D_{\text{ep}} \left.\frac{\partial C_{A,r}}{\partial r}\right|_{r=R} + D_{\text{s}} \rho_p \frac{\partial q}{\partial r} = k_L (C_A - C_{A,r}) \tag{6.9}$$

where k_L is external mass transfer coefficient of bioactives in liquid phase, and R is radius of adsorbent particle. To solve Eq. 6.4, another hypothesis should be made that the adsorption rate on active sites is usually regarded to be instantaneous [119]. Therefore, spontaneous equilibrium can be achieved on the solid–liquid interface within the pore of adsorbent. The relationship between total phenolic concentrations inside yeast at any time and radical location and their concentration in liquid phase can be expressed by the adsorption isotherm, which is written as [119]:

$$q = f(C_{A,r}) \tag{6.10}$$

Different adsorption isotherm models, such as Langmuir isotherm, Freundlich isotherm, and Sips isotherm, can be considered to fit the adsorption isotherm data [121].

Prior to solving Eq. 6.4, the external mass transfer coefficient (k_L) needs to be determined using the method of Furusawa and Smith [122]. When t approaches 0, $C_{a,r}$ and C_a values are closed to 0 and $C_{a,0}$, respectively. Consequently, the following formula can be obtained:

$$\left[\frac{\mathrm{d}\left(\frac{C_A}{C_{A,0}}\right)}{\mathrm{d}t}\right]_{t=0} = \frac{-mSk_L}{V} \tag{6.11}$$

The right side of Eq. 6.11 corresponds to the slope of concentration decay of bioactives at $t = 0$. The value of k_{L} can be determined using the data points of concentration decay curve at the early stage [123].

The diffusion model has been successfully employed to simulate the ultrasound-assisted adsorption kinetics of ibuprofen on an activated carbon fabric, as well as various adsorption processes without ultrasound treatment [96, 117, 119, 120]. It is believed that this methodology should also be qualified to model the adsorption of plant origin bioactives by different adsorbents under ultrasonic fields.

To get the solutions of diffusion model in both extraction and adsorption cases, the partial differential equations are usually discretized, so as to get the numerical solutions. The calculation of Eqs. 6.1 and 6.4 is relatively simple, since both of them are partial differential equations in one space variable x and time t. Thus, the "*pdepe*" function in MATLAB can be used to solve the equations directly without writing tedious codes [116]. The details about the application of *pdepe* function can be found in the "*help*" function of this software.

Additionally, an important issue should be pointed out that sonication can result in the disruption of solid matrix. For example, the study of Wu et al. [103] showed that the structure of macroporous resins was damaged by ultrasound during adsorption (Fig. 6.3). Liu et al. [124] also reported that the size of yeast cells decreased rapidly during ultrasound-assisted extraction of intracellular protein from yeasts. Therefore, to build more realistic models about ultrasound-assisted extraction and adsorption of bioactives from plant foods, the influence of ultrasound on solid matrix size needs

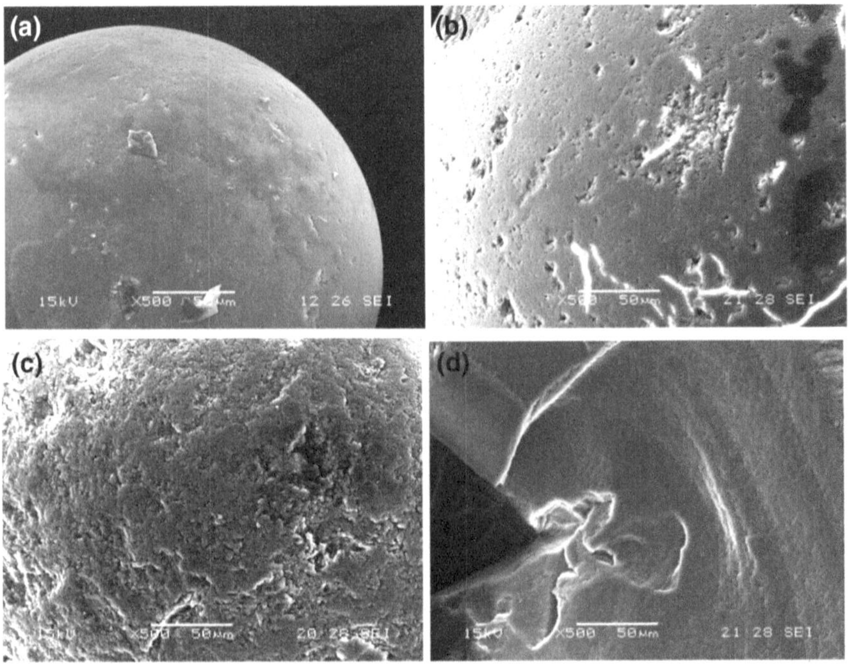

Fig. 6.3 Scanning electron micrographs (500 × magnification) for macroporous resins before and after ultrasound-assisted adsorption of blueberry anthocyanins. **a** control resins; **b** resins treated at 106 W/L; **c** resins treated at 199 W/L; **d** resins treated at 279 W/L [103]

to be incorporated into the diffusion model. To date, few studies have considered changes in particle size in the diffusion models about ultrasound-assisted extraction and adsorption processes.

If sonication changes the size of solid matrix during extraction and adsorption, the mass transfer process is then addressed as moving boundary problem. Similarly, the diffusion problem during hot-air drying of food materials also has moving boundary conditions due to volume shrinkage [125, 126]. Thus, the mathematical treatment used in moisture diffusion modeling can be taken as a reference to simulate ultrasound-assisted extraction and adsorption including sample disruption problem. The first step is to assume that the solid matrix can maintain their original geometry despite size variation. After that, the variation of particle size along with sonication needs to be established experimentally and mathematically. Following the relationship between particle size and sonication duration, the size of solid matrix at each sample time can be obtained. Therefore, the diffusion distance included the boundary conditions (Eqs. 6.3 and 6.9) can be confirmed. Through solving Eqs. 6.1 and 6.4 separately at each sampling time, the mass transfer-related parameters involved in extraction and adsorption process can be obtained. This method is recommended for the diffusional modeling of ultrasound-assisted extraction and adsorption of bioactives from plant foods in future studies.

6.7 Conclusions

In today's food industry, the green concept has penetrated into almost every corner of food and food waste processing areas. The green concept may contain different elements, such as high processing efficiency, low energy consumption, low pollution to the environment, low generation of harmful components, and high quality of final products. Proper ultrasound treatment can satisfy the food processes in all the listed aspects. In the whole processes (not limited to extraction and adsorption-based purification) to transform bioactives from plant food wastes to valuable chemicals, ultrasound has tremendous potential to benefit both processes and final products. At this stage, ultrasound has been widely used for extraction in laboratory and industry. However, fewer applications have been found in the adsorption area. To boom ultrasonic technology in the field of utilization of plant food processing wastes, more fundamental studies are needed to optimize the processing conditions, scale up the sono-reactors and processes, and evaluate the product qualities. It is expected that ultrasonic approach to process bioactives from plant food wastes to valuable chemicals can generate more "green" benefits in future applications.

Acknowledgements The authors would like to thank Prof. Farid Chemat and Dr. Ying Li for their kind invitation to join this book writing project.

References

1. Gouw V, Jung J, Simonsen J, Zhao Y (2017) Fruit pomace as a source of alternative fibers and cellulose nanofiber as reinforcement agent to create molded pulp packaging boards. Composites: Part A 99:48–57
2. Tao Y, Wu D, Zhang QA, Sun DW (2014) Ultrasound-assisted extraction of phenolics from wine lees: modeling, optimization and stability of extracts during storage. Ultrason Sonochem 21:706–715
3. Petkowicz CLO, Vriesmann LC, Williams PA (2016) Pectins from food waste: extraction, characterization and properties of watermelon rind pectin. Food Hydrocolloids 65:57–67
4. Kadam SU, Tiwari BK, Álvarez C, O'Donnell CP (2015) Ultrasound applications for the extraction, identification and delivery of food proteins and bioactive peptides. Trends Food Sci Technol 46:60–67
5. Gadkari PV, Balaraman M (2015) Catechins: sources, extraction and encapsulation: a review. Food Bioprod Process 93:122–138
6. Chemat F, Huma Z, Khan MK (2011) Applications of ultrasound in food technology: processing, preservation and extraction. Ultrason Sonochem 18:813–835
7. Tao Y, Sun DW (2015) Enhancement of food processes by ultrasound: a review. Crit Rev Food Sci Nutr 55:570–594
8. Kawsar SM, Fujii Y, Matsumoto R, Ichikawa T, Tateno H, Hirabayashi J, Yasumitsu H, Dogasaki C, Hosono M, Nitta K (2008) Isolation, purification, characterization and glycan-binding profile of a d-galactoside specific lectin from the marine sponge, Halichondria okadai. Comp Biochem Physiol B: Biochem Mol Biol 150:349–357
9. Pavlić B, Teslić N, Vidaković A, Vidović S, Velićanski A, Versari A, Radosavljević R, Zeković Z (2017) Sage processing from by-product to high quality powder: I Bioactive potential. Ind Crops & Prod 107:81–89

10. Heleno SA, Barros L, Martins A, Morales P, Fernández-Ruiz V, Glamoclija J, Sokovic M, Ferreira ICFR (2015) Nutritional value, bioactive compounds, antimicrobial activity and bioaccessibility studies with wild edible mushrooms. LWT—Food Sci and Technol 63:799–806
11. Dumitriu D, Peinado RA, Peinado J, Lerma NLD (2015) Grape pomace extract improves the in vitro and in vivo antioxidant properties of wines from sun light dried Pedro Ximénez grapes. J Funct Foods 17:380–387
12. Katalinić V, Možina SS, Skroza D, Generalić I, Abramovič H, Miloš M, Ljubenkov I, Piskernik S, Pezo I, Terpinc P (2010) Polyphenolic profile, antioxidant properties and antimicrobial activity of grape skin extracts of 14 Vitis vinifera varieties grown in Dalmatia (*Croatia*). Food Chem 119:715–723
13. Nohynek LJ, Alakomi HL, Kähkönen MP, Heinonen M, Helander IM, Oksmancaldentey KM, Puupponenpimiä RH (2006) Berry phenolics: antimicrobial properties and mechanisms of action against severe human pathogens. Nutr Cancer 54:18–32
14. Kam WYJ, Mirhosseini H, Abas F, Hussain N, Hedayatnia S, Lin H, Chong F (2018) Antioxidant activity enhancement of biodegradable film as active packaging utilizing crude extract from durian leaf waste. Food Control 90:66–72
15. Khalifa I, Barakat H, El-Mansy HA, Soliman SA (2016) Preserving apple (*Malus domestica* var. *Anna*) fruit bioactive substances using olive wastes extract-chitosan film coating. Inf Process Agric 4:90–99
16. Shekari MA, Rajabalian S (2010) Antioxidative effect of aqueous Date fruit extract in PC12 cell line. Adv Mater Res 314:123–126
17. Dai J, Mumper RJ (2010) Plant phenolics: extraction, analysis and their antioxidant and anticancer properties. Molecules 15:7313–7352
18. Holland D, Bar-Ya'Akov I (2014) Pomegranate: aspects concerning dynamics of health beneficial phytochemicals and therapeutic properties with respect to the tree cultivar and the environment. Med Aromat Plants of the World 2:225–239
19. Rice-Evans CA, Miller NJ, Paganga G (1996) Structure-antioxidant activity relationships of flavonoids and phenolic acids. Free Radic Biol Med 20:933–956
20. Oszmiański J, Wojdylo A (2005) *Aronia melanocarpa* phenolics and their antioxidant activity. Eur Food Res Technol 221:809–813
21. Haas ICDS, Toaldo IM, Burin VM, Bordignon-Luiz MT (2018) Extraction optimization for polyphenolic profiling and bioactive enrichment of extractives of non-pomace residue from grape processing. Ind Crops Prod 112:593–601
22. Boyer J, Liu RH (2004) Apple phytochemicals and their health benefits. Nutr J 3:5–15
23. Munir MT, Kheirkhah H, Baroutian S, Quek SY, Young BR (2018) Subcritical water extraction of bioactive compounds from waste onion skin. J Clean Prod 183:487–494
24. Rop O, Mlcek J, Jurikova T, Valsikova M, Sochor J, Reznicek V, Kramarova D (2010) Phenolic content, antioxidant capacity, radical oxygen species scavenging and lipid peroxidation inhibiting activities of extracts of five black chokeberry (*Aronia melanocarpa* (Michx.) Elliot) cultivars. J Med Plants Res 4:2431–2437
25. Skupień K, OszmiańSki J (2008) The effect of mineral fertilization on nutritive value and biological activity of chokeberry fruit. Agric Food Sci 16:46–55
26. Fazaeli M, Hojjatpanah G, Emamdjomeh Z (2013) Effects of heating method and conditions on the evaporation rate and quality attributes of black mulberry (*Morus nigra*) juice concentrate. Int J Food Sci Technol 50:35–43
27. Zhang L, Wang M (2017) Optimization of deep eutectic solvent-based ultrasound-assisted extraction of polysaccharides from *Dioscorea opposita* Thunb. Int J Biol Macromol 95:675–681
28. Luengo E, Condón-Abanto S, Condón S, Álvarez I, Raso J (2014) Improving the extraction of carotenoids from tomato waste by application of ultrasound under pressure. Sep Purif Technol 136:130–136
29. Durante M, Montefusco A, Marrese PP, Soccio M, Pastore D, Piro G, Mita G, Lenucci MS (2017) Seeds of pomegranate, tomato and grapes: an underestimated source of natural bioactive molecules and antioxidants from agri-food by-products. Subtropical Plant Sci 63:65–72

30. Shi Y, Kovacs-Nolan J, Jiang B, Rong T, Mine Y (2014) Antioxidant activity of enzymatic hydrolysates from eggshell membrane proteins and its protective capacity in human intestinal epithelial Caco-2 cells. J Funct Foods 10:35–45
31. Vásquez-Villanueva R, Muñoz-Moreno L, Carmena MJ, Marina ML, García MC (2018) In vitro antitumor and hypotensive activity of peptides from olive seeds. J Funct Foods 42:177–184
32. Kovacs-Nolan J, Zhang H, Ibuki M, Nakamori T, Yoshiura K, Turner PV, Matsui T, Mine Y (2012) The PepT1-transportable soy tripeptide VPY reduces intestinal inflammation. Biochem Biophys Acta 1820:1753–1763
33. González-García E, Marina ML, García MC (2014) Plum (*Prunus Domestica* L.) by-product as a new and cheap source of bioactive peptides: extraction method and peptides characterization. J Funct Foods 11:428–437
34. Adebiyi AP, Adebiyi AO, Yamashita J, Ogawa T, Muramoto K (2009) Purification and characterization of antioxidative peptides derived from rice bran protein hydrolysates. Eur Food Res Technol 228:553–563
35. Rizzello CG, De AM, Di CR, Camarca A, Silano M, Losito I, De VM, De Bari MD, Palmisano F, Maurano F (2007) Highly efficient gluten degradation by lactobacilli and fungal proteases during food processing: new perspectives for celiac disease. Appl Environ Microbiol 73:4499–4507
36. Sarmadi BH, Ismail A (2010) Antioxidative peptides from food proteins: a review. Peptides 31:1949–1956
37. Shahidi F, Zhong Y (2008) Bioactive peptides. J Aoac Int 91:914–931
38. Hernández-Ledesma B, Hsieh CC, de Lumen BO (2009) Lunasin, a novel seed peptide for cancer prevention. Peptides 30:426–430
39. Rizzello CG, Nionelli L, Coda R, Gobbetti M (2012) Synthesis of the cancer preventive peptide lunasin by lactic acid bacteria during sourdough fermentation. Nutr Cancer-an Int J 64:111–120
40. Lee SY, Sun JH (2017) Antihypertensive peptides from animal products, marine organisms, and plants. Food Chem 228:506–517
41. Megías C, Yust MdM, Pedroche J, Lquari H, Girón-Calle J, Alaiz M, Millán F, Vioque J (2004) Purification of an ACE inhibitory peptide after hydrolysis of sunflower (*Helianthus annuus* L.) protein isolates. J Agric Food Chem 52:1928–1932
42. Wu Q, Du J, Jia J, Kuang C (2016) Production of ACE inhibitory peptides from sweet sorghum grain protein using alcalase: hydrolysis kinetic, purification and molecular docking study. Food Chem 199:140–150
43. Dacoba TG, Olivera A, Torres D, Crecente-Campo J, Alonso MJ (2017) Modulating the immune system through nanotechnology. Semin Immunol 34:78–102
44. Chalamaiah M, Yu W, Wu J (2018) Immunomodulatory and anticancer protein hydrolysates (peptides) from food proteins: a review. Food Chem 245:205–222
45. Liu J, Willför S, Xu C (2015) A review of bioactive plant polysaccharides: biological activities, functionalization, and biomedical applications. Bioact Carbohydr & Dietary Fibre 5:31–61
46. Maki-Arvela P, Salmi T, Holmbom B, Willfor S, Murzin DY (2011) Synthesis of sugars by hydrolysis of hemicelluloses-a review. Chem Rev 111:5638–5666
47. Yang L, Zhang LM (2009) Chemical structural and chain conformational characterization of some bioactive polysaccharides isolated from natural sources. Cheminform 40:349–361
48. Liu J, Bai R, Liu Y, Zhang X, Kan J, Jin C (2017) Isolation, structural characterization and bioactivities of naturally occurring polysaccharide-polyphenolic conjugates from medicinal plants—a reivew. Int J Biol Macromol 107:2242–2250
49. Chawla R, Patil GR (2010) Soluble dietary fiber. Compr Rev Food Sci Food Saf 9:178–196
50. Pool-Zobel BL (2005) Inulin-type fructans and reduction in colon cancer risk: review of experimental and human data. Br J Nutr 4:S73–S90
51. Purohit AJ, Gogate PR (2015) Ultrasound-assisted extraction of -carotene from waste carrot residue: effect of operating parameters and type of ultrasonic irradiation. Sep Sci Technol 50:1507–1517

52. Cárcel JA, Garcíapérez JV, Benedito J, Mulet A (2012) Food process innovation through new technologies: use of ultrasound. J Food Eng 110:200–207
53. Ghafoor K, Yonghee C, Juyeong J, Inhee J (2009) Optimization of ultrasound-assisted extraction of phenolic compounds, antioxidants, and anthocyanins from grape (*Vitis vinifera*) seeds. J Agric Food Chem 57:4988–4994
54. Cruz L, Clemente G, Mulet A, Ahmad-Qasem MH, Barrajon-Catalan E, Garcia-Perez J (2016) Air-borne ultrasonic application in the drying of grape skin: Kinetic and quality considerations. J Food Eng 168:251–258
55. Yu L, Jie S, Liu S, Jie B, Zhang C, Yang Q (2012) Ultrasonic-assisted enzymolysis to improve the antioxidant activities of peanut (*Arachin conarachin* L.) antioxidant hydrolysate. Int J Mol Sci 13:9051–9068
56. Huang L, Liu B, Ma H, Zhang X (2014) Combined effect of ultrasound and enzymatic treatments on production of ace inhibitory peptides from wheat germ protein. J Food Process Preserv 38:1632–1640
57. Moure A, Domínguez H, Parajó JC (2006) Ultrafiltration of industrial waste liquors from the manufacture of soy protein concentrates. J Chem Technol Biotechnol Biotechnol 81:1252–1258
58. Portenlänger G, Heusinger H (1997) The influence of frequency on the mechanical and radical effects for the ultrasonic degradation of dextranes. Ultrason Sonochem 4:127–130
59. Ebringerová A, Hromádková Z (2010) An overview on the application of ultrasound in extraction, separation and purification of plant polysaccharides. Open Chemistry 8:243–257
60. Chemat F, Rombaut N, Sicaire AG, Meullemiestre A, Fabiano-Tixier AS, Abert-Vian M (2017) Ultrasound assisted extraction of food and natural products. Mechanisms, techniques, combinations, protocols and applications. A review. Ultrason Sonochem 34:540–550
61. Tiwari BK (2015) Ultrasound: A clean, green extraction technology. Trends Anal Chem 71:100–109
62. Wen C, Zhang J, Zhang H, Dzah CS, Zandile M, Duan Y, Ma H, Luo X (2018) Advances in ultrasound assisted extraction of bioactive compounds from cash crops—a review. Ultrason Sonochem 48:538–549
63. Bosiljkov T, Dujmić F, Bubalo MC (2017) Natural deep eutectic solvents and ultrasound-assisted extraction: green approaches for extraction of wine lees anthocyanins. Food Bioprod Process 102:195–203
64. Chanioti S, Tzia C (2018) Extraction of phenolic compounds from olive pomace by using natural deep eutectic solvents and innovative extraction techniques. Innovative Food Sci Emerg Technol 48:228–239
65. Jiang HL, Yang JL, Shi YP (2017) Optimization of ultrasonic cell grinder extraction of anthocyanins from blueberry using response surface methodology. Ultrason Sonochem 34:325–331
66. Wang M, Huang B, Fan C, Zhao K, Hu H, Xu XY, Pan SY, Liu FX (2016) Characterization and functional properties of mango peel pectin extracted by ultrasound assisted citric acid. Int J Biol Macromol 91:794–803
67. Albahari P, Jug M, Radić K, Jurmanovićb S, Brnčićc M, Brnčićc SR, Čepob DV (2018) Characterization of olive pomace extract obtained by cyclodextrin-enhanced pulsed ultrasound assisted extraction. LWT—Food Sci Technol 92:22–31
68. Tchabo W, Ma Y, Engmann FN (2015) Ultrasound-assisted enzymatic extraction (UAEE) of phytochemical compounds from mulberry (*Morus nigra*) must and optimization study using response surface methodology. Ind Crops Prod 63:214–225
69. Yin C, Fan X, Fan Z, Shi D, Gao H (2018) Optimization of enzymes-microwave-ultrasound assisted extraction of *Lentinus edodes* polysaccharides and determination of its antioxidant activity. Int J Biol Macromol 111:446–454
70. Hu AJ, Zhao S, Liang H, Qiu TQ, Chen G (2007) Ultrasound assisted supercritical fluid extraction of oil and coixenolide from adlay seed. Ultrason Sonochem 14:219–224
71. Santoszea L, Antunesricardo M, Gutierrezuribe JA, Garcíapérezb JV, Beneditob J (2018) Effect of ultrasound transducer design on the acoustically-assisted supercritical fluid extraction of antioxidants from oregano. Ultrason Sonochem 47:47–56

72. Yu Q, Li C, Duan ZH, Liu B, Duan WW, Shang FF (2017) Ultrasonic microwave-assisted extraction of polyphenols, flavonoids, triterpenoids, and vitamin C from *Clinacanthus nutans*. Czech J of Food Sci 35:89–94
73. Xiang Z, Wu X (2017) Ultrasonic-microwave assisted extraction of total flavonoids from *Scutellaria baicalensis*, using response surface methodology. Pharm Chem J 51:318–323
74. Chen HM, Fu X, Luo ZG (2015) Properties and extraction of pectin-enriched materials from sugar beet pulp by ultrasonic-assisted treatment combined with subcritical water. Food Chem 168:302–310
75. Allaf T, Tomao V, Ruiz K, Chemat F (2013) Instant controlled pressure drop technology and ultrasound assisted extraction for sequential extraction of essential oil and antioxidants. Ultrason Sonochem 20:239–246
76. Moradi N, Rahimi M (2018) Effect of simultaneous ultrasound/pulsed electric field pretreatments on the oil extraction from sunflower seeds. Sep Sci Technol 53:2088–2099
77. Li J, Luo ZG, Xiao ZG, Wang LM, Zhu XQ, Sun SK (2013) Optimization of extraction technology of rice bran polysaccharide by extrusion in conjunction with ultrasound. Trans Chin Soc Agric Mach 44:174–179
78. Abbott AP, Capper G, Davies DL, Rasheed RK, Tambyrajah V (2003) Novel solvent properties of choline chloride/urea mixtures. Chem Commun 9:70–71
79. Zdanowicz M, Wilpiszewska K, Spychaj T (2018) Deep eutectic solvents for polysaccharides processing: a review. Carbohydr Polym, in press
80. Cunha SC, Fernandes JO (2018) Extraction techniques with deep eutectic solvents. Trends Anal Chem 105:225–239
81. Nadar SS, Rao P, Rathod VK (2018) Enzyme assisted extraction of biomolecules as an approach to novel extraction technology: a review. Food Res Int 108:309–330
82. Zhang W, Huang J, Wang W (2016) Extraction, purification, characterization and antioxidant activities of polysaccharides from *Cistanche tubulosa*. Int J Biol Macromol 93:448–458
83. Chemat F, Fabiano-Tixier AS, Vian MA, Allaf T, Vorobiev E (2015) Solvent-free extraction of food and natural products. Trends Anal Chem 71:157–168
84. Corrales M, Toepfl S, Butz P, Knorr D, Tauscher B (2008) Extraction of anthocyanins from grape by-products assisted by ultrasonics, high hydrostatic pressure or pulsed electric fields: a comparison. Innovative Food Sci Emerg Technol 9:85–91
85. Silva RPFF, Rocha-Santos TAP, Duarte AC (2016) Supercritical fluid extraction of bioactive compounds. Trends Anal Chem 76:40–51
86. Nastić N, Švarc-Gajić J, Delerue-Matos C, Morais S, Barroso MF, Moreira MM (2018) Subcritical water extraction of antioxidants from mountain germander (*Teucrium montanum* L.). J Supercrit Fluids 138:200–206
87. Buran T, Sandhu AK, Li Z, Rock CR, Yang WW, Gu L (2014) Adsorption/desorption characteristics and separation of anthocyanins and polyphenols from blueberries using macroporous adsorbent resins. J Food Eng 128:167–173
88. Ongkowijoyo P, Luna-Vital DA, Mejia EG (2018) Extraction techniques and analysis of anthocyanins from food sources by mass spectrometry: an update. Food Chem 250:113–126
89. Shi L (2016) Bioactivities, isolation and purification methods of polysaccharides from natural products: a review. Int J Biol Macromol 92:37–48
90. Wang E, Yin Y, Xu C, Liu J (2014) Isolation of high-purity anthocyanin mixtures and monomers from blueberries using combined chromatographic techniques. J Chromatogr A 1327:39–48
91. Chen Y, Zhang W, Zhao T, Li F, Zhang M, Li J, Zou Y, Wang W, Cobbina S, Wu X, Yang L (2016) Adsorption properties of macroporous adsorbent resins for separation of anthocyanins from mulberry. Food Chem 194:712–722
92. Soto ML, Moure A, Domínguez H, Parajó JC (2011) Recovery, concentration and purification of phenolic compounds by adsorption: a review. J Food Eng 105:1–27
93. Gao X, Dai Y, Zhang Y, Fu F (2017) Effective adsorption of phenolic compound from aqueous solutions on activated semi coke. J Phys Chem Solids 102:142–150

94. Rubio FT, Maciel GM, Silva MV, Corrêa VG, Peralta RM, Haminiuk CWI (2018) Enrichment of waste yeast with bioactive compounds from grape pomace as an innovative and emerging technology: kinetics, isotherms and bioaccessibility. Innovative Food Sci Emerg Technol 45:18–28
95. Mehrabi F, Dil EA (2017) Investigate the ultrasound energy assisted adsorption mechanism of nickel(II) ions onto modified magnetic cobalt ferrite nanoparticles: multivariate optimization. Ultrason Sonochem 37:37–46
96. Ondarts M, Reinert L, Guittonneau S, Baup S, Delpeux S, Lévêque J, Duclaux L (2018) Improving the adsorption kinetics of ibuprofen on an activated carbon fabric through ultrasound irradiation: simulation and experimental studies. Chem Eng J 343:163–172
97. Milenković DD, Dašić PV, Veljković VB (2009) Ultrasound-assisted adsorption of copper (II) ions on hazelnut shell activated carbon. Utrason Sonochem 16:557–563
98. Hamdaoui O, Naffrechoux E, Tifouti L, Pétrierb C (2003) Effects of ultrasound on adsorption-desorption of p-chlorophenol on granular activated carbon. Ultrason Sonochem 10:109–114
99. Jing GH, Zhou ZM, Lei S, Dong MX (2011) Ultrasound enhanced adsorption and desorption of chromium (VI) on activated carbon and polymeric resin. Desalination 279:423–427
100. Korkut O, Sayan E, LaċN O, Bayrak B (2010) Investigation of adsorption and ultrasound assisted desorption of lead (II) and copper (II) on local bentonite: a modelling study. Desalination 259:243–248
101. Karimi M, Entezari MH, Chamsaz M (2010) Sorption studies of nitrate ion by a modified beet residue in the presence and absence of ultrasound. Ultrason Sonochem 17:711–717
102. Kostić M, Mitrović J, Radović M, Đorđević M, Petović M, Bojić D, Bojić A (2016) Effects of power of ultrasound on removal of Cu(II) ions by xanthated *Lagenaria vulgaris* shell. Ecol Eng 90:82–86
103. Wu Y, Han Y, Tao Y, Fan S, Chu DT, Ye X, Ye M, Xie X (2018) Ultrasound assisted adsorption and desorption of blueberry anthocyanins using macroporous resins. Ultrason Sonochem 48:311–320
104. Li Z, Xu KF, Li XB, Xi HX, Hua B, Li LS (2006) Effect of ultrasound on desorption kinetics of phenol from polymeric resin. Ultrason Sonochem 13:225–231
105. Chandrasekhar J, Madhusudhan MC, Raghavarao KSMS (2012) Extraction of anthocyanins from red cabbage and purification using adsorption. Food Bioprod Process 90:615–623
106. Zhao Z, Dong L, Wu Y, Lin F (2011) Preliminary separation and purification of rutin and quercetin from *Euonymus alatus*, (*Thunb.*) Siebold, extracts by macroporous resins. Food Bioprod Process 89:266–272
107. Goyal N, Jain SC, Banerjee UC (2003) Comparative studies on the microbial adsorption of heavy metals. Adv Environ Res 27:311–319
108. Wang J, Chen C (2006) Biosorption of heavy metals by Saccharomyces cerevisiae: a review. Biotechnol Adv 24:427–451
109. Bagheri AR, Ghaedi M, Asfaram A, Jannesar R, Goudarzi A (2017) Design and construction of nanoscale material for ultrasonic assisted adsorption of dyes: application of derivative spectrophotometry and experimental design methodology. Ultrason Sonochem 35:112–123
110. Aliakbarian B, Sampaio FC, Faria JT, Pitangui CG, Lovaglio F, Casazza AA, Converti A, Perego P (2018) Optimization of spray drying microencapsulation of olive pomace polyphenols using response surface methodology and artificial neural network. LWT-Food Sci and Technol 93:220–228
111. Patil D, Akamanchi KG (2017) Ultrasound-assisted rapid extraction and kinetic modelling of influential factors: extraction of camptothecin from *Nothapodytes nimmoniana* plant. Ultrason Sonochem 37:582–591
112. Tao Y, Zhang Z, Sun DW (2014) Kinetic modeling of ultrasound-assisted extraction of phenolic compounds from grape marc: influence of acoustic energy density and temperature. Ultrason Sonochem 21:1461–1469
113. Dil EA, Ghaedi M, Asfaram A, Bazrafshan AA (2018) Ultrasound wave assisted adsorption of congo red using gold-magnetic nanocomposite loaded on activated carbon: optimization of process parameters. Ultrason Sonochem 46:99–105

114. Kumar A, Kumaresan T, Pandit AB, Joshi JB (2006) Characterization of flow phenomena induced by ultrasonic horn. Chem Eng Sci 61:7410–7420
115. Gogate PR, Sutkar VS, Pandit AB (2011) Sonochemical reactors: important design and scale up considerations with a special emphasis on heterogeneous system. Chem Eng J 166:1066–1082
116. Tao Y, Zhang Z, Sun DW (2014) Experimental and modeling studies of ultrasound-assisted release of phenolics from oak chips into model wine. Ultrason Sonochem 21:1839–1848
117. Ocampo-Perez R, Leyva-Ramos R, Alonso-Davila P, Rivera-Utrilla J, Sanchez-Polo M (2010) Modeling adsorption rate of pyridine onto granular activated carbon. Chem Eng J 165:133–141
118. Yao C, Chen T (2017) A film-diffusion-based adsorption kinetic equation and its application. Chem Eng Res Des 119:87–92
119. Ocampo-Perez R, Leyva-Ramos R, Mendoza-Barron J, Guerrero-Coronado RM (2011) Adsorption rate of phenol from aqueous solution onto organobentonite: surface diffusion and kinetic models. J Colloid Interface Sci 364:195–204
120. Ponnusami V, Rajan KS, Srivastava SN (2010) Application of film-pore diffusion model for methylene blue adsorption onto plant leaf powders. Chem Eng J 163:236–242
121. Kim TY, Lee JW, Cho SY (2015) Application of residual brewery yeast for adsorption removal of Reactive Orange 16 from aqueous solution. Adv Powder Technol 26:267–274
122. Furusawa T, Smith JM (1973) Fluid-particle and intraparticle mass transport rates in slurries. Ind Eng Chem Fundam 12:197–203
123. Dotto GL, Pinto LAA (2012) Analysis of mass transfer kinetics in the biosorption of synthetic dyesonto *Spirulina platensis* nanoparticles. Biochem Eng J 68:85–90
124. Liu D, Zeng XA, Sun DW, Han Z (2013) Disruption and protein release by ultrasonication of yeast cells. Innovative Food Sci Emerg Technol 18:132–137
125. García-Pérez J, Ozuna C, Ortuño C, Cárcel JA, Mulet A (2011) Modeling ultrasonically assisted convective drying of eggplant. Drying Technol 29:1499–1509
126. Tao Y, Zhang J, Jiang S, Xu Y, Show PL, Han Y, Ye X, Ye M (2018) Contacting ultrasound enhanced hot-air convective drying of garlic slices: mass transfer modeling and quality evaluation. J Food Eng 235:79–88

Chapter 7
Natural Terpenes as Building Blocks for Green Chemistry

Mohamed Touaibia, Chahrazed Boutekedjiret, Sandrine Perino and Farid Chemat

Abstract This chapter presents a complete picture of current knowledge on recent developments of extraction techniques of essential oils' source of terpenes and their use not only in food ingredients as aromas, antioxidants or antimicrobials but also as synthons and solvents for green chemistry. The modern applications of terpenes and their original procedures are summarized in terms of their applications, benefits and future trends.

Keywords Terpene · Extraction · Synthon · Solvent · Antioxidant · Antimicrobial

7.1 Introduction

Natural terpenes have always been part of people's everyday lives, whether as fragrances, flavours, poisons or even drugs [1]. More recently, natural terpenes have had a considerable impact on chemistry and especially on the organic synthesis of molecules of interest [1–5].

Whether in their acyclic form (as for citronellol), monocyclic (as for limonene), or bicyclic (as for β-pinene, Fig. 7.1), terpenes have been widely used in organic synthesis of various natural and non-natural products with different applications [5–11].

A rich literature dismounts the great potential of terpenes as synthons for classical organic synthesis of other more complex terpenes or other molecules [5–15].

M. Touaibia (✉)
Département de Chimie et Biochimie, Université de Moncton, Moncton, NB, Canada
e-mail: mohamed.touaibia@umoncton.ca

C. Boutekedjiret
Laboratoire des Sciences et Techniques de l'Environnement (LSTE), Ecole Nationale Polytechnique, 16200 Alger, Algeria

S. Perino · F. Chemat
Université d'Avignon et des Pays de Vaucluse, INRA, UMR408, GREEN Extraction Team, 84000 Avignon, France

Y. Li and F. Chemat (eds.), *Plant Based "Green Chemistry 2.0"*,
Green Chemistry and Sustainable Technology,
https://doi.org/10.1007/978-981-13-3810-6_7

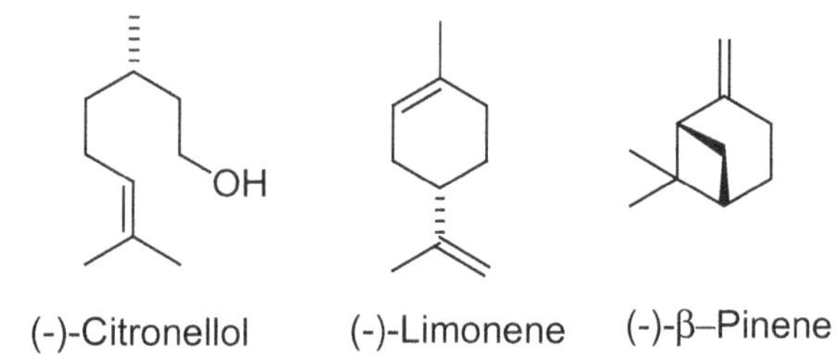

Fig. 7.1 Structure of (-)-citronellal, (-)-limonene and β-pinene

Even considering the enormous advances in synthesis developed, the use of terpenes as starting materials for green chemistry continues today. In this chapter, we have focused exclusively on the use of terpenes as building blocks in environmentally friendly processes by respecting one or more green chemistry principles [16].

This chapter presents a complete picture of current knowledge on recent developments of extraction techniques of essential oils' source of terpenes and their use not only in food ingredients as aromas, antioxidants or antimicrobials but also as synthons and solvents for green chemistry. The modern applications of terpenes and their original procedures are summarized in terms of their applications, benefits and future trends.

7.2 Essential Oils: Extraction

Terpenes are major ingredients of essential oils (EOs) considered as secondary metabolites synthesized for combating infectious or produced in response to stress conditions. EOs are extracted from various aromatic plants generally localized in temperate to warm countries like Mediterranean and tropical countries where they represent an important part of the traditional pharmacopoeia. They are liquid, volatile, limpid and rarely coloured, lipid soluble and soluble in organic solvents with a generally lower density than water. They can be synthesized by all plant organs, i.e. buds, flowers, leaves, stems, twigs, seeds, fruits, roots, wood or bark, and are stored in secretory cells, cavities, canals, epidermic cells or glandular trichomes. EOs are defined as products extracted from natural plants by physical means only such as distillation, cold press and dry distillation.

7.2.1 Steam Distillation

Steam distillation is one of the ancient and official approved methods for isolation of EOs from plant materials. The plant materials charged in the alembic are subjected to the steam without maceration in water. The injected steam passes through the plants from the base of the alembic to the top. The vapour laden with essential oils flows through a "swan-neck" column and is then condensed before decantation and collection in a Florentine flask (Fig. 7.2). EOs that are lighter or heavier than water form two immiscible phases and can be easily separated. The principle of this technique is that the combined vapour pressure equals the ambient pressure at about 100 °C so that the volatile components with the boiling points ranging from 150 to 300 °C can be evaporated at a temperature close to that of water. Furthermore, this technique can be also carried out under pressure depending on the EOs' extraction difficulty.

7.2.2 *Hydrodiffusion*

In reverse of steam distillation, steam injected in this system is from the top of the alembic to the bottom. The vapour mixture with EOs is directly condensed below the plant support through a perforated tray. The way of separating EOs is the same as that in other distillation methods. This method can reduce the steam consumption and the distillation time; meanwhile, a better yield can be obtained in comparison with steam distillation.

7.2.3 *Hydrodistillation*

Hydrodistillation is one of the oldest and easiest methods. It is a variant of steam distillation, which is standardized by the French Pharmacopoeia for the extraction of EOs from dried plants and the quality control of essential oils in the laboratory. Instead of the steam input, the plant materials in HD are directly immersed in water. This solid–liquid mixture is then heated until boiling under atmospheric pressure in an alembic, where the heat allows the release of odorous molecules in plant cells. These volatile aroma compounds and water form an azeotropic mixture, which can be evaporated together at the same pressure and then condensed and separated in a Florentine flask due to their immiscibility and density difference. Moreover, a cohobation system can recycle the distilled water through a siphon so as to improve the yield and quality of EOs. It is important to mention that the recovered EOs are different from the original essence due to the long treatment duration.

7.2.4 *Cold Expression*

This technique is an extraction without heating for EOs of citrus family. This method involves the rind portion of the fruit, or the outermost waxy layer of the fruit's peel. It does not require a heat source, so it is often referred to as "cold pressing". The fruit is placed in the expression machinery where it is passed across rasping cylinders that

Fig. 7.2 Traditional and industrial alembic and Florentine flask

cut up the surface of the peel. The small sacs containing essential oil break open and release their contents. Water is sprayed over the fruit to collect the essential oil. At this point, the essential oil is still mixed with some watery materials from the fruit's flesh and some solid particles from the rind. To isolate the essential oil, this mixture is filtered and centrifuged, allowing separation of the pure essential oil product from any water or solid particles.

7.2.5 *Turbodistillation*

Turbodistillation has been used until nowadays for the extraction of essential oils from hard matrixes (such as wood, bark and seeds) due to a mechanical stirrer which breaks down agglomerates and chunks of hard matrices and homogenizes the medium at the same time.

7.3 Terpenes as Synthons for Green Chemistry

Due to the availability of different functional groups on terpenes (i.e. double bonds, hydroxyl or carboxyl groups), which can be used to introduce new functionality and then further be modified, a wealth of possible molecules can be prepared from these natural products.

Oxidative and reductive transformations of olefins can be complicated by competitive reactions such as isomerization, skeletal rearrangements and nucleophilic addition.

De Oliveira et al. [17] have efficiently converted camphene **1** to epoxy camphene **2a**, glycol camphene **2b** and aldehyde camphene **2c** using a free-metal reoxidant protocol based on $PdCl_2/H_2O_2/CH_3CN$ system (Scheme 7.1). The water is formed during the reaction as well as the water of H_2O_2 solution is resulted in lowering of the selectivity of oxidation products. With the same system, β-pinene **3** was converted into epoxy-β-pinene **2a**, pinocarveol **2b** and pinocarvone **2c**. Compound **2b** was always the major product. Unlike the oxidation of camphene **1**, which had a conversion rate of more than 90%, that of β-pinene did not exceed 58% even with an excess of H_2O_2.

No oxidation products were detected when the system was used with α-pinene and 3-carene. The endocyclic double bonds as well as the bulky two methyl groups must slow down or block this oxidation.

Hydroformylation of terpenes, to obtain natural aroma aldehydes, has been achieved by Vieira et al. [18] in a water/toluene biphasic system. As shown in Scheme 7.2, the hydroformylation of β-citronellene (**5**) was performed in a water/toluene biphasic system using $[Rh(COD)(OMe)]_2$ as the catalyst precursor and a water-soluble tris(3-sulfonatophenyl)phosphine trisodium salt (TPPTS) ligand. In this green hydroformylation, the linear aldehyde **6a**, the major product, as

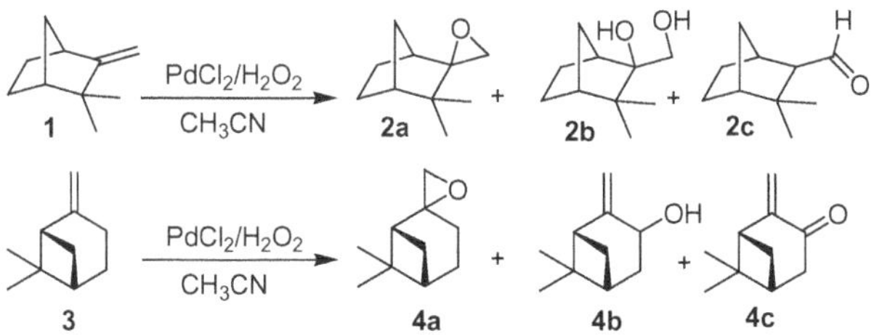

Scheme 7.1 Oxidation of camphene (**1**) and β-pinene (**3**)

well as the branched isomer **6b** was isolated from the organic phase. The aqueous phase containing the catalyst can be separated and reused.

Scheme 7.2 Hydroformylation of β-citronellene (**5**)

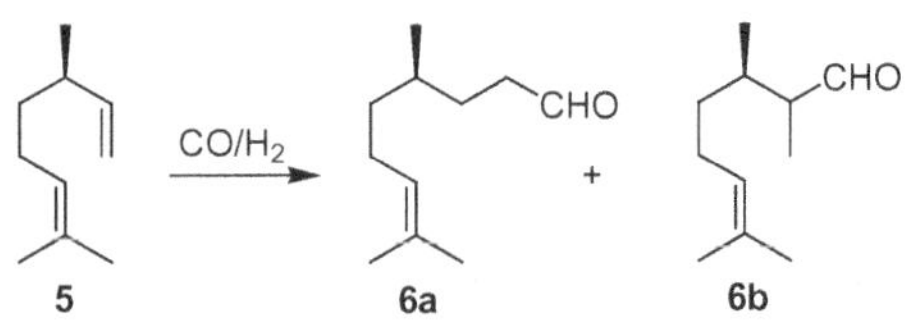

Under the same condition, linalool **7** and nerolidol **9** gave hemiacetals **8a** and **10a** as the major products (Scheme 7.3). The hemiacetals result from the intramolecular cyclization of the corresponding hydroxyl–aldehydes.

Himachalene isomers (**11** and **12**), extracted from the grinded wood of Atlantic cedar, were used as starting material to obtain himachalene epoxides which have been identified as minor components of the essential oils [19]. As shown in Scheme 7.4, himachalene epoxides were obtained by Loubidi et al. [20] through a solvent-free oxidation using tert-butyl hydroperoxide as oxidant and $[MoO_2(SAP)]_2$ as a catalyst. Epoxides **13a** and **13b** were isolated in a 33/67 ratio.

Scheme 7.3 Hydroformylation of linalool (**7**) and nerolidol (**9**)

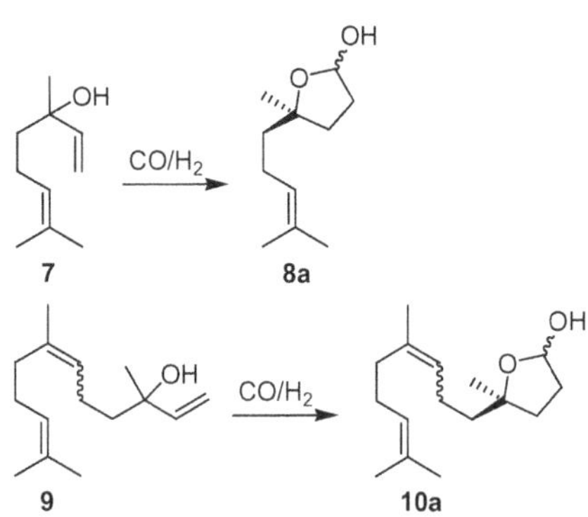

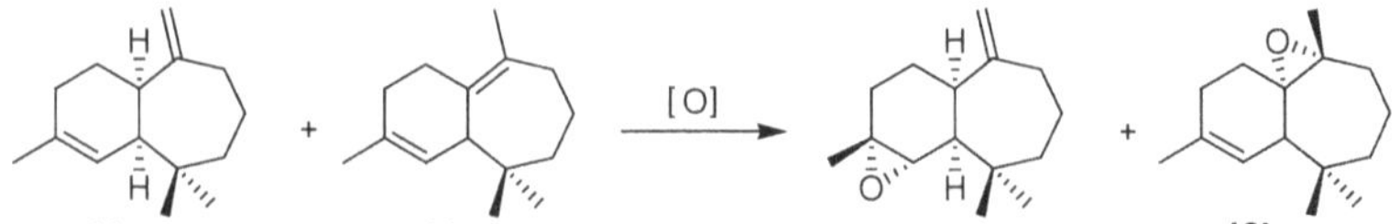

Scheme 7.4 Epoxidation of himachalenes (**11**) and (**12**)

Rodilla et al. [21] have used hydrogen peroxide as oxidant and hydrotalcites [22] as catalyst for a Baeyer–Villiger oxidation of carvone **14** to obtain compound **17** as well as epoxides **15** and **16** in moderate to low yield (Scheme 7.5). Opening and lactonization occurred when isolated epoxide **2** was oxidized with *meta*-Chloroperbenzoic acid and hydrotalcites to obtain a racemic mixture of compounds **19** and **20**.

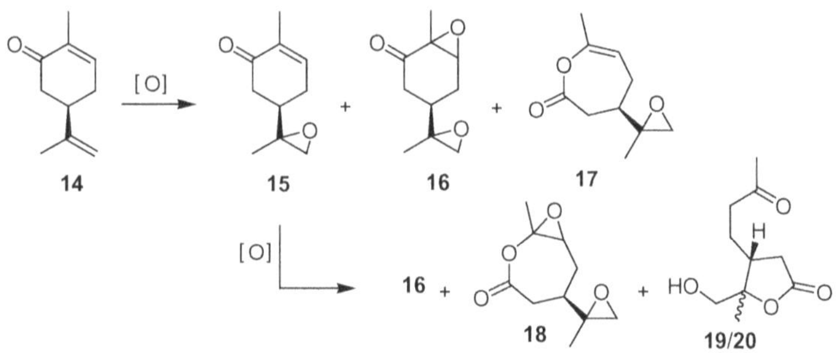

Scheme 7.5 Oxidation of carvone **14**

For the synthesis of bioactive compounds such as natural terpenols derivatives, Zlotin et al. [23] have developed an efficient protocol for the orthoester Johnson—Claisen rearrangement using commercially available ionic liquid [bmim][PF6] as recyclable catalyst.

Linalool **7**, tetrahydronerolidol **22a**, isophytol **22b** and myrtenol **24** gave corresponding rearrangement's products (**21**, **23**, **25**) in moderate to good yields (Scheme 7.6).

As shown in Scheme 7.6, reaction with triethyl orthoacetate (R = H) and linalool **7** yields compound **21a** which contain geranylacetic acid structural unit founded in various bioactive molecules [24].

Many terpenic compounds are industrially produced from more abundant precursors under homogeneous acid catalysis with mineral acids as catalysts. Cotta et al. [25] have developed an alternative β-pinene **3** and linalool **7** heterogenous acid catalysis by using tungstophosphoric heteropoly acid $H_3PW_{12}O_{40}$ (HPW) supported on silica and benzaldehyde (R=H) to obtain oxabicyclo[3.3.1]nonene (Scheme 7.7).

The same oxabicyclo compound was obtained by a series of monoterpenes and monoterpenic alcohols with benzaldehyde or cuminaldehyde (R=Me) under the

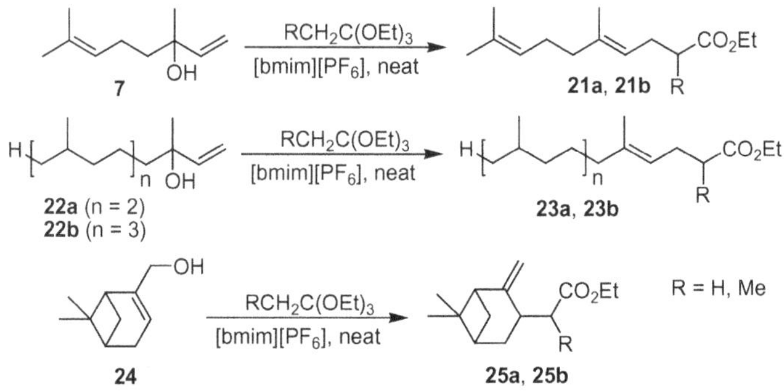

Scheme 7.6 Johnson–Claisen rearrangements of linalool **7**, tetrahydronerolidol **22a**, isophytol **22b** and myrtenol **24**

same conditions and diethyl carbonate or dimethyl carbonate as green solvents (Scheme 7.7) [26].

This cycloaddition represents a promising approach to the synthesis of oxygen-containing heterocyclic compounds from natural terpenes. Besides fragrant properties, oxabicyclo[3.3.1]nonene ethers have been described as ligands for estrogen receptors which are involved in many physiological processes [27].

Starting from Eucalyptol **27**, Luska et al. [28] have developed bifunctional catalysts composed of an acid-functionalized supported ionic liquid phase and ruthenium nanoparticles (RuNPs@SILP) for the hydrodeoxygenation of eucalyptol to produce *p*-menthane **28** in excellent yield (Scheme 7.8). Hydrocarbons as menthane derived from renewable biomass can be a green alternative of fossil fuels.

Some natural hydrocarbons with specific architectures, including pinanes, are of great interest for the pharmaceutical, bioenergy, fine chemistry and flavour industry

Scheme 7.7 Cycloaddition of β-pinene (**3**) and linalool (**7**) with benzaldehyde and cuminaldehyde

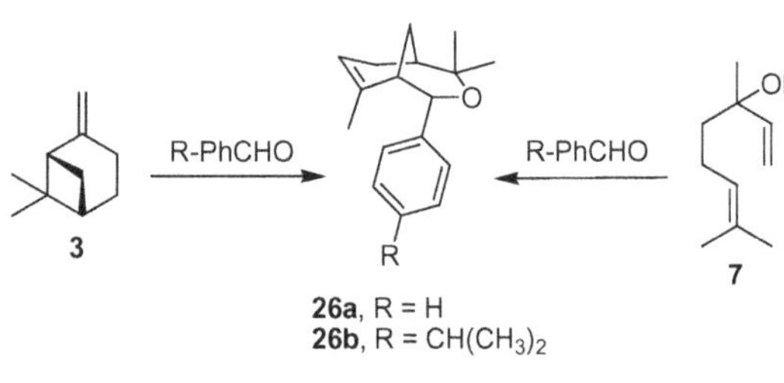

Scheme 7.8 Hydrodeoxygenation of Eucalyptol **27** using RuNPs@SILP

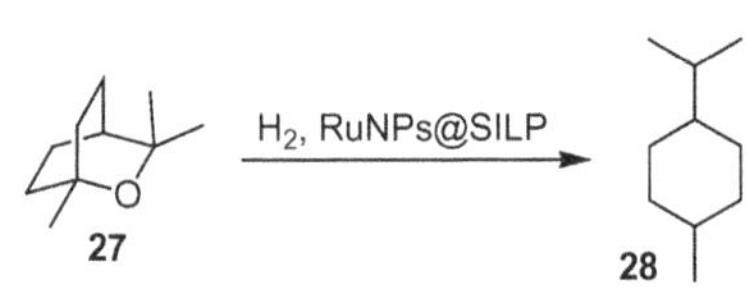

[29, 30]. The low abundance of the natural *cis*- and *trans*-pinane isomers in essential oils obliges the industry to use the synthetic products.

Under neat conditions, hydrogenation of pinene was investigated over palladium (Pd) by Selka et al. [31]. Pd catalyst on carbon, alumina and silica supports yields to a quantitative conversion of pinene to *cis*- and *trans*-pinanes (**31** and **32**) and a higher selectivity for the *cis* isomer (72–89%) (Scheme 7.9). Recycling of the Pd on the three tested supports varied greatly. Pd/C and Pd/alumina were successfully recycled thirteen and fourteen times, respectively. Pd/silica could be used only once to convert pinenes into pinanes. ^{1}HNMR and GCMS analyses show clearly the fast isomerization of β-pinene **3** to α-pinene **29** occurs after only few minutes. The hydrogenation was exclusively heterogeneous due to the extremely low leaching rate of palladium in pinenes and pinanes as determined by ICP/MS. Remaining very effective during all thirteen cycles, Pd/C was the best catalyst/support of the three tested supports for the environmentally friendly hydrogenation of α- and/or β-pinenes (Scheme 7.9).

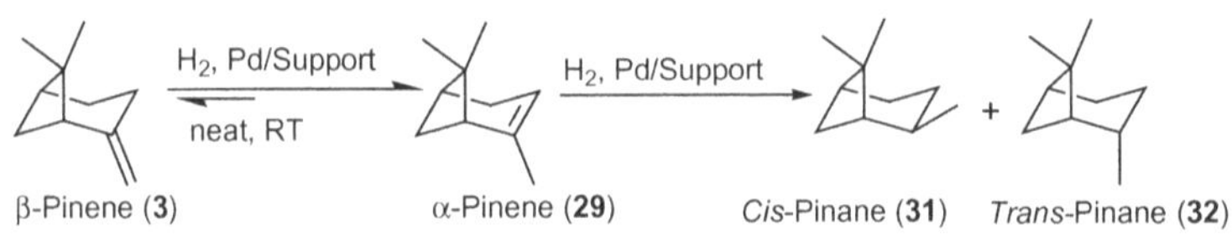

Scheme 7.9 Hydrogenation of pinenes (**3**, **29**) under neat conditions

Under the same conditions, a complete pinene conversion and comparable *cis*-pinane selectivity (71%) were obtained when turpentine oil was used as starting material. The pinane obtained from α/β pinenes was tested as a forestry-based alternative solvent for the extraction of natural products. The *cis*-rich pinane extracted a higher amount of carotenoids from carrots compared to n-hexane and showed comparable potential to the petrochemical solvent to extract oil from rapeseeds and aromas from caraway seeds [32].

cis-Pinane was obtained by Liu et al. [33] through the hydrogenation of α-pinene (**29**) on Ruthenium (Ru) nanoparticles stabilized by amine-functionalized magnetite nanoparticles. The selectivity towards *cis*-pinane **31** was almost quantitative (97%). Fe_3O_4/1,6-hexanediamine/Ru catalyst was completely recoverable with the application of an external magnetic field and the catalytic efficiency even after ten repeated cycles.

Polyoxyethylene–polyoxypropylene–polyoxyethylene triblock copolymer (P123)-Ru micellar catalyst was used by Yu et al. [34] for the hydrogenation of α-pinene **29** in water. The selectivity for *cis*-pinane **31** reached 98.9%. The isolated catalyst aqueous phase was used seven times with no treatment, and its catalytic activity and selectivity were almost unchanged.

7.4 Terpenes as Green Solvents

Solvents are essential for various industrial applications and fields, such as chemicals, pharmaceutical, food ingredients, nutraceuticals, perfume, cosmetic and biofuel industries [35]. They are generally liquid with boiling point below 100 °C and with ability to dissolve, dilute and extract solutes from inert matrices without causing chemical modifications. The commonly used solvents come from petrochemical industry with an economical risk impact due to volatility of petroleum prices and the decrease of the world reserve and also a high risk for some industries for their sustainability due to regulations such as REACH but also the societal demand for more sustainable products.

"Green Chemistry" had a positive effect on discovering and using alternative solvents resulting from bio-refineries and synthesized according to principles of green chemistry and green engineering [36]. Bio-based solvents, such as organic acid esters, fatty acid esters, alcohols and terpenes, could be economic and ecological alternatives to petroleum solvents. More especially, terpenes mainly used for their fragrant properties in various fields, such as perfumes, cosmetics, pharmaceuticals, food ingredients and nutraceuticals, could be green solvents as organic molecules produced naturally by the plants, are found in essential oils and oleoresins of aromatic herbs, spices and fruits.

7.4.1 Terpenes: Solvation Properties

Terpenes are isolated from vegetable raw materials by steam distillation, hydrodistillation or cold pressing of citrus peel oils. Industrially, they can be obtained from the spirits of turpentine isolated by distillation from wood resins or black liquors from paper industry. The relevant properties of some terpenes and *n*-hexane are listed in Table 7.1.

These last years, an increasing interest was given to the use of terpenes as a green solvent for substitutions of petrochemical solvents. However, it is necessary to evaluate and compare the performances of solubilisation of these compounds compared to those of petrochemical solvents. Quantification of the behaviour solvent can be carried out by several tests such as Kauri-Butanol index, aniline point, coefficient of division octanol/water, or by predictive methods of calculation such as Hansen solubility parameter method [37], or COSMO-RS approach. The interest to use these predictive methods is that they allow having information as a decision tool to reduce experiment tests.

Table 7.1 Relevant properties of n-hexane and some terpenes

	n-Hexane	*d*-Limonene	*α*-Pinene	*p*-Cymene	Pinane
Chemical structure	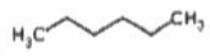	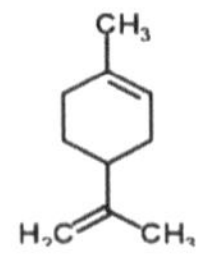	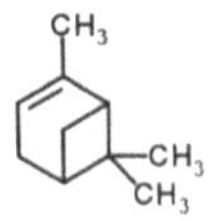	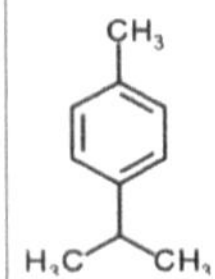	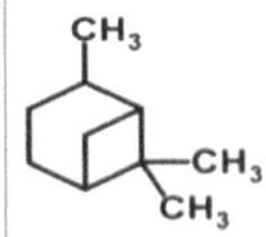
Molecular formula	$C_{10}H_{14}$	$C_{10}H_{16}$	$C_{10}H_{16}$	$C_{10}H_{14}$	$C_{10}H_{18}$
Properties					
Molar weight (g/mol)	86.17	136.23	136.23	134.22	138.2
Density (g/cm^3)	0.675	0.834	0.879	0.861	
Solubility in pure water, 25 °C (mg/ml)	0.11	0.012	0.069	0.025	
Molar volume (cm^3)	127.5	163.2	154.9	155.7	
Boiling point (°C)	68.54	175	158	174	157.0
Vapour pressure, 25 °C (mmHg)	150.9	1.54	3.49	1.65	
Enthalpy of vaporization (kJ/mol)	28.85	39.49	37.83	39.34	37.8
Rate evaporation, 25 °C	8.30	0.25	0.41	0.14	
Flash point (°C)	−23	48.3	32	47.2	36.0
Viscosity 25 °C (Cp)	0.31	0.83	1.32	0.83	0.83
Surface tension (dyne/cm)	20.3	25.8	25.3	28.5	
Dielectric constant, 20 °C	1.87	2.44	2.58	2.34	
Polarizability (cm^3)	11.83	17.98	17.42	17.94	

(continued)

Table 7.1 (continued)

	n-Hexane	*d*-Limonene	α-Pinene	*p*-Cymene	Pinane
Hansen Parameters					
δ_d ($MPa^{1/2}$)	14.9	17.2	17	18.5	16.1
δ_p ($MPa^{1/2}$)	0	1.8	1.3	2.6	0
δ_h ($MPa^{1/2}$)	0	4.3	2	1.9	0

7.4.1.1 Theory of Hansen Solubility Parameters (HSP)

The theory of the solubility parameters was developed in the beginning of the XXe by Hildebrand who postulated that a substance can dissolve another if the forces of interns' cohesion of the two involved species are comparable [38]. For that, he defines a global solubility parameter $\boldsymbol{\delta}$ which takes account of the internal cohesive energy of the system. From there, two substances will be miscible if their solubility parameters are identical or very close. However, the experiments showed that the use of this only parameter was not always precise. That was explained by the fact that this theory does not take account of the interactions related to the polar associations and the hydrogen bonds.

On the basis of the insufficiencies of Hildebrand theory, Hansen made improvements by breaking up the total cohesive energy of the system into the sum of cohesive energies corresponding to the dispersive interactions of London, the polar interactions of Keesom and the hydrogen bonds. This decomposition makes it possible to define three parameters of solubility: $\boldsymbol{\delta_d}$, $\boldsymbol{\delta_p}$ and $\boldsymbol{\delta_h}$ connected to the global solubility parameter $\boldsymbol{\delta}$ defined by Hildebrand by the following relation:

$$\delta^2 = \delta_{\mathrm{d}}^2 + \delta_{\mathrm{p}}^2 + \delta_{\mathrm{h}}^2 \tag{7.1}$$

Consequently, solubility condition is extended to the three parameters, and thus so that two substances are miscible, it is necessary that their three parameters of solubility are equal or closes.

In order to determine the solubility between solvent and solute, the relative energy difference (RED) number has been calculated using Eq. (7.2):

$$\mathrm{RED} = \frac{R_{\mathrm{a}}}{R_0} \tag{7.2}$$

where R_0 is the radius of a Hansen solubility sphere, and R_{a} is the distance of a solvent from the centre of the Hansen solubility sphere, given by Eq. (7.3):

$$R_{\mathrm{a}}^2 = 4(\delta_{\mathrm{d1}} - \delta_{\mathrm{d2}})^2 + \left(\delta_{\mathrm{p1}} - \delta_{\mathrm{p2}}\right)^2 + (\delta_{\mathrm{h1}} - \delta_{\mathrm{h2}})^2 \tag{7.3}$$

where index 1 refers to the solute and index 2 refers to the solvent. In general, the parameter follows the classical like dissolves like" rule, than greater affinity between solute and solvent correspond to a smaller R_a. Consequently, good solvents exhibit RED numbers smaller than 1, while RED numbers progressively higher than 1 correspond to inappropriate solvents.

Developed in the year 1960 for the paintings and varnishes industries, the HSP theory has since that largely used in many spheres of activity: pharmaceutical, detergency, cosmetic, plant health [39]. This theory was used for the optimization of extraction processes in order to reduce the number of tests and to choose suitable solvent to extract the organochlorinated pesticides from the ground. The experimental essays carried out with solvents selected by the theory showed that the hexane/methanol and hexane/acetone mixtures were the most powerful [40]. Srinivas et al. [41] exploited the Hansen theory to extract in subcritical medium compounds of pharmaceutical and nutraceutic interest (betulin, niacin, flavonoids). The solubility parameters of the three molecules were calculated, and the solubility spheres were traced what made it possible to study the miscibility of each compound with water and ethanol at various temperatures. The best conditions of extraction then could be given for each system.

In the field of natural products extraction, Hansen solubility methodology was used in many studies. Filly et al. [42] evaluate the performance of nine alternative solvents (α-pinene, MeTHF, ethyl acetate, methyl acetate, ethyl lactate, butanol, isopropanol, ethanol and CO_2 supercritical fluid) for extracting aromas from blackcurrant buds (*Ribes nigrum* L) compared to that of *n*-hexane. This study shows that terpenes in blackcurrant extracts and *n*-hexane display similar Hansen solubility parameters values, whereas oxygenated compounds have different HSPs due to their hydroxyl group or carboxyl group, which resulted in a higher δ_h value. α-pinene solvent also showed solubility parameters with aroma compounds similar to those of hexane. The values of the (RED) numbers lie between 0 and 1 what makes it to consider that the all used solvents are good for each terpene. Comparison of experimental and theoretical results shows that Hansen's method describes α-pinene as a good alternative for aroma extraction, but experimentally the flavour profile is very different from that of the hexane extract. This difference can be explained by the difficulty to evaporate the solvent.

Hansen solubility methodology was used by Lie et al. [43] to evaluate the efficiency of many alternative solvents for the oil extraction from rapeseed. In this study, five alternative solvents were investigated (*d*-limonene, α-pinene, *p*-cymene, ethanol, isopropanol) and compared to *n*-hexane. Experimental results show that the most effective solvents were *d*-limonene and isopropanol, whereas of a point of view selectivity *p*-cymene, isopropanol and *n*-hexane were the better. A higher solubility potential of the terpenes is due to their high boiling points which allows a lower viscosity involving a better diffusion of solvent in the solid phase and a good dissolution of the solute in the solvent. Simulation results indicate that terpenes are the most solvent for the triglyceride, fatty acids, tocopherol and sterol extraction from rapeseed oil. Indeed, the computed values of their solubility parameters are comparable. In this study, predictive and experimental results were in accordance and proved that terpenes could be preferential solvents for the extraction of oils from rapeseed as compared to hexane and alcohols. Among the terpenes, *p*-cymene is the most solvent with a higher lipid yield and a good selectivity.

Breil et al. [44] simulate with Hansen solubility simulations the solubilisation of lipids extracted from oleaginous yeasts biomass using several alternative bio-based solvents (hexane, isopropanol, ethanol, ethyl acetate, ethyl lactate, DMC, *p*-cymene, α-pinene, *d*-limonene, MeTHF and CPME). GC analysis showed that triglycerides (TAG), diglycerides (DAG) and free fatty acids (FFA) are the main classes of lipids present in the oil. The solubility potential in the used bio-based solvents of these compounds is evaluated by calculating the RED number. For the extraction of TAGs, the RED values are lower than 1 for *p*-cymene, *d*-limonene, α-pinene, MeTHF and CPME, which means that these solvents are those best suited for the extraction of TAGs. Ethyl acetate, *d*-limonene, MeTHF and CPME with a RED < 1 are the most suitable solvents for DAGs' and FFAs' solubilisation.

The Hansen solubility methodology remains a good approach for the selection of a new solvent; however in certain cases, the results obtained do not agree with the experimentation. This could be due to that this theory takes into account only the interactions relating to hydrogen bonds, van der Waals forces and dipolar interactions. However, the interactions between a solvent and a solute utilize other properties of which account would have to be held. One of the methods which make it possible to make improvements to the Hansen theory is a Conductor-like Screening Model for Realistic Solvent (COSMO-RS) approach. This method integrates more parameters than Hansen theory what brings additional information on the interactions between the solute and the solvent as well as the mechanisms of solvation.

7.4.1.2 Conductor-like Screening Model for Realistic Solvent (COSMO-RS) Theory

COSMO-RS is a predictive method for thermodynamic equilibrium of the fluids and the liquid mixtures which use a statistical thermodynamic approach based on the results of quantum chemistry calculations. Developed in 1991 by Klamt, COSMO (Conductor-like Screening Model) is based on a solvation according to a "dielectric continuum" [45]. This model was then extended to real solvents (COSMO-RS) by only holding account interactions within the liquid due to local contacts of surface of charge densities σ and σ' [46]. COSMO-RS approach develops in two steps: the first step corresponds to calculations of quantum chemistry (COSMO) for each insulated molecule, whereas the second stage relating to extension (RS) makes it possible thanks to thermodynamic calculations to reach the electrostatic interactions between the various species [47].

COSMO-RS theory was used in various fields where the choice of solvent is an important step in a process such as in pharmaceutical formulation [48], in the liquid–liquid extraction [49] and in the cosmetic field [50]. This predictive approach was also applied by Moity et al. in a comparative study of the solvent properties of the conventional organic solvents and new green solvents [51]. One of the fields where the search for alternative solvents for petrochemical solvents is of topicality is the extraction of the natural products. In this context, several studies were undertaken using the predictive COSMO-RS method as tools of assistance to the evaluation of the solvation properties and the comprehension of the solvation mechanism of the studied systems. Filly et al. [52] applied a COSMO-RS approach to evaluate the performance of eight alternative solvents (α-pinene, MeTHF, ethyl acetate, ethyl lactate, butanol, isopropanol, ethanol, dimethyl carbonate) in comparison with hexane. GC-FID analysis shows that carvone witch causes typical freshness character of caraway aromas and limonene are the main compounds in caraway extract, but differences are observed in the relative composition of the extracts. COSMO-RS study indicates that σpotential of limonene is similar to that of *n*-hexane and α-pinene and that of carvone is similar to that of ethyl acetate, dimethyl carbonate, MeTHF or ethyl lactate with different indices. Thus, α-pinene appears as an alternative solvent and most promising for limonene extraction and ethyl acetate especially for carvone. Experimentally, the results are in agreement with COSMO-RS prediction for the carvone extraction, but for limonene, COSMO-RS prediction was wrong; indeed, limonene content in α-pinene extract is considerably lower. This difference can be explained by the difficulty to evaporate the solvent.

COSMO-RS simulation has also been used to evaluate the efficiency of bio-based solvents (2-methyltetrahydrofuran (MeTHF), cyclopentyl methyl ether (CPME), dimethyl carbonate (DMC), isopropanol (IPA), ethanol (EtOH), ethyl acetate (EtOAc), *p*-cymene and *d*-limonene) for the substitution of hexane for vegetable oil extraction [53]. Calculation of the relative solubility of typical triglycerides (TAGs), tocopherols and sterols of rapeseed oil in various solvents indicates that *p*-cymene and *d*-limonene and three other solvents (MeTHF, CPME, ethyl acetate) are equivalent to *n*-hexane for TAGs and better than *n*-hexane regarding the solubility of tocopherols. Experimentally, the yields obtained by *p*-cymene and *d*-limonene which represent around 80% of the amount extracted with *n*-hexane are slightly lower than the other solvents. The statistical study (ANalysis Of Variance with a Student test) showed that *p*-cymene and *d*-limonene significantly give a lower yield than hexane ($p = 0.0255$ and $p = 0.0047$, respectively). This study shows a divergence between the experimental and theoretical results particularly for terpenes which appeared to be good theoretically but experimentally give the lowest extraction yield. COSMO-RS predictive method was also used as a tool of investigation in research of renewable sources of production of biofuels [44]. In this context, extraction of lipids from oleaginous yeast biomass was investigated using bio-based solvents. Experimentation for extraction and solvation of lipids with bio-based solvents was correlated with theoretical study in order to make a preliminary evaluation for the replacement of hexane. COSMO-RS was used to evaluate the relative solubility of the main lipid contents in the yeast *Yarrowia lipolytica* IFP29 in the selected bio-solvents. Experimental results show that extraction yield is not affected by solvent. Fatty acid profiles obtained by GC analyses of *p*-cymene, *d*-limonene and α-pinene extracts are similar in terms of selectivity to that of hexane. In the other hand, HPTLC analyses show that the main classes of lipids present in the oil were triglycerides (TAG), diglycerides (DAG) and free fatty acids (FFA).

COSMO-RS simulation indicates that the best solubility of TAGs was obtained for hexane. However, the relative solubility of *p*-cymene, *d*-limonene and α-pinene is similar to that of hexane. Consequently, these terpenes can be considered as suitable solvents to replace hexane for the extraction of TAGs of oil from *Yarrowia lipolytica* yeast. Regarding the solubilisation of DAGs, FFAs and phospholipid families, the behaviour of the three terpenes is similar to that of hexane; indeed, the low values of relative solubility indicate that these three solvents are not suitable for the solubilisation of the considered families of lipids. Experimental results obtained in this study are in agreement with those predicted by a theoretical approach; therefore, *p*-cymene, *d*-limonene and α-pinene could constitute an alternative to replace hexane for lipid extraction from oleaginous yeast biomass.

The same results on the behaviour of *d*-limonene, *p*-cymene and hexane were observed for lipid extraction in salmon fish, but in this study, both experimental and theoretical approaches diverge [54]. Indeed, the theoretical predictions showed that *p*-cymene and limonene are the best solvents to dissolve TAGs, whereas the analyses of the extracts highlighted low contents of these compounds. This is probably due to the degradation of the latter by the high boiling points of these terpenes used for the extraction. Similar observations were noted for some unsaturated fatty acids which theoretically should be well solubilized by limonene and particularly by *p*-cymene but which are found in low content in the extracts because of oxidation reaction induced by the high extraction temperatures.

COSMO-RS approach was also used to simulate the extraction capacity of the *cis*-rich pinane (*cis*/*trans*: 7/3) an alternative solvent investigated to substitute hexane for the extraction of bioactive compounds: carotenoid from carrots, oil from rapeseeds and aromas from caraway seeds [32]. Results obtained by a predictive method show a relatively similar behaviour of *cis*-rich pinnae and hexane. Indeed, the solubility of α-and ß-carotene is similar for both solvents (60–100%), whereas lycopene is slightly more soluble in pinane than in hexane. The solubility of lutein is low for the two solvents. Pinane and hexane showed a higher probability of solubility for TAGs from rapeseed oil and a low probability of solubility for DAGs. Concerning the aromas from caraway seeds, the probability of solubility in *cis*-rich pinane is higher and similar to that in *n*-hexane. In this study, the experimental and predictive approaches were in the agreement which allowed considering that pinane could be an interesting sustainable solvent for *n*-hexane substitution in the extraction of natural products.

Even if sometimes the differences are observed between simulations and the experiments probably because of other parameters which are not taken into account such as the physicochemical properties of the solvents (viscosity, density, volatility, specific heat, surface tension, etc.), or the diffusion phenomena intervening during the extraction process, these studies illustrates that theoretical simulations can be seen as a powerful screening tool employed as a guide to look for green alternative solvents to substitute petroleum solvents.

7.4.2 Application of Terpenes as a Solvent for Natural Product Extraction

Terpenes as an alternative solvent for cleaning and degreasing properties have been used since 1990 [55]. Limonene could constitute an alternative to the traditional solvents (halogenous hydrocarbons, petrochemical solvents, etc.) used in the industries and household cleaning. A few years later, Liu and Mamidipally [56, 57] tried out this solvent in the extraction of oils from rice bran and found that it was possible to use it instead of hexane.

Because of their relatively high cost compared with that of the traditional solvents, the use of the terpenes was not developed; but with the hardening of the laws and the environmental regulations on the use of the petrochemical solvents, the growing interest in this family of compounds has emerged. Thus, several researches were carried out these last years to evaluate their solvent capacity.

Various examples of the use of terpenes to extract natural products are available in the literature. The most commonly used terpene as solvent is probably *d*-limonene. It is used for fats and oils' determination [58, 59], for moisture determination as a substitute of toluene [60], for the extraction of carotenoids as a substitute of dichloromethane for extraction of lycopene [61] and recently for extraction of carotenoids from orange peels instead of hexane [62]. All these studies showed that there is no significant difference as well qualitatively as quantitatively between each extract obtained with petroleum solvents and limonene allowing us to conclude that this latter is effective and valuable as a substitute solvent.

α-pinene is another monoterpene susceptible to be an interesting alternative solvent. It is tested in the extraction of peanuts, soya, sunflower and olive oils [63], and also of extraction of oil from microalgae (*Chlorella vulgaris*) [64] using Soxhlet extraction. From a quantitative point of view, the yields are better than those obtained with hexane, which is probably due to the polarity of α-pinene which is slightly higher than that of hexane. Qualitatively, gas chromatography coupled to mass spectrometry (GC-MS) and gas chromatography (GC) analyses of the extracts showed that fatty acids extracted by both solvents are equivalent in terms of compounds identified and relative proportions, and comparable to that cited in the literature [65, 66]. Pinene was also used for oil extraction from rapeseeds [43], and for lipid extraction from oleaginous yeast biomass [44]. In these studies, the yields are also better than those obtained with hexane, and fatty acid profiles of the extracts are similar for both solvents.

In addition to hexane which is very much used as solvent in the extraction processes and for which is necessary to find a solvent of replacement, they are other hazardous petroleum solvents which is necessary to substitute. Toluene is one of these solvents; it is used in the Dean–Stark distillation, a reference method for moisture determination in food products containing volatile compounds [67].

Veillet et al. [60] used limonene in the place of toluene for moisture determination in food products, and tests were carried out by Bertouche et al. [68], for moisture determination of several matrices (coriander and caraway seeds, onion, garlic, carrot, leek, olive and oregano), using Dean–Stark method with α-pinene as solvent instead of toluene. The results show that the moisture values obtained with the two solvents are comparable. In addition, the processing time using α-pinene is shorter with a profit of about 60% and thus a reduction in the energy consumption.

p-cymene is a bio-based solvent among terpenes family occurs in tree leaf oils [69]. It is used as intermediate in the chemical industry and as a solvent in different fields: for dyes and varnishes, heat transfer medium, additive in fragrances and musk perfumes, and as a masking odour for industrial products. This terpene was tested by Dejoye et al. for the extraction of oils from microalgae [64]. This study gives good results in terms of gravimetric determination and fatty acid composition and shows that there was no real variation in the lipid profile when extracted by *d*-limonene, α-pinene, *p*-cymene or *n*-hexane.

Simultaneous Distillation and Extraction Process (SDEP) is a procedure developed by Dejoye Tanzi et al. [70], for lipid extraction from wet microalgae (*Nannochloropsis oculata* and *Dunaliella salina*) using three terpenes (limonene, α-pinene and *p*-cymene) as solvent. The comparison of SDEP procedure with Soxhlet extraction using hexane, and Bligh and Dyer method showed that lipid yields obtained by SDEP procedure with three terpenes were higher than Soxhlet and slightly lower than Bligh and Dyer reference method. The fatty acid composition of oils obtained by SDEP method is almost similar to those obtained by both conventional Soxhlet extraction and Bligh and Dyer method, and with that found in the literature.

A *cis*-rich pinane (cis/trans: 7/3) a stable terpene derived from pine tree by-product was tested for the extraction of carotenoid from carrots, oil from rapeseeds and aromas from caraway seeds and compared with hexane [32]. This solvent was obtained by catalytic hydrogenation of α/ß-pinenes or turpentine oil under neat conditions over a Pd/C catalyst. Results show that *cis*-rich pinane gives highest carotenoid yield (95.4%) than hexane (78.1%), extraction yields and fatty acid profiles of rapeseed oil are similar for both solvents, and the aromas composition of caraway seeds oil, particularly for carvone and limonene the main components of the oil, is comparable for the two extracts.

Always in optics to find an alternative solvent for petroleum solvents, a new solvent was patented by Chemat et al. [71], the para-menthane, a saturated monocyclic terpenoide obtained by catalytic hydrogenation. The patent deposited by Chemat et al. relates the use of the para-menthane as a solvent of extraction and solubilisation of natural bio-compounds such as lipids, antioxydants, flavours and fragrances and colorants. The tests of solubilisation of several natural bio-compounds namely beta-carotene, carvone, linalool, limonene, estragol and colza oil gave the same results as those obtained by hexane. In the extraction field, this solvent was used for the extraction of colza seeds oil and was compared with hexane. GC-FID analyses of the extracts showed that the fatty acid profiles are similar for both solvents. Para-menthane was also used for the extraction of the odorous compounds from basil sheets. The profiles as well as the proportions in terpenic compounds obtained by GC-FID analysis are comparable with those obtained by hexane.

Through the reported research, we can see that terpenes could be an interesting sustainable solvent and constitute a real alternative for the substitution of petrochemical solvents in the field of natural products' extraction.

7.5 Essential Oils and Terpenes as «Actifs» Ingredients

7.5.1 Essential Oils and Terpenes as Antioxidants

Lipid peroxidation involves the oxidative deterioration of unsaturated fatty acids and the changes such as membrane fragmentation, disruption of membrane-bound enzyme activity result from this process. Reactive oxygen species (ROS) and more particularly hydrogen peroxide cause lipid peroxidation and DNA oxidative damage in cells. Antioxidant application is the simplest technique for reducing lipid peroxidation. Vitamins, phenolic compounds and EOs are naturally occurring antioxidants. EOs are considered as potential resources of natural bioactive molecules, which have been numerously investigated for their antioxidant properties. It is not surprising in view of the presence of phenol groups. It is well known that almost all phenols can function as antioxidants of lipid peroxidation because they trap the chain-carrying lipid peroxyl radicals. Plant phenolics can act as reducing agents, hydrogen-donating antioxidants and singlet-oxygen quenchers. EOs are liposoluble, non-toxic antioxidants and minimize rancidity at a low dose without affecting the quality of the products. Numerous antioxidant methods for the evaluation of antioxidant activity are reported in the literature and consider the chemical complexity of EOs. Several studies have been carried out in order to elucidate the activity of the components. Factors such as reducing power, total phenolic content, b-carotene bleaching activity, Trolox equivalent antioxidant capacity and 1,1-diphenyl-1-picrylhydrazyl free radical scavenging ability have been evaluated to investigate the antioxidant property. For instance, phenolic compounds such as eugenol, thymol, α-tocopherol and carvacrol are very effective free radical scavengers. This phenomenon might be explained by their ability to donate a hydrogen atom from their phenolic hydroxyl groups [72].

Kulisic et al. [73] determined the antioxidant activity of oregano essential oil using three different methods: the beta-carotene bleaching (BCB) test, free radical scavenging activity and the thiobarbituric acid reactive species (TBARS) assay. These authors observed that oregano essential oil showed a strong antioxidant activity.

Zaouli et al. [74] have compared two *Rosmarinus Officinalis* L. varieties and the incidence of antioxidant activity. The oils' antioxidant activity determined by 1,1-diphenyl-1-picrylhydrazyl (DPPH) assay, ferric reducing (FRAP) assay and b-carotene bleaching test was relatively high. The highest activity was found in oil containing a higher content of camphor and α-pinene. Among the oxygenated terpenes, α-terpineol and 1,8-cineole had a high hydrogen-donating capacity towards the DPPH radical. In many cases, the antioxidant activity could be attributed to synergistic effects with the different compounds contained in the essential oil. In

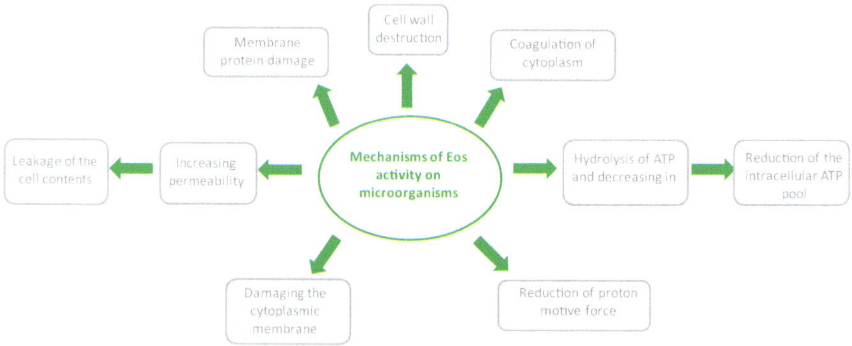

Fig. 7.3 Mechanisms of antimicrobial activity

Melaleuca species, EO containing 1,8 cineole and terpinen-4-ol exhibited stronger antioxidant activity than those with high methyl eugenol or 1,8 cineole contents.

7.5.2 Essential Oils as Antimicrobials

Since the EOs' antibacterial property was firstly investigated by De La Croix in 1881, many researches on EOs' chemical composition and their antimicrobial activities have been reported. The antimicrobial activity of EOs can be completed associated with their active constituents. About 90–95% of the whole EO comprises volatile components consisting of terpenes, sesquiterpene hydrocarbons and their oxygenated derivatives. The non-volatile portion constitutes only 5–10% of the whole EO that includes hydrocarbons, fatty acids, sterols, carotenoids, etc. The most active antimicrobial compounds present in EOs can be divided into four groups: terpenes (limonene, p-cymene, etc.), terpenoids (thymol, α-tocopherol, etc.), phenylpropenes (vanillin, etc.) and other compounds such as allicin or isothiocyanates. The fundamental mechanisms of EOs antimicrobial activity are shown in Fig. 7.3. The mechanism of action of EOs has not entirely understood yet and cannot be attributed to a single mechanism [75].

The antibacterial activity is pertained to the hydrophilic or lipophilic character of the EOs' components, the microorganism type and its cell wall structure. It is assumed that EOs are more effective against gram-positive bacteria compared to gram-negatives ones though there are some exceptions. The gram-negative bacteria *Aeromonas hydrophila* and *Campylobacter jejuni* have been found to be particularly sensitive to the EOs' action. The distinction between gram-positive and gram-negative bacteria is based on a difference in the cell wall composition. The structure of wall in gram-positive bacteria is homogenous with a thickness varying from 10 to 80 nm; it is rich in teichoic acid and osmaines but poor in lipids (<2%). On the contrary, the wall structure in gram-negative bacteria is more complex with a thickness of about 10 nm, and it is rich in lipids (10–22%) and contains less osmaines. The compounds with the highest antibacterial efficacy are phenols such as thymol, carvacrol, followed by aldehydes, ketones, alcohols [76].

For example, the mixture between oregano and lavender essential oils exhibited good antibacterial properties against *E. coli* and *S. aureus*. Yet oregano proved to be better due to its phenolic compounds content which also provided antioxidant capacity. The addition of oregano and lavender essential oils to a biopolymer film as natural bacteriostatics could be an interesting election. This mixture led to an increase in their antimicrobial and antioxidant efficacy. These properties can be attributed to the high content of terpenic compounds (α-pinene, β-pinene, 1,8-cineol, menthol, linalool) or phenolic compounds such as carvacrol, eugenol and thymol. It is common knowledge that essential oils are characterized by changes in their chemical composition, depending on the state of development of the plant, the part used for the extraction, the geographical location and the physical and chemical characteristics of the soil and climate [77].

The antifungal activity of EOs has been demonstrated in several studies. A study has been realized with twenty EOs' effects against the most important moulds regarding the spoilage of bakery products. The results showed that only cinnamon leaf, rosemary, thyme, bay and clove EOs exhibited antifungal activity against all fungus, which suggest a possibility of using EOs as alternatives to synthetic chemicals in the preservation of bakery products. The more active constituents (carvacrol and thymol) in oregano essential oil performed highest fungicidal activity so that this essential oil was identified as the most active against various fungi. Results from the antifungal tests demonstrated that the six major compounds exhibited strong fungicidal activity. In addition, carvacrol was the most toxic against the wood-rot fungi, with the IC values against *T. hirsuta* and *L. sulphureus* of 33.6 and 17.2 μg/mL, respectively. Furthermore, phenol and aldehyde groups were more toxic than alcohol. The results show that the six plant essential oils and their major chemical constituents have potential for the development of natural wood preservatives [78].

Another study was to find an alternative to synthetic fungicides currently used in the control of postharvest fungal diseases of citrus. To be done, antifungal activities of essential oils from four Thymus species were investigated against *Penicillium digitatum*, *Penicillium italicum* and *Geotrichum citri-aurantii*. In the in vitro mycelial growth assay, complete inhibition of the three pathogens was obtained by EOs with a high content of carvacrol, *p*-cymène et γ-terpinene. The effect of essential oils on spore germination varied significantly between tested Thymus species. The high content of carvacrol displayed the highest bioactivity, inhibiting completely the spore germination of *G. citri-aurantii*, *P. digitatum* and *P. italicum*. These findings suggest that essential oils obtained from Thymus may be useful and effective agents for control of citrus fungal pathogens [79].

Xie et al. [80] have evaluated the fungicidal activity of six plant essential oils (*Origanum vulgare, Cymbopogon citratus, Thymus vulgaris, Pelargonium graveolens, Cinnamomum zeylanicum* and *Eugenia caryophyllata*) against the wood-rot fungi, *Trametes hirsuta* and *Laetiporus sulphureus*. Antifungal tests revealed that six essential oils showed good antifungal activities, in which *O. vulgare* was the most toxic to the wood-rot fungi, with the IC values against *T. hirsuta* and *L. sulphureus* of 79.1 and 36.9 μg/mL, respectively. The six major compounds from six essential oils were tested individually for their fungicidal activities against two wood-rot fungi.

7.6 Future Trends

Essential oils as major source of terpenes are one of the most promising themes that can strongly contribute to green chemistry, not only in research laboratories, but also in various industries and at the teaching level from primary schools to universities. Essential oils are widely used in the perfume, cosmetic, pharmaceutical, agricultural and food industries. Essential oils are source of valuable terpenes recognized to possess biological activities, including antibacterial, antifungal, antiviral, anti-mycotic, anti-toxigenic, anti-parasitic and insecticidal properties. We are totally convinced that it is the starting point for future of a new "green chemistry based on terpenes and essential oils" between research, industry and education.

References

1. Breitmaier E (2006) Terpenes: flavors, fragrances, pharmaca, pheromones. Wiley-VCH, Weinheim, Germany
2. Rude MA, Schirmer A (2009) New microbial fuels: a biotech perspective. Curr Opin Microbiol 12:274–281
3. Wilbon PA, Chu F, Tang C (2013) Progress in renewable polymers from natural terpenes, terpenoids, and rosin. Macromol Rapid Commun 34:8–37
4. Blaser HU (1992) The chiral pool as a source of enantioselective catalysts and auxiliaries. Chem Rev 92:935–952
5. Agar D (2005) Handbook of chiral chemicals, 2nd edn. CRC Press, Taylor & Francis, Boca Raton, FL
6. Hanessian S (1983) Total synthesis of natural products: the chiron approach. Pergamon Press, Oxford
7. Nakata M (2008) Monosaccharides as chiral pools for the synthesis of complex natural compounds. In: FraserReid BO, Tatsuta K, Thiem J (eds) Glycoscience, 2nd edn, vol. 2: Chapter 4.4. Springer-Verlag, Berlin, pp 957–994
8. Chida N, Sato T (2012) Chiral pool synthesis: chiral pool syntheses starting from carbohydrates. In: Comprehensive chirality vol. 2. Elsevier, pp 207–239
9. Paek SM, Jeong M, Jo J, Heo YM, Han YT, Yun H (2016) Recent advances in substrate-controlled asymmetric induction derived from chiral pool α-amino acids for natural product synthesis. Molecules 21:951
10. Ho TL (1992) Enantioselective synthesis. natural products from chiral terpenes. Wiley, New York
11. Money T, Wong MKC (1995) The use of cyclic Monoterpenoids as Enantiopure starting materials in natural product synthesis. Stud Nat Prod Chem 16:123–288
12. Maimone TJ, Baran PS (2007) Modern synthetic efforts toward biologically active terpenes. Nat Chem Biol 3:396–407
13. Jansen DJ, Shenvi RA (2014) Synthesis of medicinally relevant terpenes: reducing the cost and time of drug discovery. Future Med. Chem. 6:1127–1148

14. Qiao T, Liang G (2016) Recent advances in terpenoid synthesis from China. Sci. China: Chem. 59:1142–1174
15. Urabe D, Asaba T, Inoue M (2015) Convergent strategies in total syntheses of complex terpenoids. Chem Rev 115:9207–9231
16. Poliakoff M, Fitzpatrick JM, Farren TR, Anastas PT (2002) Green chemistry: science and politics of change. Science 297:807–810
17. De Oliveira AA, Da Silva ML, Da Silva MJ (2009) Palladium-catalysed oxidation of bicycle monoterpenes by hydrogen peroxide in acetonitrile solutions: a metal reoxidant-free and environmentally oxidative process. Catal Lett 130:424–431
18. Vieira CG, de Freitas MC, de Oliveira KC, de Camargo Faria A, dos Santos EN, Gusevskaya EV (2015) Synthesis of fragrance compounds from renewable resources: the aqueous biphasic hydroformylation of acyclic terpenes. Catal Sci Technol 5:960–966
19. Miyazawa M, Teranishi A, Ishikawa Y (2003) Components of the essential oil from *Petasites japonicas*. Flavour Frag. J. 18:231–233
20. Loubidi M, Agustin D, Benharref A, Poli R (2014) Solvent-free epoxidation of himachalenes and their derivatives by TBHP using $[MoO_2(SAP)]_2$ as a catalyst. C R Chimie 17:549–556
21. Rodilla JM, Neves PP, Pombal S, Rives V, Trujillano R, Díez D (2016) Hydrotalcite catalysis for the synthesis of new chiral building blocks. Nat Prod Res 30:834–840
22. Bookin AS, Drits VA (2001) Layered double hydroxides: present and future. Crystal structure and X-ray identification of layered double hydroxides. In: Rives V (ed), Chapter 2. Nova Sci. Pub. Co, Inc., New York pp 39–92
23. Kryshtal GV, Zhdankina GM, Ignat'ev NV, Schulte M, Zlotin SG (2016) The orthoester Johnson–Claisen rearrangement of allylic terpenols in the presence of acidic ionic liquid. J Flu Chem 183:23–29
24. Zlotin SG, Kryshtal GV, Zhdankina GM, Sukhanova AA, Kucherenko AS, Smirnov BB, Tartakovsky VA (2014) Kinetic resolution of racemic (cyclohexyl)(geranyl)acetic acid. Mendeleev Commun 24:257–2259
25. Cotta RF, da Silva Rocha KA, Kozhevnikova EF, Kozhevnikov IV, Gusevskaya EV (2017) Coupling of monoterpenic alkenes and alcohols with benzaldehyde catalyzed by silica-supported tungstophosphoric heteropoly acid. Catal Today 289:14–19
26. Cotta RF, da Silva Rocha KA, Kozhevnikova EF, Kozhevnikov IV, Gusevskaya EV (2017) Heteropoly acid catalysts in upgrading of biorenewables: cycloaddition of aldehydes to monoterpenes in green solvents. Catalysis B Environ 217:92–99
27. Saha P, Gogoi P, Saikia AK (2011) Synthesis of oxabicyclo[3.3.1]nonenes and substituted tetrahydropyrans *via* (3,5)-oxonium-ene reaction. Org Biomol Chem 9:4626–4634
28. Luska KL, Migowski P, El Sayed S, Leitner W (2016) Bifunctional ruthenium nanoparticle-SILP Catalysts (RuNPs@SILP) for the hydrodeoxygenation of eucalyptol under batch and continuous flow conditions. ACS Sustain Chem Eng 4:6186–6192
29. Mimoun H (1996) Catalytic opportunities in the flavor and fragrance industry. CHIMIA Int J Chem 50:620–625
30. Li Y, Fabiano-Tixier AS, Chemat F (2014) Essential oils as reagents in green chemistry, Springer
31. Selka A, Levesque N, Foucher D, Clarisse O, Chemat F, Touaibia M (2017) A Comparative study of solvent-free and highly efficient pinene hydrogenation over Pd on carbon, alumina, and silica supports. Org Process Res Dev 21(1):60–64
32. Yara-Varon E, Selka A, Fabiano-Tixier AS, Canela-Garayoa R, Balcells M, Bily A, Touaibia M, Chemat F (2016) Solvent from forestry biomass. Pinane a stable terpene derived from pine trees by product to substitute n-hexane for the extraction of bioactive compounds. Green Chem 18:6596–6608
33. Liu Y, Li L, Liu S, Xie C, Yu S (2016) Magnetically recyclable Ru immobilized on amine-functionalized magnetite nanoparticles and its high selectivity to prepare *cis*-pinane. J Mol Cat A Chem 424:269–275
34. Hou S, Xie C, Zhong H, Yu S (2015) Mild water-promoted ruthenium nanoparticles as an efficient catalyst for the preparation of *cis*-rich pinane. RSC Adv 5:89552–89558

35. Chemat F, Vian MA (eds) (2014) Alternative solvents for natural products extraction. Springer, Berlin Heidelberg, Berlin, Heidelberg
36. Ernenwein C, Fréville V, Pezron I (2011) Les agrosolvants en extraction. In Chemat F (ed) Book chapter, Eco-Extraction du végétal. Procédés innovants et solvants alternatifs. Dunod, Paris
37. Roire J (1989) Les solvants. EREC, Puteaux
38. Hildebrand J, Scott RL (1950) The solubility of nonelectrolytes. Reinhold, New York
39. Hansen CM (2007) Hansen solubility parameters: a user's handbook, 2nd edn. CRC Press, Boca Raton, FL
40. Lang YH, Cao ZM, Jiang X (2005) Prediction of solvents extraction-the organochlorine pesticides in soil using solubility parameters. Talanta 66:249–252
41. Srinivas K, King JW, Monrad JK, Howard LR, Hansen CM (2009) Optimization of subcritical fluid extraction of bioactive compounds using Hansen solubility parameters. J Food Sci 74:342–354
42. Filly A, Fabiano-Tixier A-S, Lemasson Y, Roy C, Fernandez X, Chemat F (2014) Extraction of aroma compounds in blackcurrant buds by alternative solvents: theoretical and experimental study. C R Chimie 17:1268–1275
43. Li Y, Fine F, Fabiano-Tixier A-S, Abert-Vian M, Carre P, Pages X, Chemat F (2014) Evaluation of alternative solvents for improvement of oil extraction from rapeseeds. C R Chimie 17:242–251
44. Breil C, Meullemiestre A, Vian M, Chemat F (2016) Bio-based solvents for green extraction of lipids from oleaginous yeast biomass for sustainable aviation biofuel. Molecules 21:196
45. Klamt A, Schüürmann G (1993) COSMO: a new approach to dielectric screening in solvents with explicit expressions for the screening energy and its gradient. J Chem Soc Perk T2:799–805
46. Klamt A (1995) Conductor-like screening model for real solvents: a new approach to the quantitative calculation of solvation phenomena. J Phys Chem 99:2224–2235
47. Sicaire A-G, Filly A, Abert Vian M, Fabiano-Tixier AS, Chemat F (2016) COSMO-RS assisted solvent screening for green extraction of natural products. In: Handbook of Green Chemistry. Wiley
48. Pozarska A, da Costa Mathews C, Wong M, Pencheva K (2013) Application of COSMO-RS as an excipient ranking tool in early formulation development. Eur J Pharm Sci 49:505–511
49. Garcia-Chavez LY, Hermans AJ, Schuur B, de Haan AB (2012) COSMO-RS assisted solvent screening for liquid–liquid extraction of mono ethylene glycol from aqueous streams. Sep Purif Technol 97:2–10
50. Benazzouz A, Moity L, Pierlot C et al (2014) Hansen approach versus COSMO-RS for predicting the solubility of an organic UV filter in cosmetic solvents. Colloids Surf Physicochem Eng Asp 458:101–109
51. Moity L, Durand M, Benazzouz A et al (2012) Panorama of sustainable solvents using the COSMO-RS approach. Green Chem 14:1132
52. Filly A, Fabiano-Tixier A-S, Fernandez X, Chemat F (2015) Alternative solvents for extraction of food aromas: experimental and COSMO-RS study. LWT-Food Sci Technol 61:33–40
53. Sicaire A-G, Vian MA, Fine F, Carre P, Tostain S, Chemat F (2015) Experimental approach versus COSMO-RS assisted solvent screening for predicting the solubility of rapeseed oil. Oil seeds Fats Crops Lipids 22(4):D404
54. Tixier A-S, Chemat F (2017) A green analytical chemistry approach for lipid extraction: computation methods in the selection of green solvents as alternative to hexane. Anal Bioanal Chem 409:3527–3539
55. Toplisek T, Gustafson R (1995) Cleaning with D-limonenes: a substitute for chlorinated solvents. Precis Clean 3:17–20
56. Mamidipally PK, Liu SX (2004) First approach on rice bran oil extraction using limonene. Eur J Lipid Sci Tech 106:122–125
57. Liu SX, Mamidipally PK (2005) Quality comparison of rice bran oil extracted with d-limonene and hexane. Cereal Chem 82:209–215

58. Virot M, Tomao V, Ginies C, Visinoni F, Chemat F (2008) Green procedure with a green solvent for fats and oils' determination Microwave—integrated Soxhlet using limonene followed by microwave clevenger distillation. J Chromatogr A 1196–1197:147–152
59. Virot M, Tomao V, Ginies C, Chemat F (2008) Total lipid extraction of food using *d*-limonene as an alternative to *n*-hexane. Chromatographia 68:311–313
60. Veillet S, Tomao V, Ruiz K, Chemat F (2010) Green procedure using limonene in the Dean-Stark apparatus for moisture determination in food products. Anal Chim Acta 674:49–52
61. Chemat-Djenni Z, Ferhat MA, Tomao V, Chemat F (2010) Carotenoid extraction from tomato using a green solvent resulting from orange processing waste. J Essent Oil Bearing Plants 13:139–147
62. Boukroufa M, Boutekedjiret C, Chemat F (2017) Development of a green procedure of citrus fruits waste processing to recover carotenoids. Res-Efficient Technol 3:252–262
63. Bertouche S, Tomao V, Hellal A, Boutekedjiret C, Chemat F (2013) First approach on edible oil determination in oilseeds products using α-pinene. J Essent Oil research 25(6):439–443
64. Dejoye Tanzi C, Abert Vian M, Ginies C, El maataoui M, Chemat F (2012) Terpenes as green solvents for extraction of oil from microalgae. Molecules 17:8196–8205
65. Petkov G, Garcia G (2007) Which are fatty acids of the green alga Chlorella. Biochem Syst Ecol 35:281–285
66. Plaza M, Santoyo S, Jaime L, Avalo B, Cifuentes A, Reglero G, García-Blairsy G, Señoráns FJ, Ibáñez E (2011) Comprehensive characterization of the functional activities of pressurized liquid and ultrasound-assisted extracts from Chlorella vulgaris. LWT Food Sci Technol 46:245–253
67. AOCS Official Method Ja 2a-46 (1993) American Oil Chemist' Society, Champaign
68. Bertouche S, Tomao V, Ruiz K, Hellal A, Boutekedjiret C, Chemat F (2012) First approach on moisture determination in food products using α-pinene as an alternative solvent for dean–stark distillation. Food Chem 134:602–605
69. Yatagai M, Sato T, Takahashi T (1984) Terpenes of leaf oils from Cupressaceae. Biochem Syst Ecol 13:377–385
70. Dejoye Tanzi C, Abert Vian M, Chemat F (2013) New procedure for extraction of algal lipids from wet biomass: a green clean and scalable process. Biores Technol 134:271–275
71. Chemat F, Rombaut N, Fabiano-Tixier AS, Birtic S, Roller M, Bily A (2017) New solvent for extraction and solubilisation, WO2017064424 (A1)
72. Ruberto G, Baratta MT (2000) Antioxidant activity of selected essential oil components in two lipid model system. Food Chem 69:167–174
73. Kulisic T, Radonic A, Katalinic V, Milos M (2004) Use of different methods for testing activity of oregano essential oil. Food Chem 85:633–640
74. Zaouali Y, Bouzaine T, Boussaid M (2010) Essential oils composition in two Rosmarinus officinalis L. varieties and incidence for antimicrobial and antioxidant activities. Food Chem Toxicol 48:3144–3152
75. Burt S (2004) Essential oil: their antibacterial properties and potential applications in foods. A Rev Int J Food Microbiol 94:223–253
76. Sacchetti G, Maietti S, Muzzoli M, Scaglianti M, Manfredini S, Radice M, Bruni R (2005) Comparative evaluation of 11 essential oils of different origin as functional antioxidants, antiradicals and antimicrobials in foods. Food Chem 91:621–632
77. Bajalan I, Rouzbahani R, Ghasemi Pirbalouti A, Maggi F (2017) Antioxidant and antibacterial activities of the essential oils obtained from seven Iranian populations of *Rosmarinus officinalis*. Ind Crops Prod 107:305–311
78. Solorzano-Santos F, Miranda-Novales MG (2011) Essential oils from aromatic herbs as antimicrobial agents. Curr Opin Biotechnol 23:136–141
79. Nevas M, Korhonen AR, Lindstrom M (2004) Antibacterial efficiency of Finnish spice essential oils against pathogenic and spoilage bacteria. J Food Protect 67:199–202
80. Xie Y, Wang Z, Huang Q, Zhang D (2017) Antifungal activity of several essential oils and major components against wood-rot fungi. Ind Crops Prod 108:278–285

Chapter 8
Polyphenols as Natural Antioxidants: Sources, Extraction and Applications in Food, Cosmetics and Drugs

Muhammad Kamran Khan, Larysa Paniwnyk and Sadia Hassan

Abstract Antioxidants are able to scavenge free radicals that cause degradation of food, pharmaceutical and cosmetic products during processing and storage. Natural ingredients in food products are everywhere continually gaining popularity, and the use of plant extracts in cosmetic formulations are on the rise. Additionally, chemists, pharmacists and nutritionists are focusing on the development of new nutritional applications that target not only to disease prevention but also in improving skin health, appearance of the food product and overcoming any off flavors. This approach is the beginning of the new era where the products intended to promote well-being and health will be available to meet the consumers' demands. Polyphenols are the most abundant natural antioxidants in nature commonly found in both edible and non-edible parts of plants, and they have been reported to have multiple biological effects, including antioxidant, antimicrobial, anti-viral, anti-inflammatory, analgesic and antipyretic activities. The extraction of polyphenols from plant sources has drawn increasing attention as a low-value source of antioxidants. Although the development in analytical techniques has played a significant role in the detection of a wide range of polyphenols, success still depends on the method of extraction. In literature, several combinations of solvent, temperature, extraction time and agitation speed have been presented in order to achieve maximum phenolic yields, but these conventional proposals are considered as burden on the environment due to the use of high temperatures for long periods of time and secondly due to the effect on the heat-sensitive components of polyphenols. Alternatively, the application of novel techniques such as ultrasound, pulsed electrical field, high-pressure CO_2 and others is more welcom-

M. K. Khan (✉) · S. Hassan
Institute of Home and Food Sciences, Faculty of Life Sciences, Government College University, Faisalabad 38000, Pakistan
e-mail: mk.khan@gcuf.edu.pk

S. Hassan
e-mail: sadiahassan88@gmail.com

L. Paniwnyk
School of Life Sciences, Faculty of Health and Life Sciences, Coventry University, Coventry CV1 5FB, UK
e-mail: apx122@coventry.ac.uk

Y. Li and F. Chemat (eds.), *Plant Based "Green Chemistry 2.0"*,
Green Chemistry and Sustainable Technology,
https://doi.org/10.1007/978-981-13-3810-6_8

ing due to their reduced usage of organic solvents, low operational temperatures, short processing times and better quality and yields with high selectivity toward targeted compounds. The extracts rich in phenolic compounds are attractive ingredients for food, cosmetics and pharmaceutical products due to their beneficial biological properties importantly their antioxidant potential. This chapter is aimed at discussing the extraction of polyphenols using environment-friendly techniques and to use these polyphenol-rich extracts in various food, cosmetic and pharmaceutical products.

Keywords Polyphenols · Natural antioxidant · Solvent extraction · Novel techniques · Food products · Cosmetics · Pharmaceuticals

8.1 Introduction

The word antioxidant can be explained as: "any substance capable of delaying or inhibiting the oxidation of that substrate even if present in low concentrations." Antioxidant compounds have the ability to prevent the effects of very damaging and reactive, free radicals through oxidative reactions taking place in food. The probable mechanisms regarding antioxidant activity were discovered for the first time when it was known that a substance with antioxidative potential one that itself having ability to become readily oxidized [67]. Oxidation is a chemical reaction process producing free radicals by transferring electrons from one to other oxidizing molecules. These free radicals containing unpaired electrons in the outer orbits are very reactive species. The formation of free radicals initiates further chain reactions. Antioxidants interact with these free radicals by removing free radical intermediates in order to stop these chain reactions and also oxidize themselves to prevent further oxidation reactions [132]. There are a variety of multiple types of free-radical-scavenging antioxidants found in plants and animals and include ascorbic acid, tocopherol and different enzymes (catalase, superoxide dismutase and peroxidases) [72]. If these antioxidants are not situated at proper position and time in the right concentration, they can act as pro-oxidants which can cause oxidative stress in itself or even prevent the action of antioxidants [150].

Currently, the comparative status of the antioxidant and pro-oxidant activities is not fully determined; therefore, more research is required to identify their modes of action. Some antioxidants (types of micronutrients) cannot be produced in the body like ascorbic acid, tocopherol and β-carotene; therefore, they should be complemented through normal diet [187]. Some important natural antioxidants are commercially available like tocopherols, ascorbic acid and rosemary extract [34]. For designing different new functional foods, nutraceutical products and drugs, the research and characterization regarding new bioactive constituents have been increased. Different studies showed contentious outcomes of these exogenous antioxidants; depending upon their various forms, concentration and matrix, all these factors might define and impact the stability between useful and harmful properties of these compounds [31]. The awareness and interest regarding phenolic and other substances having antioxidant potential have grown in recent years among agricultural and food scientists, nutritionists, food industry professionals and consumers [109]. As a result, different plant polyphenols, as value-adding substances, are now being used as integral components in cosmetic, food and pharmaceutical industries [92]. This aim of this chapter is to render a perspective of plant polyphenols, along with different aspects related to structure, sources, antioxidant mechanism, potential health benefits, environment-friendly extraction techniques as well as their application in different food, cosmetic and pharmaceutical products.

8.2 Mechanism of Antioxidants' Activity

The general mechanism of antioxidant activity consists of three different steps, namely chain breaking, preventive and synergetic action [132]. Antioxidants are the compounds have the ability to interrupt the oxidation process either by hindering free radical formation or by disturbing their propagation. Propagation of free radicals can be inhibited by antioxidants in several ways such as (1) scavenge the species causing initiation of peroxidation, (2) chelate metal ions to resist the generation of reactive species, (3) quench $^{\bullet}O_2^-$ which prevents peroxide formation, (4) disrupt oxidative chain reactions, (5) reduce localized O_2 concentrations [34]. Chemical conformation and physical position within a food differentiate chain-breaking antioxidants in their antioxidative effectiveness (proximity to membrane phospholipids, emulsion interfaces or in the aqueous phase) [190]. The reaction of phenolic antioxidants (P–H) with $ROO^{\bullet}$ generates ROOH and a comparatively inert phenoxyl radical ($P^{\bullet}$).

$$ROO^{\bullet} + P\text{–}H \rightarrow ROOH + P^{\bullet}$$

Further, $P^{\bullet}$ can experience chain termination reactions with $ROO^{\bullet}$ generating the non-radical products [64].

$$ROO^{\bullet} + P^{\bullet} \rightarrow \text{Non-radical products [164]}$$

Therefore, the phenolic compounds compete with the substrate (RH) for the chain-carrying peroxyl radicals, thus acting as chain-breaking antioxidants [16]. Accessibility of an antioxidant to peroxy radicals is influenced by its chemical effectiveness, solubility in oil and the amphiphilic character. Amphiphilic character is vital for efficiency of an antioxidant in membrane, micellar and emulsion systems [207]. Different factors affect the efficiency of antioxidants including energy of activation, rate constants, oxidation–reduction reactions, volatility, heat sensitivity and solubility of antioxidant [190]. Furthermore, both inhibitor and chain propagation reactions are exothermic reactions. Increased energies of A:H and R:H bond detachment increase the activation but decrease the effectiveness of the antioxidant. On the other hand, the decrease of bond energies increases the efficiency of antioxidants. Antioxidants having ability to interpose the free radical chain reaction are considered more efficient. During the oxidation process, antioxidants become radicals by giving their $H^{\bullet}$

to the free radicals and become a radical themselves. Both resonance delocalization of the electron within the aromatic ring and formation of quinone structures stabilize the radical intermediates. Synthetic antioxidants and natural botanical extracts containing phenolics also react in this manner [34].

8.3 Classification of Antioxidants

Antioxidants produced by the body to deactivate free radicals so defending the body against different diseases are called endogenous antioxidants. Endogenous defense systems of antioxidants are classified as either enzymatic or non-enzymatic systems. Exogenous antioxidants are externally supplied to the body through food having an important part in body protection against diseases [11]. Naturally originated as well as chemically produced antioxidants can be used as preservatives having a difference in their performance activity. Both types of antioxidants natural or synthetic donate electrons in order to prevent their breakdown. In oxidation prevention processes, synthetic antioxidants showed better performance than natural antioxidants, but natural antioxidants demonstrated reluctance regarding the donation of hydrogen atoms during oxidation prevention. Natural antioxidants are costly than the synthetic ones, due to the cost of derivation processes from different natural sources [195].

8.3.1 Synthetic Antioxidants

Synthetic antioxidants are artificially synthesized by using different techniques. Principally, they all are polyphenolic compounds and work by capturing free radicals to stop chain reactions. All polyphenolic derivatives generally comprise more than one hydroxyl or methoxy groups [150]. Due to improved activity and large accessibility of synthetic antioxidants, they are extensively applied as additives to avoid rancidification in foods. They are also applied in pharmaceutical industry, as preservatives in cosmetics, and also used in foods to stabilize fats, oils or lipids [173]. According to their mode of action, these antioxidants are characterized into two main groups, i.e., primary and secondary antioxidants. Primary antioxidants avert the development of free radicals during oxidation process and divided into three groups which include free radical terminators, oxygen scavengers and chelating agents [41]. Most of synthetic antioxidants are composed of radical terminators which discontinue free radical chains to avoid lipid oxidation. Main examples of radical terminators are butylated hydroxytoluene (BHT), butylated hydroxyanisole (BHA), tertiary butyl hydroquinone and gallates (dodecyl gallate, propyl gallate and octyl gallate) [72]. Oxygen scavengers perform as reducing agents for instance sulfites, glucose oxidase and ascorbyl palmitate. Chelating agents bind the lipid oxidation catalysts (heavy metals) which result in either precipitation of the metal or occupation of all its coordination sites. Important examples of chelating agents are polyphosphatases and ethylene EDTA. Secondary antioxidants break down the hydroperoxides developed through lipid oxidation into stable end products with main examples being thiodipropionic acid and dilauryl thiodipropionate [196]. The various properties of important synthetic antioxidants discussed in this section are summarized in Table 8.1 while their permissible levels in food products are briefly described in Table 8.2.

8.3.2 Natural Antioxidants

Food tissues are always susceptible to oxidative stress caused by free radicals, reactive oxygen species and pro-oxidants. Pro-oxidants responsible for oxidative stress are produced either by exogenous factors (heat and light) or endogenous factors (H_2O_2 and transition metals). To control the formation of free radicals, lipid oxidation catalysts, oxidation intermediates and secondary breakdown products, the food tissues have established their antioxidant systems [40]. Natural antioxidants are synthesized by plants and become available when we eat food while synthetic antioxidants are added to food products to enhance their shelf life. Natural antioxidants all differ regarding their composition, physical and chemical characters and mechanism and site of action and may be of higher or lower molecular weight with polyphenols being a major structural group [196].

8.4 Polyphenols

Polyphenols' secondary metabolites have more than 8000 phenolic structures and are the most abundant and extensively dispersed group of substances in the plant kingdom [109]. As the largest group among phytochemicals with the ability of preventing and treating different health disorders related to oxidative stress, they have attracted more consideration. They possess interesting benefits to health, such as antioxidant, immune modulatory actions, anticancer, antibacterial activity and wound healing [153]. These health benefits are provided by different ways like by removing free radicals, protecting and regenerating other dietary antioxidants and by chelating other pro-oxidant metals. Phenolics are mainly in esterified or glycosylated form so differ dramatically in their nature and content among plants [109].

Polyphenols significantly affect food properties like bitterness, astringency, color, flavor, odor and oxidative stability [145]. As compared to new developments in instrumentation, chemical complexity of polyphenols and their frequent occurrence in plants makes extraction, separation, identification and analysis of polyphenols more remarkable. The development of a common protocol is resisted by complex glycosylation and polymerization patterns of polyphenols [161]. The main antioxidative phenolics are divided into four groups such as phenolic, acids, phenolic diterpenes and volatile oils. Phenolic acids normally trap free radicals to perform as antioxidants while flavonoids scavenge the free radicals along with chelating the metals [169].

Table 8.1 Physical properties of the important synthetic antioxidants

Name	Chemical structure/family/melting temp	Morphology/solubility	Chemical structure	Adopted from
Butylated hydroxyanisole (BHA)	• Monohydric phenol • Radical scavenger • 50–52 °C	• White waxy flakes • Fat soluble • Water insoluble	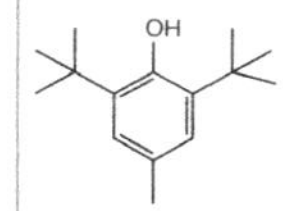	[13, 119, 177, 196]
Butylated hydroxytoluene (BHT)	• Monohydric phenol • Radical scavenger • 69–70 °C	• White crystalline compound • Fat soluble • Water insoluble	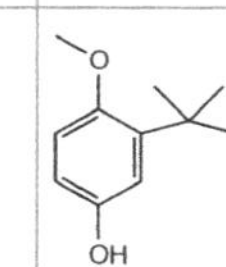	
Tertiary butyl hydroquinone (TBHQ)	• Di-phenolic • Radical scavenger • 126–128 °C	• Beige colored powder • Fat soluble	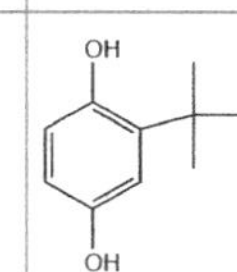	
Propyl gallate (PG)	• Phenol • Radical scavenger • 146–148 °C	• White crystalline powder • Sparingly water soluble	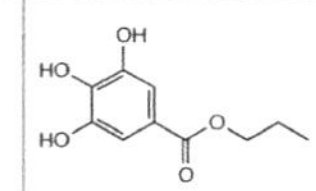	

(continued)

Table 8.1 (continued)

Name	Chemical structure/family/melting temp	Morphology/solubility	Chemical structure	Adopted from
Dodecyl gallate (DG)	• Phenol • Radical scavenger • 94–96 °C	• White to creamy white crystalline solid • Water insoluble	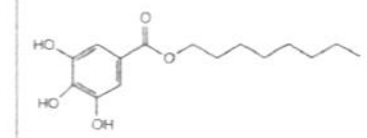	
Octyl gallate (OG)	• Di-phenolic • Radical scavenger • 91–92 °C	• White to creamy white crystalline solid • Water insoluble	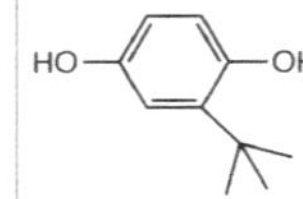	
Ethylene diaminetetra acetic acid (EDTA)	• Metal chelator		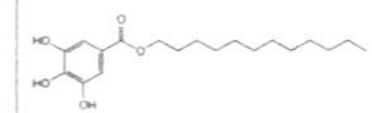	

Table 8.2 Typical synthetic antioxidants used as preservatives, their legal limits in foods and their common sources

Name	Limits in food	Food items	References
Butylated hydroxyanisole (BHA)	<200 mg/kg	Cereals, chewing gum, potato chips, vegetable oils, biscuits, cakes, pastries, sugar, honey, spices, meat products, milk products, etc.	[133, 134, 196, 208]
Butylated hydroxytoluene (BHT)	<100 mg/kg		
Tertiary butyl hydroquinone (TBHQ)	Not allowed	Milk, milk products like cheese, meat and meat products, chewing gum, fish and fish products, seafood, sugar, honey, spices, etc.	
Propyl gallate (PG)	<200 mg/kg	Vegetable oil, meat products, potato sticks, chicken soup base, chewing gum, sugar, honey, spices, milk products, etc.	
Dodecyl gallate (DG)	<200 mg/kg	Oils and fats, cereals, snack foods, dairy produce, meat products, etc.	
Octyl gallate (OG)	<200 mg/kg	Oils and fats, cereals, snack foods, dairy products, sugar, honey, meat products, etc.	
Ethylene diaminetetra acetic acid (EDTA)		Salad dressing, margarine, sandwich spreads, mayonnaise, processed fruits and vegetables, canned shellfish, soft drinks	

The number and location of free –OH groups on the flavonoid skeleton influence the free radical-scavenging efficiency of natural polyphenolics [120].

8.4.1 *Classification of Polyphenols*

Polyphenols include a large group of substances, all having more than one phenolic hydroxyl group which is bounded with one or more benzene rings [49]. The presence of hydroxy-substituted benzene ring within the structure of phenolic compounds is their common feature. Based on the number of phenol rings contained, and the structural elements that bind these rings to one another, polyphenols can be classified into different groups. They are composed of simple compounds, like phenolic acids and as complex compounds like tannins (hydrolysable or condensed). All different subgroups are differentiated on the basis of their degree of hydrogenation and replacement of the heterocycle [212]. Depending on the basic chemical structure, polyphenols can be divided into at least 10 different categories. Flavonoids establish one of the major groups, and further can be subdivided into 13 different categories, having more than 5000 compounds [168]. According to their difference in the type of heterocycle present, flavonoids can be further divided into six subgroups including flavones, flavonols, flavanones, flavanols, anthocyanidins and isoflavonoids [200, 216].

As non-flavonoid groups, phenolic acids are the most common comprising two unique constitutive carbon frameworks such as hydroxycinnamic and hydroxybenzoic structures [100]. Hydroxybenzoic acids are present either in the free form or in the combined form as esters of glycosides [49]. Stilbenes belonging to a comparatively small group of non-flavonoids also found in a wide range of plant sources. Among stilbenes, trans-resveratrol present largely in grapes is studied extensively. Another group of non-flavonoid phenolics is lignans structurally composed of phenylpropanoid units coupled by a bond between the positions in the propane side chains. Lignans are mainly present in the free form, but their glycoside derivatives are only present in minor form [200]. Phenolic compounds esterified with sugars or organic acids result in a complex spectrum of over 5000 compounds naturally found in plants [127]. Table 8.3 shows the basic chemical structures and typical representatives of the most investigated polyphenolic classes.

8.5 Plant Sources

Major naturally occurring sources of antioxidants are fruits [17], vegetables, whole grains [87], and herbs and spices [7].

8.5.1 Legumes, Nuts and Oilseeds

In literature, antioxidant activities of many legumes have been reported, such as yellow and green peas, lentils, chickpea, fava beans, common beans, beach bean, and yellow and black soybeans [206]. Different oil seeds like sesame, sunflower, cold pressed oils and olives are composed of polyphenolic compounds like flavonoids and phenolic acid [216]. Pistachio nuts are a rich source of phenolic compounds containing anthocyanins mainly in their skins [122]. Similarly, in red peanut skins 7–9% procyanidins are present and 50% of which are composed of low molecular weight phenolic oligomers [3]. Walnut seeds contain monomers like ellagic acid, gallic acid and methyl gallate [118]. Antioxidants present in flaxseed, sunflowers,

Table 8.3 Classification of some common flavonoids and non-flavonoid phenolic compounds

Class	Compounds	Main structure
Flavonoids		
Flavone	• Apigenin • Luteolin	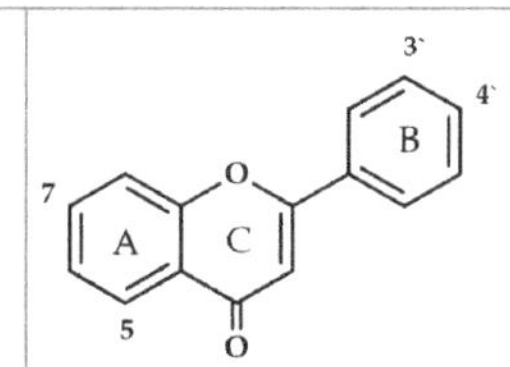
Flavonol	• Quercetin • Kaempferol	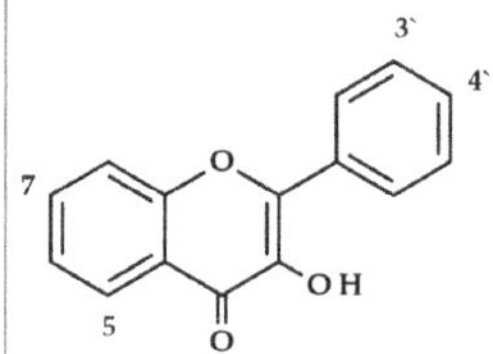
Flavanone	• Naringenin • Hesperetin	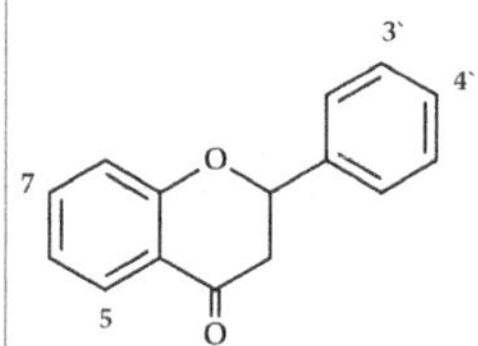
Flavanol	• Catechin • Gallocatechin	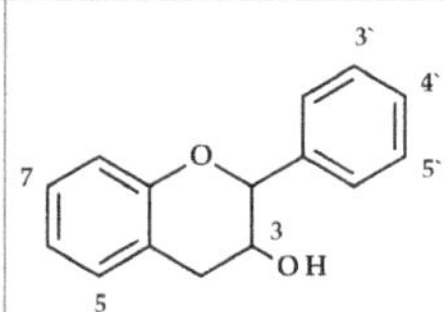
Anthocyanidin	• Cyanidin • Plargonidin	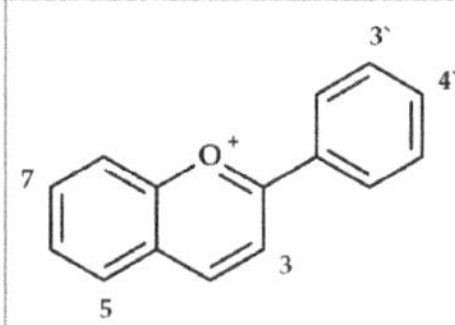
Isoflavonoid	• Daidzein • Genistein	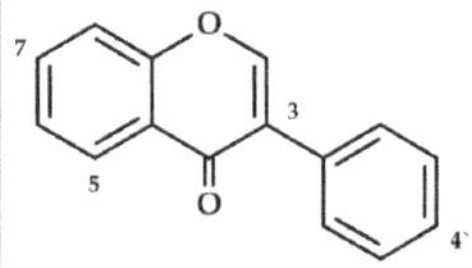

(continued)

Table 8.3 (continued)

Class	Compounds	Main structure
Non-flavonoid phenolic compounds		
Hydroxycinnamic acid (HCA)	• Ferulic acid • Caffeic acid • Chlorogenic acid • Rosmarinic acid	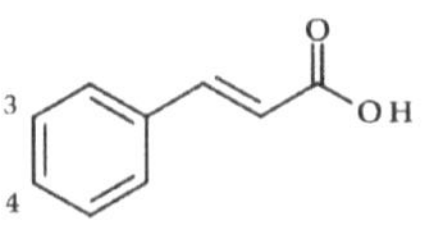
Hydroxybenzoic acid [9]	• p-HBA • Gallic acid	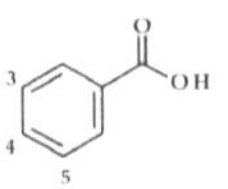
Stilbene	• Resveratrol • Piceatannol	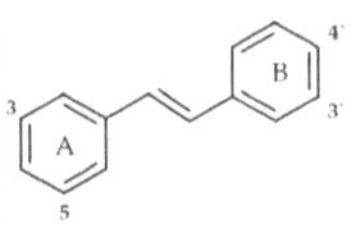
Lignan	• Secoisolariciresinol • Isotaxiresinol	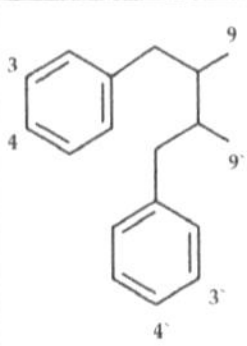

soybean, cottonseed, and canola are representative examples of the antioxidants from oilseeds. Isoflavones and phenolic acids as the main antioxidants are present in soy concentrates and aqueous extracts of soybeans [3].

8.5.2 *Cereals*

Pigmented cereals like black rice, red rice and purple rice are composed of the highest level of phenolic contents [53]. In cereals among bound polyphenols, phenolic acids are the major constituent, among which ferulic acid is more dominant (wheat and rye). Apart from this, vanillic and *p*-coumaric acids also have an important role, although they are present in low concentration. The presence of avertramidin has been reported in oat, while in buckwheat rutin was reported as the main polyphenol [201]. Moreover, buckwheat seeds and hulls contain 387 and 1314 mg/100 g flavonoids and 47 and 77 mg/100 g rutin, respectively. Cereals are also a good source of catechins, and its higher contents are present in buckwheat, following oats and rye, and lowest amount present in wheat [3]. Rice flour also contains phenolic acids up to 86 mg/100 g, similar to wheat flour and oats. The main phenolic component in both rice and wheat flour is ferulic acid establishing at least 89% of total phenolic acids [212].

8.5.3 Fruits and Vegetables

Mostly fruits are a rich source of carotenoids and polyphenolic compounds [63]. Among assessed edible and wild fruits, grape, olive, blueberry, sweetsop, mango and citrus showed the highest content of polyphenols [54, 106]. Grape is an important source of polyphenols with its antioxidant activity depending upon the berries color with the level of anthocyanins and flavonoids increasing alongside an increased color intensity [85]. In tropical fruits, phenolic compounds are mostly concentrated in their peels to protect these fruits against UV radiation (pineapple, passion fruit, acerola, pomegranate); as a result, these fruits are also a good source of phenolic compounds [47, 108]. Wastes and by-products resulting from food production are an interesting source of polyphenols such as grape or apple pomace. The grape skins, seeds and stems and waste products produced during juice processing, are rich source of flavonoids, phenolic acids and stilbenes [91]. Similarly, seeds of red grapes contain catechin and epicatechin with grape skin extract showing a presence of quercetin, rutin and resveratrol [37]. Apple pomace is a good source of catechins, hydroxycinnamates, phloretin glycosides, quercetin glycosides and procyanidins, which are mainly localized in the peels [199]. Major polyphenols stated in pineapple peels include gallic acid, catechin, epicatechin and ferulic acid [108].

In daily diets, vegetables are also an important source of polyphenols. Potato is a good source of polyphenolic compounds, and several studies have demonstrated antioxidant activity of phenolic compounds present in potato [5]. Cabbage also contains more than twenty compounds of quercetin and kaempferol [157]. Moreover, onions and onion scales have been analyzed for their antioxidant activity in lipid oxidation models and in radical-scavenging assays [3]. Vegetables and fruits are also chief sources of flavonoids [113]. The content of quercetin glycoside and kaempferol glycoside in the outer leaves of lettuce was analyzed as containing 237 and 250 mg/kg of fresh weight, respectively [112]. The best sources of antioxidants among vegetables are tomatoes, red pepper, Brassica vegetables, onion, garlic and red beet. Two derivatives of quercetin including luteolin and apigenin were found in red pepper [123]. Some unique antioxidants were discovered in spinach while aqueous spinach leaf extract showed the presence of derivatives of flavonoids and *p*-coumaric acid [179].

8.6 Extraction Techniques

Extraction is the process to separate medicinally important constituents from plants with the help of different selective solvents by following standard protocols [21]. There is a growing interest for isolation of these phytochemicals from plant materials, with the aim to produce a safe, natural and low-cost alternative to synthetic compounds [6]. Polyphenols differ broadly in their structures due to different plant sources; therefore, it is not possible to use a single standard extraction protocol to extract targeted components from a wide variety of sources [36]. Different extraction techniques have been currently developed, which differ in their mechanism, plant material, solid/solvent ratio, time, temperature, pressure, solvent type and pH [6, 66]. Traditional techniques used for plant extraction are maceration, percolation, digestion and soxhlet extraction [21]. These procedures are simple, but have numerous drawbacks such as low final yield, long extraction times, requirement for large amounts of plant material and solvents [135].

Therefore, recently innovative extraction techniques have been developed and assessed for their application, examples being ultrasound-assisted, microwave-assisted, supercritical fluid, accelerated solvent and pressurized liquid extractions [30, 48, 129, 135, 146]. According to different studies, these procedures have several advantages such as higher yield, short extraction times, faster kinetics, decreased consumption of organic solvent and positive influence on the environment [176]. The objective of all these extraction techniques is the separation of soluble plant constituents, leaving the remained insoluble cellular residues. The most commonly applied extraction techniques along with their advantages and limitations are given in Table 8.4.

8.6.1 Maceration

Maceration is the most commonly and effectively used technique for extraction of heat-sensitive phenolic compounds from plants [46, 193]. Regarding working conditions, it is carried out at room temperature by soaking plant materials in stoppered glass containers followed by constant mixing for 10–30 min, or even several hours, according to the nature of the plant matrix. After soaking, the mixture is strained through filtration. This technique is based on the principle of polyphenol diffusion into an appropriate solvent extraction medium [82]. The method releases the soluble phytochemicals by softening and breaking the plant's cell wall. In this traditional technique, heat is manipulated through the principles of convection and conduction [21]. Different factors affect the maceration process such as the characteristics of the plant material, particle size and internal diffusion, which may be the restrictive steps [82].

According to the previous studies, smallest particles result in increased active surface area and also improved contact of the matrix with the extraction solvent [46]. Extraction solvents also have a critical role, so their selection is conducted according to the physical and chemical properties of the target compounds [35]. Most commonly used solvents are water, ethanol, methanol and their mixtures [46, 193]. In mixture form, one solvent can increase the solubility of the phenolic constituents, while the other solvent can recover desorption [135]. In a binary solvent system, the small water quantity has significant influence, as it creates a polar medium, so breaking of hydrogen bonds enabling the improved extraction of polyphenols [192]. According to the different studies, a mixture of alcohol and water was used to extract polyphenols from aromatic plants mainly in a glycoside form [44, 193]. The polyphenol yield

Table 8.4 Advantages and limitations of different extraction techniques to get polyphenols from plant sources

Extraction technique	Working principle	Advantages	Limitations	References
Maceration	Diffusion into an appropriate extraction medium leads to softening and breaking down of the plant's cell wall to release soluble phytochemicals	• Convenient for thermo-sensitive components • Useful for various plant materials, solvents and pH • Simple easy and cheap method	• Long extraction time • Large quantity of solvent • Lower polyphenol yield	[46, 83, 193]
Heat-assisted extraction	Thermal energy causes disruption of cellular structures, inducing increment of cell membrane permeability and breakdown of polyphenol lipoproteins interactions, which can enhance the solubility of polyphenols	• Enhance polyphenol solubility and mass transfer due to high temperature • Faster kinetics • Efficient extraction and higher polyphenol yield • Simple apparatus and method	• Degradation of thermo-sensitive phenolic compounds • High solvent consumption • Limited choice of plant material and extraction medium	[76, 131, 135]
Soxhlet extraction	Solvent heated to reflux, condenses and floods into the chamber housing the sample thimble	• Requires a smaller quantity of solvent	• Exposure to hazardous and flammable liquid organic solvents • Solvents for extraction need to be of high-purity that might add to cost	[21, 138]
Ultrasound-assisted extraction	Ultrasound waves cause the degradation of cell walls, and release of cell contents, which increases penetration of solvent into plant material, and enhances mass transfer, which increases the polyphenol yield	• Increased extraction yield • Improved extract quality • Fast kinetics • Low-priced instrument, wide range of solvents can be used • Simple technique, ease of use	• Degradation of phenolic compounds • Effects of ultrasound waves on polyphenol yield and extraction kinetics depend on the characteristics of plant material • The presence of a larger quantity of plant particles contributes to the ultrasound waves' attenuation and results in the restriction of the active part of ultrasound inside the zone located near the ultrasonic emitter	[76, 75, 93, 105, 205]

(continued)

Table 8.4 (continued)

Extraction technique	Working principle	Advantages	Limitations	References
Microwave-assisted extraction	Radiation interacts with polarizable materials and causes heating near the surface of the material. Dipole rotation of the molecules disrupts hydrogen bonding; enhancing the migration of dissolved ions promotes solvent penetration into the matrix	• Significant reduction of extraction time • Reduction of solvent consumption • Increased extraction yield • Simplicity and economy of extraction process	• Degradation of thermo-sensitive polyphenol constituents • Lower yield of non-polar and volatile target compounds • Effects of microwaves on polyphenol yield depend on the polarity of solvents • Possibility of the extraction of a large quantity of ballast substances	[18, 50, 90, 130, 181]
Pressurized liquid extraction and accelerated solvent extraction	Organic solvents at high pressures and temperatures above their normal boiling point accelerate the diffusivity of solvents, which in result increases the extraction kinetics. The high pressure helps the extraction cells to be filled faster and forces liquid into the solid matrix	• Less consumption of solvent during extraction • Procedure is less time consuming • Handling of sample is reduced	• Not suitable for thermolabile phenolic compounds • High temperature can have deleterious effects on structure and functional activity of phenolic compounds	[2, 21, 33, 90]
Supercritical extraction	Separating one component (the extractant) from another (the matrix) using supercritical fluids as the extracting solvent	• Rapid extraction • Small amount of organic solvent or no solvent • No solvent residue • Preserves thermally labile compounds • Tunable solvent (supercritical fluid) density • Selective extraction • Inexpensive to operate/run	• High setup cost • Technical knowledge of properties of supercritical fluid required (e.g., phase behavior, crossover region)	[2, 94]

(continued)

Table 8.4 (continued)

Extraction technique	Working principle	Advantages	Limitations	References
Solid–liquid extraction	Initial extraction with different solvents (methanol, ethanol, acetone) and further fractionation by liquid–liquid extraction, (with ethyl acetate or hexane,) and followed by column chromatography or solid-phase extraction.	• Well-known procedure • Wide experience in the extraction field	• Long extraction time • Large consumption of toxic and inflammable solvents • Problem with evaporation • Multiple extractions • Laborious, time consuming and expensive • Filtration required after extraction	[68, 114]
Ultrafiltration	Filtration of solutions under pressure through a semipermeable membrane having pores that allow the solvent and small molecules to pass through and the larger molecules to be retained	• Longer life • Higher flux • Resistance to temperature, pH and organic solvents • Better defined cleaning protocols	• Fouling results in a decline of performance	[29, 139]
Enzyme-assisted extraction	Enzymes disintegrate the plant cell wall matrix and facilitate polyphenol extraction	• Easy and rapid • Enhance the disintegration of plant cell wall	• Application of enzymes in industrial processes has high cost	[102, 171, 184]

increased by increasing the solid/solvent ratio, so saturation of the extraction medium can be prevented by using an increased solvent volume [49]. Among disadvantages of this technique, one is the requirement of large amount of plant material contributing to an increased viscosity of the extraction medium, which results in a decreased diffusion rate [95].

Other techniques like infusion and decoction are also based on the same principle of maceration, and in both techniques, plant matrix is soaked in cold or boiled water. However, in infusion the maceration period is shorter, while in decoction, the sample is boiled in a specific water volume ratio (e.g., 1:4 or 1:16) for a definite time period [21]. For extraction of thermostable constituents, e.g., from hard plants' materials, decoction is the only suitable technique as compared to maceration and infusion and also produces more oil-soluble compounds. Another method based on a similar central principle is called percolation. Dried powdered samples are placed in the percolator with the addition of boiling water and maceration carried out for 2 h. The percolation process is normally performed at a moderate rate (e.g., 6 drops/min) and as the extraction is completed with further evaporation being conducted to concentrate the extract [136]. However, as compared to novel extraction techniques, maceration is an old-fashioned and time-consuming method with a low final yield of phenolic compounds [83].

8.6.2 Heat-Assisted Extraction (HAE)

Considering the industrial requirements such as higher polyphenol yield and shorter time duration, heat-assisted extraction is suggested as the most convenient technique for polyphenol extraction [76, 131]. This extraction method is performed at high-temperature-controlled environment within glass containers with constant mixing. The thermal energy disrupts the cellular structures, enhances the permeability of the cell membrane and also breaks down the interaction of polyphenols with lipoproteins, causing the improvement of mass transfer and polyphenol solubility, thus enhancing the extraction efficiency [135]. High temperature also results in solvent penetration into the plant matrix and faster kinetics which in turn reduces the viscosity of the extraction medium [131]. Increased temperature of solvents also reduces the surface tension which in turn improves the wetting of the plant matrix, resulting in increased extraction yield [197]. Highest polyphenol contents were obtained from *Ficus carica*, *Origanum vulgare*, *Artemisia dracunculus* and *Thymus serpyllum* by using a mixture of ethanol and water [36, 131].

According to the previous studies, application of high temperature decreased the polarity of solvents, thus making water and ethanol appropriate solvents for extraction of polar, moderately polar and non-polar organic compounds [135]. Several previous studies proved that a mono-solvent system caused very low polyphenol yield; therefore, it is not suggested as an appropriate solvent for extraction [131]. Studies have also discovered that solid/solvent ratio and temperature had a prominent effect on phenolic yield as the increased ratio also increased polyphenol contents [83,

136]. Furthermore, due to temperature sensitivity, enzymatic degradation, oxidation of polyphenols and polymerization among insoluble constituents, the increased temperature and long exposure duration can decrease the extracted yield [197].

8.6.3 Soxhlet Extraction (SE) or Hot Continuous Extraction (HCE)

In this technique, the ground sample is packed into a porous bag or "thimble" which is made of filter paper or cellulose, and then, it is placed in thimble chamber of the soxhlet apparatus. In the bottom flask of apparatus, extraction solvents is heated which vaporized into thimble, then condenses in the condenser and trickles back. Once the liquid contents reach the siphon arm, they are emptied into the bottom flask, and the process repeats itself as a continuous process [21]. Using this technique extraction from *Azadirachta indica* leaf powder with methanol (~1:5 w:v) showed that abundant phytochemicals, mostly non-polar compounds, were recovered [77]. A study of *Centella asiatica* showed that optimization of the extraction process at the following conditions, i.e., 25 °C temperature, sample/solvent ratio (1:45), agitation speed 200 rpm and process duration 1.5 h, showed optimum metal chelating activities [10]. In powdered *Clitorea ternate* flowers, this technique resulted in a 2.2% yield w/w and also removed lypodial constituents using petroleum ether at 60–80 °C [15]. Further extraction using ethanol ensured the occurrence of alkaloids and saponins, but the important component of *C. ternate* flowers, the anthocyanin, was absent, and the possible reason was their oxidation and degradation during extraction. This technique is considered not environmental approachable due to long extraction times compared to advanced extraction methods such as supercritical fluid extraction (SFE) [4]. The soxhlet extraction is also limited to have a solid sample to be used, and many extraction conditions such as solvent-sample ratio, temperature and agitation speed are required to be optimized for this method [10].

8.6.4 Solid–Liquid Extraction (SLE)

SLE is the easiest method and is also among the most common techniques of extraction [189]. This method is a multiphase process; in the first phase, the sample is extracted by using different solvents which is then followed up by a cleanup stage. The sample is then further subjected to fractionation through liquid–liquid extraction (LLE), generally using ethyl acetate or hexane. After LLE, the process is followed by column chromatography or solid-phase extraction. Normally, this technique directly extracts the fresh or freeze-dried material using a suitable solvent through any extractor, homogenizer or by ultrasonic bath for a specific time. SLE utilizes an unsteady-state mass transfer involving continuous changes of concentration of the solute and the solid [2]. The chemical nature and polarity of the phenolic compounds affect their solubility. Phenolic compounds have the ability to interact with other compounds like carbohydrates and proteins to form complexes which also disturb their solubility and efficiency of the extraction procedures. Sometimes, extra measures are required for the removal of unwanted impurities from the extracts because mostly undesirable non-phenolic compounds are also extracted (chlorophylls, fats, terpenes, waxes and others). This requirement of additional measures of impurities removal from extracts also increases the extraction cost [69].

8.6.5 Ultrasound-Assisted Extraction (UAE)

This extraction process can be carried out through using either an ultrasound bath or ultrasound probe involving the use of ultrasound frequencies ranging between 20 and 2000 kHz. Ultrasound waves destruct plant tissues and pass through plant cells as mechanical vibrator inducing expansion and compression cycles during movement through the extraction medium and result in local increased temperatures and negative pressures [126]. The surface contact between solvents, samples and cell permeability increases through the mechanical effect of acoustic cavitation created through ultrasound [185]. The mechanical effects degrade the cell walls, release cell contents also cause diffusion of solvent into plant matrix, resulting in increased mass transfer and enhanced polyphenol yield [55, 75]. Polyphenol yield is not significantly affected by the particle size of the plant matrix because ultrasound waves are able to reduce the particle size almost 5%, along with resultant cell damage and alteration of microstructure irrespective of initial size of plant particles, resulting in increased solvent penetration into matrix and higher release of phenolic constituents [30, 55]. Concerning the selection of the extraction medium, pure alcohol or water alone cannot fully extract polyphenols alone while an alcohol/water mixture is demonstrated to be the most appropriate solvent for maximum recovery of polyphenols from different sources like *Jatropha dioica*, *Eucalyptus camaldulensis*, *Origanum majorana* and *Thymus* species [209].

According to previously conducted studies, increased solid/solvent ratio also increased polyphenol content in plant extracts [126]. Increase of solid/solvent ratio prevents the saturation of the extraction medium and also increases the phenolic content in the extracts. The large amount of plant matrix also weakens the ultrasonic waves while their active part becomes limited to a small zone [209]. Moreover, the prolonged extraction time can destroy extract quality and also damage extracted antioxidants because of increased temperature and formation of free radicals [75]. As compared to conventional extraction techniques, several studies stated that high yield of extracted polyphenols is obtained using ultrasound-assisted extraction for shorter time durations [75, 105]. Consequently, UAE can be suggested as an effective and fast technique of extracting polyphenols from plant materials [75, 193].

8.6.6 Microwave-Assisted Extraction (MAE)

Microwave-assisted extraction has been developed as a modern practice with the purpose for improving the extraction efficiency, providing short extraction time, low solvent consumption and high polyphenol yield. This technique uses microwave energy to enable separation of analytes from the plant matrix into the solvent [27]. Microwave radiations interact with dipoles of polar and polarizable materials and result in heating of the materials near the surface that heat is distributed through conduction. Microwaves cause dipole rotation of molecules and as a result disturb the hydrogen bond, increase the movement of dissolved ions and enhance the solvent diffusion into the matrix [156]. Non-polar solvents result in poor heating because in this case energy is transported through dielectric absorption [80]. Microwaves cause effective and consistent heating of the sample which offer a speedy energy distribution to a total volume of extraction medium and plant matrix. Water presence in plant cells causes absorbance of microwave energy resulting in internal superheating which can disrupt the cell and enhance the polyphenol recovery [82]. This extraction process can be conducted in microwave oven or microwave reactor with microwave reactors guaranteeing a better control of temperature and pressure in the sample.

According to a study conducted on *Achillea millefolium* and *Myrtus communis*, increased concentration of ethanol in water medium also increased the polyphenol yield through high polar distribution and dielectric constant [48, 130]. Increased dielectric constant through water addition to ethanol permits more absorption of microwave energy. On the other hand, higher water concentration can reduce the heat expansion into the extraction medium, leading to reduced extraction efficiency [130]. Microwave radiation destroying plant cells also increase the diffusion coefficient leading to release of polyphenols and quick saturation of the extraction medium. Hence, increased solvent volume induces improved polyphenol recovery from plant matrix. Microwave extraction can be regarded as a discriminating technique that is suitable for polar molecules and solvents having a high dielectric constant [42].

8.6.7 Pressurized Liquid Extraction (PLE) and Accelerated Solvent Extraction (ASE)

PLE is a novel extraction technique using organic solvents above their normal boiling point at high pressures and temperatures. In this method, the solid sample is packed in an extraction cell and further extraction is conducted by using a suitable solvent at high temperature (40–200 °C) and pressure (500–3000 psi) for short time duration (5–15 min). Further compressed gas is used to collect the sample extract into a collection vial [2].

ASE is similar to SLE and is a type of pressurized solvent extraction technique. To maintain the solvents in their liquid form, ASE is also conducted at higher temperatures (50–200 °C) and pressure (1450–2175 psi). High temperature enhances the solvents diffusion, which in result accelerates the extraction kinetics [33, 90]. High pressure fills the extraction cells quicker and induces the liquid into the plant matrix. ASE is primarily applied for thermally stable organic substances to extract them more quickly. Some studies reported the application of ASE in the area of nutraceuticals and natural products [90]. As compared to maceration and soxhlet extraction techniques, ASE is an effective method of liquid solvent extraction.

To avoid the sample aggregating and blockage of system, tubing an inert material such as sand is used to pack the sample within the stainless steel extraction cell [160]. The packed extraction cell is composed of cellulose filter paper and sand layers in between which layers of sand–sample mixture are placed in between cellulose filter paper and sand layers. This extraction process can be completed in less than an hour, and for each individual sample, it is possible to control the temperature and pressure. ASE also mainly relies upon different solvent types like all other solvent-based methods. *Bixa orellana* extraction for 5 min at 50 °C with cyclohexaneacetone solution in ratio of 6:4 v/v showed maximum yield of bixin [160]. In another study, *Rheum palmatun* was extracted with 80% aqueous methanol using ASE technique, which showed maximum flavonoids recoveries (~94%), signifying the appropriateness of this technique regarding quality control evaluation [186].

8.6.8 Supercritical Extraction (SCE)

Supercritical fluid, also known as dense-gas at its critical point, contains the physical characters of both gas and liquid. For polyphenol extraction, SCE is also an alternative effective method. The critical point is stated as being the maximum temperature and pressure of substance at which it can exist as vapor liquid at equilibrium. Above the critical point, high temperatures and pressures form a single homogeneous fluid called a supercritical fluid (SCF). SCF is a heavy liquid having a gaseous equivalent diffusion power, these properties establish SCF as an efficient and selective extraction solvent [9]. A number of SCFs are used for extraction such as ethane, butane, pentane, nitrous oxide, ammonia, trifluoro-methane and water, but most frequently applied SCF is supercritical CO_2. Supercritical CO_2 decreases the danger of degradation reactions due to the absence of light and air therefore regarded as a safe, non-toxic technique having low critical temperature ($T_c = 31.1$ °C). By altering the temperature, pressure or by addition of different modifiers, strength of SC-solvents can be easily changed which can minimize the extraction timing [160]. For the extraction of polar polyphenols, CO_2 is not a good solvent because of its non-polar nature. To improve the solvating power of CO_2 and increase the yield of extracted polyphenols, different organic solvents can be used as co-solvents such as methanol, ethanol and acetone. Addition of different organic solvents in CO_2 mixture maximizes its critical temperature [2].

Supercritical CO_2 extraction method by using ethanol and methanol as a co-solvent has been applied in different materials like grape seeds, wine and pistachio hulls [84]. Quercetin, catechin, epicatechin and resveratrol are soluble in supercritical fluid having 5–30% ethanol, and anthocyanins are also soluble to some extent [148] while hydroxycinnamic acid and derivatives of coumaric acid show a little solubility in supercritical CO_2 extraction without any co-solvent [117]. Supercritical fluid can be used for extracting heat stable components having high boiling points at comparatively low temperatures while at low temperature, heat-sensitive components can also be extracted with minimum damage. Regardless this technique can maintain the structural integrity of phenolic compounds, but on the other hand, it also requires a high capital investment, equipment, and higher pressure. Additionally, intense recycling measures are required for solvent compression to decrease the energy cost [2].

8.6.9 Ultrafiltration (UF)

To separate and purify phenolic compounds, another technique using membrane separation is used known as ultrafiltration (UF) [111]. Normally, this technique is used for juice clarification during their processing in industries. As compared to traditional filtration techniques, this method is more efficient regarding energy conservation, environmental and processing benefits [62]. This filtration technique is affected by particle size of the compounds and the driving source. For transport across the membrane, the driving source is the pressure difference between the membranes, and generally, it works at 2–10 bar [71]. Molecules having molecular weight ranging between >3000 Da and >100 kDa can be separated through this technique; therefore, this system can easily separate polyphenols [2]. This technique has been applied to the juice of *Echinacea* herb in order to separate and concentrate the polyphenols by using commercially accessible membranes (polyethersulfone and regenerated cellulose) through ethanol [139, 158]. This technique presented high extraction rates and selectivity with shorter extraction duration and labor savings. Phenolic compounds having low molecular weight can also be easily and rapidly extracted from compound mixture through following ultrafiltration by using semipermeable membranes [39].

8.6.10 Enzyme-Assisted Extraction (EAE)

Different enzymes can be used to extract phenolic compounds like pectinase, cellulase, hemi-cellulase and glucanase [159] as the enzymes enhance extraction through the disintegration the plant cell wall which results in release of the cell wall complex polyphenols [170]. Phenolic extraction from apple pomace was enhanced through pectinase [78]. Similarly, in barley application of alpha-amylase and cellulase has improved the phenolic release [214] while commercially available enzymes such as Viscozyme, Termamyl and Lallzyme were also used in combination in barley grain extraction [25]. Application of non-starch polysaccharide hydrolyzing enzymes in vegetables and fruits can also release bound phenolics situated in the cell wall [61]. As compared to conventional extraction techniques like ultrafiltration, enzyme-assisted extractions have several advantages; it is an easy and rapid technique to increase the cell wall disintegration leading to enhanced separation and concentration of polyphenolic compounds [2].

8.7 Application of Polyphenols in Food Products

According to literature, polyphenols have contributed different properties in food products like bitterness, astringency, color, flavor, odor and oxidative stability [1]. Recently use of polyphenols as food additives in different products like baked products, noodles and pasta has enhanced which in result has increased polyphenol economy globally [98, 152, 162, 174] particularly in functional beverage sector (44%) and functional foods (33%). In developed countries, demand of polyphenol-enriched beverages like juices, energy drinks and enhanced water has grown [204]. It has been recommended that fruit juices [43], powders [203] and extracts rich in polyphenols have the ability to be applied as functional ingredients in food industry. For instance, in dairy sector, yogurt is a well-known fermented product, in spite of its importance in diet and nutritional composition; it is not regarded as a good source of phenolic compounds. Therefore, plant-based additives had been applied for improving phenolic level of yogurt [142]. In the other study, acidified ethanol extracts of four different grape varieties and grape callus as functional components were used to enrich the yoghurt [86]. Similarly, in another study fortification of stirred dahi with polyphenol extract of strawberry at the rate of 0.5 mg /mL showed sevenfold higher antioxidant activity as compared to control [182]. Addition of *Allium sativum* significantly enhanced the antioxidant activity in cow milk yogurt ($26.4 \pm 0.7\%$ to $37.9 \pm 0.8\%$) and camel milk yogurt ($15.4 \pm 1.3\%$ to $26.1 \pm 0.8\%$) [180]. Yoghurt having the addition of fennel and chamomile extract had also showed increased antioxidant activity as compared to synthetic additives [38]. During nine months of the storage at −18 °C, addition of *Rosmarinus officinalis* extract in chicken nuggets changed TBARS values 4.07–5.88 mg malonaldehyde/kg sample [188]. The influence of various polyphenol-enriched extracts from natural herbs like jojoba, jatropha, ginseng and ginger was analyzed in burgers regrading preventing lipid oxidation. These herbal extracts exhibited decreased TBARS values during storage and thus prevented lipid oxidation [79]. Similarly, rosemary lyophilized extract added in chicken burger also showed enhanced antioxidative activity which may be due to the presence of different phenolic compounds such as acid gallic, catechin, *p*-coumaric acid, ferulic acid, *p*-cinnamic acid, rutin, quercetin and kaempferol [155]. Similarly, lignans or resins as less polar phenolic compounds may be used in fats and oils to increase their stability during heating and storage. More polar phenolic substances are beneficial to stabilize food dispersions [144]. Chlorogenic acid and several other polyphenol substances are widely used as additives in beverage, tea products and other food products [81]. Most effective usage of plant extracts having polyphenols is in beverage (water- or tea-based functional drinks), dairy products and other novel product like "smoothies," while green tea extract and rooibos extract are most commonly used extracts in these beverages [204].

Extracts from grape seed, pomegranate by-products, rosemary, oregano and various other spices have been evaluated for their antioxidant potential in meat and poultry products [101]. The antioxidant effect of water and methanolic extract of garlic was evaluated in ground meat during refrigerated storage. Results showed greater total phenolic content, DPPH radical-scavenging activity and reducing power for methanolic extract [151]. Study showed very effective results of these extracts against lipid oxidation, while antioxidant potential was increased due to synergistic effects of different antioxidative factors when extracts were used in combination [22]. Addition of rosemary and oregano extracts (400 ppm) in beef burger showed better results in preventing lipid oxidation without affecting their sensory properties as compared with 200 ppm of BHT/BHA [191]. In another study, extracts of barley husks used in frozen salmon were evaluated on lipid oxidation during prolonged frozen storage. The results showed that barley husk extract efficiently increased oxidative stability and decreased the lipid hydrolysis [51]. Various studies evaluated the ability of cinnamon oil in the preservation of rainbow trout fresh fillets [12, 143]. Different phenolic compounds like catechin, caffeic acid, ferulic acid and tannic acid exhibited their preventive effect against lipid oxidation in fish mince by lowering peroxide and TBARS values [121]. Similarly, tea polyphenols and rosemary extract were used as natural preservatives in air-packaged whole crucian carp stored at 4 ± 1 °C. The results showed significantly decreased values of TVB-N and TBA in treated groups of samples [107].

However, it is not possible that every commercially accessible plant extract fulfills the regulatory requirements regarding their usage in food products. For any added ingredient, the form of food matrix depends on several factors including traditional use of the plant as a food, presenting form to the consumer (i.e., a pill or a candy) and physiological functions, along with techniques used for extracting ingredient [73]. Besides these legal perspectives, other technological aspects like stability and solubility are also significant for effective usage in food products. In the food matrix to avoid undesirable precipitation, plant extracts should be stable without hindering product physical features like color or taste [204]. Polyphenolic compounds can cause an astringent or bitter taste, and brown coloring in food products. For example, grape seed extracts due to having dark brown color and lower water solubility are difficult to be added in functional food [202]. Hence, during product development it is important to inspect the sensorial effects of bioactive substances on foods before their usage. Several methods can be used to control these problems such as microencapsulation of bioactive extracts could be a possible system to mask their unpleasant flavor [183]. Due to having concerns regarding harmful health influences of synthetic antioxidants, now consumers mainly preferred products comprising natural additives [59, 163].

8.8 Applications in Cosmetics

The importance of polyphenol components as functional ingredients has involved significant attention to cosmetic and drug industries in past years; as a result, a number of cosmeceutical and skin care companies are established for providing products enriched with polyphenol extracts [216]. Molecules having significant biological activities must be able to be released from their formulations and influence the skin and overwhelm the effect of stratum corneum and penetrate into dermis. The properties of a substance, e.g., lipophilicity and molecular weight, are important to check the activity of these released substances and skin permeation [23]. To confirm the deliverability and stability of the formulation components to the target areas of skin, the formulation has to be microbiological, physically and chemically stable. "Beauty pills" is a well-known term in cosmetic industries and can be depicted as the usage of supplements rich with nutraceuticals, also called nutricosmetics, a novel idea in the field of cosmetic and derived from the combination of nutraceuticals and cosmeceuticals [60]. Nutricosmetics is very wide but blurry area of research. Initially, nutricosmetic products were designed to boost the excessive fluid drainage in order to decrease the cellulite appearance. Now, a wide range of different dietary supplements and formulations have been marketed according to different requirements of the skin, like dermatologic and dermo-cosmetic ones. These supplements are supposed to target aging, decreasing the wrinkles through controlling free radicals production by solar radiation [14].

However, the cosmeceuticals are the combination of cosmetic and pharmaceutical industries [65]. Albert Kligman first devised the term "cosmeceutical" in 1984 at National Scientific Meeting of the Society of Cosmetic Chemists. They are referred to as the products that are directly applied on skin to decorate having ability to change the skin condition, and they are not regarded drugs or cosmetics. Consumers are more attracted by cosmeceuticals and nutraceuticals because it has the largest market of different products and concepts presented with different marketing strategies. According to DeFelice [52] during '80s, the revolution in nutraceutical field was started when different studies based on clinical trials regarding benefits of orally administrated calcium, fiber, and fish oil were published in medical journals. Similarly, several other studies regarding possible health benefits of other food compounds were conducted. In cosmeceuticals field, a number of different research works were conducted to assess the antioxidant activities of plant extracts used in topical lotions and creams. This constituent is added in sunscreen formulations to enhance the photo-protection given by UV filters [88]. Several studied showed that distillate of *Pinus pinaster* is rich in epicatechin, catechin, ferulic acid and caffeic and procyanidins thus proved as nutricosmetics in reducing melisma [140] and UV-linked damage [172]. While in another study, oral administration of extract from grape seed enriched with procyanidins showed reduction of hyperpigmentation in women suffering from melisma [210]. Similarly, the oral administration of epicatechin and catechin, cocoa flavonols showed photoprotective influences [74]. Different studies strongly proved the anti-inflammatory, anti-mutagenic and free radical scavenger functions performed by phenolic compounds [104]. Epigallocatechin-3-gallate from green tea has been evaluated to assure safety for DNA damage induced by UV and suppression of immune system through minimizing oxidative stress and hindering NF-KappaB (pro-inflammatory constituent induced by UV radiation) [124]. Addition of pycnogenol hindered the NF-KappaB factor in a dose-dependent manner as evaluated through an in vitro study [172].

In order to ensure proper delivery of phenolic ingredients into target areas of skin, their precipitation should be avoided in the formulations used as transmission vehicles. Some polyphenols show rapid precipitation in water due to their poor solubilization, but some surfactants like polyoxyethylenesorbitan monolaurate or block copolymers enhance their solubility, as revealed for flavonoids naringenin, quercetin and rutin [110]. Most suitable way of topical formulations is the emulsions due to their improved solubility efficiency for both hydrophilic and lipophilic components. Emulsions are of heterogeneous nature having two immiscible components (oil and water), in which one phase is distributed within other one and stabilized through emulsifiers [216]. It was demonstrated that according to polarity of phenolic compounds, their solubilization was affected by non-ionic surfactant micelles, while non-polar substances showed less solubilization as compared to polar [165]. Addition of phenolic compounds into emulsions affects their properties like rheology, stability, specifically, even reduced viscosity might be observed [56]. There are some studies explaining the interaction between polyphenols and proteins [103], by involving reversible complexation results from non-covalent forces, or irreversible aggregation by establishing covalent bonds. However, polyphenols also demonstrated light properties related to surface-activation. For example, accumulation of catechin at the surface of interaction of air/water can reduce surface tension. Similarly, gallic acid, catechin and quercetin in oil-in-water emulsions reduced surface tension, by altering the droplet sizes of emulsions, like catechin-induced bigger size oil droplets, while quercetin enhanced the dispersion state of emulsion [57]. Some flavonoids mainly rutin showed stabilizing influences on emulsions [115]. Oil–water interface could absorb the flavonoids while insoluble particles performed as stabilizers by developing Pickering emulsions [116]. This capability of stabilization is influenced by pH value, for example at lower pH values rutin formed tetradecane emulsions having smaller droplet size [115]. Polyphenols can enhance the stability of emulsion against oxidation and deposition due to their antioxidant features [57] and due to their adsorption on the oil–water interface [19].

As anti-aging agent, polyphenols act dermis and the epidermis as targeted skin areas. The epidermis contains polyphenol binding sites which act as binding sites for resveratrol where its antioxidant/anti-aging properties can act [26]. The release of substances from the formulation is a part of complex process of permeation through skin. As the release rate will be faster, more earlier substance will be permeated into the skin. In most of the cases, coefficients of permeation and exoneration can correlate [217]. For example, the liberation of rutin from semisolid systems is dependent on the emulsion ingredient. Isopropanol and urea retard the exoneration of rutin from formulations. In contradiction, the existence of propylene glycol enhances the rutin exoneration rate [23]. On the other hand, the penetration of active ingredients through the skin depends upon vehicle formulations. For example, emulsions having decreased oil contents frequently assist the progressive release of phenolic compounds along with increased skin permeation [96, 217]. This behavior might be due to an emulsion of lower viscosity having decreased level of oil as compared to emulsions having increased oil contents. The higher viscosity might cause inhibition of phenolic diffusion within emulsions. Application of formulations having more water contents causes hyperhydration of the skin which could also be a reason of increased skin permeability [32]. Thermodynamically stable micro-emulsions can be also used as vehicles to improve the penetration rates of polyphenols in skin [97, 215]. The skin penetration of hydrophilic chlorogenic acid, developed to O/W micro-emulsions, was enhanced in contrast with aqueous vehicles and W/O micro-emulsions (water-in-oil) [97]. The formulation of hydrophobic quercetin to a W/O micro-emulsion significantly enhanced the intradermal transmission while quercetin penetrated deeply into skin layers. An advancement of skin permeation can be accomplished using nano-emulsions and nanocarriers [58, 175]. In contrast, some nanocarriers having different components, for example ethanol-containing vesicles composed of soya phosphatidylcholine, were evaluated more effective for improving skin penetration of resveratrol [175]. Addition of active compounds in cosmeto-textiles is used as a substitute to penetrate active compounds into the skin. In cosmetics industry, this delivery procedure is increasingly used [8].

8.9 Applications in Pharmaceuticals

Several steps such as conjugation ability with other phenolic compounds, degree of glycosylation and acylation, molecular size and solubility affect the absorption and metabolism activities of phenolics which in turn determine their possible health benefits [178]. During passage through the small intestine wall into the circulation system and consequent transportation in portal vein to the liver, at several positions all these steps occur [149]. Removal of phenolic metabolites from plasma is very rapid; therefore, it is vital to consume plant products on daily basis, suggested intakes of polyphenols of 1 g/day, in order to provide high concentrations of metabolites in blood [35]. Different factors are involved which affect the bioavailability of bioactive compounds such as their preparatory methods, gastrointestinal (GI) digestion, their absorption and further metabolic reactions in body. Intestinal enzymes and colonic microflora hydrolyze the dietary polyphenols during the absorption process. After absorption through mechanisms of methylation, sulfation or glucuronidation, these bioactive compounds become conjugated in the intestinal cells and liver. This is the method that phenolic compounds reach and accumulate in the target tissue to induce different biological properties. Finally, the excretion of phenolic derivatives is conducted through the bile and urine [24]. For various polyphenolic compounds like procyanidins, quercetin and flavanols' rapid absorption was analyzed. Results showed that 2 or 3 h after ingesting these compounds the plasma concentrations were at peak [166]. Primarily, it was considered that antioxidant properties of phenolics were responsible for their defensive benefits by dropping the level of free radicals in the body. But, later on there are developing indications suggesting that metabolites of dietary phenolics appear in the circulatory system and are responsible for inducing regulatory beneficial effects in cells. Phenolic metabolites induce actions on various parts of the intracellular signaling cascades which are important for proper cellular functioning [45].

Pomegranate polyphenols were studied to reduce the cardiovascular coincidences in high-risk patients [125]. This effect may be due to inhibition of platelet functions as in decreasing platelet stimulations pomegranate extract, rich in polyphenols, showed an efficient response. The major component in this regard was hydrolysable tannins such as ellagitannins. These studies confirmed that the ability of polyphenols for inhibiting platelet functioning may be linked with cardiovascular health benefits [154]. Phenolic acids, hydrolysable tannins and flavonoids perform as protective agents of DNA against free radicals thus having anti-carcinogenic and anti-mutagenic effects. These compounds inactivate carcinogens, constrain the enzymes that stimulate pro-carcinogen and also activate the enzymes responsible for xenobiotics detoxification. In lymphocytes, flavonoids also showed protective effect in synergism against oxidative damages of DNA [20]. According to Block et al. [28], risks of colon cancer can be reduced by consuming vegetable rich diet. Through in vitro studies, it was demonstrated that chlorogenic and caffeic acids have possible inhibiting effects against the formation of mutagenic and carcinogenic N-nitroso compounds, thus having antioxidant effects [128]. According to Katiyar and Elmets [89], detrimental effects promoting progress of skin cancers can be enhanced through overexposure of the skin to ultraviolet (UV) radiation. Prevention of photo-carcinogenesis was observed through proanthocyanidins present in grape seeds. Dietary supplements of proanthocyanidins given to SKH-1 hairless mice showed the reduction of UVB-induced skin tumor. According to results, dietary proanthocyanidins are useful against harmful UV-directed health effects of skin [89].

Efficiency of willow bark extract for treating painful mobility disorders like back pain and arthritis was observed due to having salicin and its derivatives as pro-drugs of salicylates. Furthermore, salicins and their metabolites were deficient in acetylating the potential of acetylsalicylic acid indicating an alternative action mechanism [137]. According to Katiyar and Elmets [89], green tea is used as a widespread beverage and several studies have also stated the anti-inflammatory benefits and activity of green tea. The main polyphenol responsible for this antioxidant activity is epicatechin while the chemopreventive component (-)-epigallocatechin-3-gallate is responsible for the anti-inflammatory and anti-carcinogenic effects in green tea [213]. In another study, Osakabe N et al. orally administered male rats with Alpha-tocopherol, sucralfate and cimetidine. Afterward, cacao liquor water-soluble crude polyphenols were subjected which decreased the actions of enzymes responsible for oxygen radicals. It was observed that antiulcer mechanism of cacao not only includes radical-scavenging potential but also has the modulation of leukocyte function [147]. According to Vauzour et al. [194], flavonoids have potential to overwhelm neuro-inflammation and also have ability to improve memory, learning and cognitive functions. Flavonoids can enhance neuroprotective actions within the brain, having provided protection for neurons against injuries brought about via neurotoxins. Ingesting flavonoid-rich foods can reduce neurodegeneration leading to avoiding or reversing the age-dependent loses of cognitive ability. Flavonoids may be used as an important precursor component in the formulation of drugs used for enhancing brain activity [70].

Flavonoids, catechins and their derivatives are regarded as therapeutic agents which help as neuroprotective molecules in neurodegenerative diseases like Parkinson's and Alzheimer's diseases [99]. Yang et al. determined that tea polyphenols consist of catechins, proanthocyanidins, flavonols and theaflavins which can inhibit influenza virus infections and these results are projected for developing polyphenol-based anti-viral drugs [211]. Against microbial infections, phenolics serve as a natural defense mechanism so this property can be used to increase the shelf life in food processing [149]. Nikitina et al. assessed that phenolic compounds of plants belonging to families Geraniaceae and Rosaceae have been evaluated for their activity against gram-positive and gram-negative bacteria related to genera Azotobacter, Bacillus and Pseudomona [141]. The hypoglycemic effect of bumelia sartorum extracts investigated by Ruela et al. [167] showed hypoglycemic activity. Their extracts which are rich in polyphenolic compounds can be used for treating different diseases, such as diabetes. These extracts also have the ability of decreasing blood glucose in normoglycemic mice and can also inhibit the activity of sarco/endoplasmic reticulum Ca (2+)-ATPase which is one of the mechanisms involved in reducing glucose level. As a result, this inhibition can increase cytoplasmic Ca^{2+} concentration, resulting in an increased insulin release [167]. Type-A procyanidine polyphenols extracted from cinnamon bark showed disease-modifying protective potential against inflammation and rheumatoid arthritis in animal models [198].

8.10 Future Perspectives

Antioxidants are almost ubiquitous in commonly consumed food products, as they are generally preexisting compounds in the form of natural antioxidants, or are added during processing as synthetic antioxidants. As natural antioxidants have been shown to have significant benefits in preventing cancer and heart disease, many food advertisers have taken note and begun publicizing this fact. These beneficial properties have put natural antioxidants on the forefront of recent food advertising, and public levels of awareness concerning natural antioxidants and their positive effects have increased significantly. Therefore, antioxidants are widely supplemented in diets for the maintenance of proper health and prevention of various pathological diseases

Unlike natural antioxidants, however, synthetic antioxidants have been shown to have potential toxic effects on the health of consumers. Nowadays, pharmacists, chemists, physicians and nutritionists all are working together for developing new nutritional formulations contributing not only to disease prevention, but also positively affect skin health and contribute to many industrial uses, such as for the preservation of food and cosmetics.

References

1. Adebooye OC, Alashi AM, Aluko RE (2018) A brief review on emerging trends in global polyphenol research. J Food Biochem, e12519
2. Ajila C, Brar S, Verma M et al (2011) Extraction and analysis of polyphenols: recent trends. Crit Rev Biotechnol 31:227–249
3. Akbarirad H, Ardabili AG, Kazemeini S et al (2016) An overview on some of important sources of natural antioxidants. Int Food Res J 23
4. Akram A, Younis A, Akhtar G et al (2017) Comparative efficacy of various essential oil extraction techniques on oil yield and quality of *Jasminum sambac* L. Sci Int 5:84–95
5. Akyol H, Riciputi Y, Capanoglu E et al (2016) Phenolic compounds in the potato and its byproducts: an overview. Int J Mol Sci 17:835
6. Ali IBEH, Bahri R, Chaouachi M et al (2014) Phenolic content, antioxidant and allelopathic activities of various extracts of *Thymus numidicus* Poir. organs. Ind Crops Prod 62:188–195
7. Alok S, Jain SK, Verma A et al (2014) Herbal antioxidant in clinical practice: a review. Asian Pac J Trop Biomed 4:78–84
8. Alonso C, Martí M, Martínez V et al (2013) Antioxidant cosmeto-textiles: skin assessment. Eur J Pharm Biopharm 84:192–199
9. Ameer K, Shahbaz HM, Kwon JH (2017) Green extraction methods for polyphenols from plant matrices and their byproducts: a review. Compr Rev Food Sci Food Saf 16:295–315
10. Amid A, Salim R, Adenan M (2010) The factors affecting the extraction condition for neuroprotective activity of *Centella asiatica* evaluated by metal chelating activity assay. J Appl Sci 10:837–842
11. Anbudhasan P, Surendraraj A, Karkuzhali S et al (2014) Natural antioxidants and its benefits. Int J Food Nutr Sci 3:225–232
12. Andevari GT, Rezaei M (2011) Effect of gelatin coating incorporated with cinnamon oil on the quality of fresh rainbow trout in cold storage. Int J Food Sci Technol 46:2305–2311
13. André C, Castanheira I, Cruz J et al (2010) Analytical strategies to evaluate antioxidants in food: a review. Trends Food Sci Technol 21:229–246
14. Anunciato TP, da Rocha Filho PA (2012) Carotenoids and polyphenols in nutricosmetics, nutraceuticals, and cosmeceuticals. J Cosmet Dermatol 11:51–54
15. Anuradha M, Pragyandip P, Richa K et al (2010) Evaluation of neuropharmacological effects of ethanolic extract of *Clitorea ternatea* flowers. Pharmacologyonline 1:285–292
16. Apak R, Özyürek M, Güçlü K et al (2016) Antioxidant activity/capacity measurement. 1. Classification, physicochemical principles, mechanisms, and electron transfer (ET)-based assays. J Agric Food Chem 64:997–1027
17. Arshiya S (2013) The antioxidant effect of certain fruits: a review. J Pharm Sci Res 5:265–268
18. Asofiei I, Calinescu I, Trifan A et al (2016) Microwave-assisted batch extraction of polyphenols from sea buckthorn leaves. Chem Eng Commun 203:1547–1553
19. Atarés L, Marshall LJ, Akhtar M et al (2012) Structure and oxidative stability of oil in water emulsions as affected by rutin and homogenization procedure. Food Chem 134:1418–1424
20. Awaad AS, Al-Jaber NA (2010) Antioxidant natural plant. RPMP Ethnomed Source Mech 27:1–35
21. Azwanida N (2015) A review on the extraction methods use in medicinal plants, principle, strength and limitation. Med Aromat Plants 4(2167–0412):1000196
22. Babuskin S, Babu PAS, Sasikala M et al (2014) Antimicrobial and antioxidant effects of spice extracts on the shelf life extension of raw chicken meat. Int J Food Microbiol 171:32–40
23. Baby AR, Haroutiounian-Filho CA, Sarruf FD et al (2009) Influence of urea, isopropanol, and propylene glycol on rutin in vitro release from cosmetic semisolid systems estimated by factorial design. Drug Dev Ind Pharm 35:272–282
24. Bahadoran Z, Mirmiran P, Azizi F (2013) Dietary polyphenols as potential nutraceuticals in management of diabetes: a review. J Diabetes Metab Disord 12:43

25. Bartolomé B, Gómez-Cordovés C (1999) Barley spent grain: release of hydroxycinnamic acids (ferulic and *p*-coumaric acids) by commercial enzyme preparations. J Sci Food Agric 79:435–439
26. Bastianetto S, Dumont Y, Duranton A et al (2010) Protective action of resveratrol in human skin: possible involvement of specific receptor binding sites. PLoS ONE 5:e12935
27. Blicharski T, Oniszczuk A (2017) Extraction methods for the isolation of isoflavonoids from plant material. Open Chem 15:34–45
28. Block G, Patterson B, Subar A (1992) Fruit, vegetables, and cancer prevention: a review of the epidemiological evidence. Nutr Cancer 18:1–29
29. Borneman Z, Gökmen V, Nijhuis HH (2001) Selective removal of polyphenols and brown colour in apple juices using PES/PVP membranes in a single ultrafiltration process. Sep Purif Technol 22:53–61
30. Both S, Chemat F, Strube J (2014) Extraction of polyphenols from black tea—Conventional and ultrasound assisted extraction. Ultrason Sonochem 21:1030–1034
31. Bouayed J, Bohn T (2010) Exogenous antioxidants—double-edged swords in cellular redox state: health beneficial effects at physiologic doses versus deleterious effects at high doses. Oxid Medi Cell Longev 3:228–237
32. Bouwstra JA, Honeywell-Nguyen PL, Gooris GS et al (2003) Structure of the skin barrier and its modulation by vesicular formulations. Prog Lipid Res 42:1–36
33. Brachet A, Rudaz S, Mateus L et al (2001) Optimisation of accelerated solvent extraction of cocaine and benzoylecgonine from coca leaves. J Sep Sci 24:865–873
34. Brewer M (2011) Natural antioxidants: sources, compounds, mechanisms of action, and potential applications. Compr Rev Food Sci Food Saf 10:221–247
35. Brglez Mojzer E, Knez Hrnčič M, Škerget M et al (2016) Polyphenols: extraction methods, antioxidative action, bioavailability and anticarcinogenic effects. Molecules 21:901
36. Bucić-Kojić A, Planinić M, Tomas S et al (2011) Effect of extraction conditions on the extractability of phenolic compounds from lyophilised fig fruits (*Ficus carica* L.). Pol J Food Nutr Sci 61:195–199
37. Butkhup L, Chowtivannakul S, Gaensakoo R et al (2016) Study of the phenolic composition of Shiraz red grape cultivar (*Vitis vinifera* L.) cultivated in north-eastern Thailand and its antioxidant and antimicrobial activity. S Afr J Enol Vitic 31:89–98
38. Caleja C, Barros L, Antonio AL et al (2016) Fortification of yogurts with different antioxidant preservatives: a comparative study between natural and synthetic additives. Food Chem 210:262–268
39. Cassano A, Conidi C, Ruby-Figueroa R et al (2018) Nanofiltration and tight ultrafiltration membranes for the recovery of polyphenols from agro-food by-products. Int J Mol Sci 19:351
40. Charles DJ (2012) Antioxidant properties of spices, herbs and other sources. Springer Science & Business Media
41. Chen S-K, Hsu C-H, Tsai M-L et al (2013) Inhibition of oxidative stress by low-molecular-weight polysaccharides with various functional groups in skin fibroblasts. Int J Mol Sci 14:19399–19415
42. Chupin L, Maunu S, Reynaud S et al (2015) Microwave assisted extraction of maritime pine (*Pinus pinaster*) bark: impact of particle size and characterization. Ind Crops Prod 65:142–149
43. Coïsson J, Travaglia F, Piana G et al (2005) Euterpe oleracea juice as a functional pigment for yogurt. Food Res Int 38:893–897
44. Costa P, Gonçalves S, Valentão P et al (2012) Thymus lotocephalus wild plants and in vitro cultures produce different profiles of phenolic compounds with antioxidant activity. Food Chem 135:1253–1260
45. Crozier A, Jaganath IB, Clifford MN (2009) Dietary phenolics: chemistry, bioavailability and effects on health. Nat Prod Rep 26:1001–1043
46. Ćujić N, Šavikin K, Janković T et al (2016) Optimization of polyphenols extraction from dried chokeberry using maceration as traditional technique. Food Chem 194:135–142
47. da Silva LMR, de Figueiredo EAT, Ricardo NMPS et al (2014) Quantification of bioactive compounds in pulps and by-products of tropical fruits from Brazil. Food Chem 143:398–404

48. Dahmoune F, Nayak B, Moussi K et al (2015) Optimization of microwave-assisted extraction of polyphenols from *Myrtus communis* L. leaves. Food Chem 166:585–595
49. Dai J, Mumper RJ (2010) Plant phenolics: extraction, analysis and their antioxidant and anticancer properties. Molecules 15:7313
50. Dandena A, Zuka L, Kostova M (2014) Natural antioxidants in black chokeberry marc extracts depending on the extraction method. In: 9th baltic conference on food science and technology "food for consumer well-being", p 324
51. De Abreu DP, Losada PP, Maroto J et al (2010) Evaluation of the effectiveness of a new active packaging film containing natural antioxidants (from barley husks) that retard lipid damage in frozen Atlantic salmon (*Salmo salar* L.). Food Res Int 43:1277–1282
52. DeFelice SL (1995) The nutraceutical revolution: its impact on food industry R&D. Trends Food Sci Technol 6:59–61
53. Deng G-F, Lin X, Xu X-R et al (2013) Antioxidant capacities and total phenolic contents of 56 vegetables. J Funct Food 5:260–266
54. Deng G-F, Shen C, Xu X-R et al (2012) Potential of fruit wastes as natural resources of bioactive compounds. Int J Mol Sci 13:8308–8323
55. Deng Y, Zhao Y, Padilla-Zakour O et al (2015) Polyphenols, antioxidant and antimicrobial activities of leaf and bark extracts of *Solidago canadensis* L. Ind Crops Prod 74:803–809
56. Di Mambro VM, Fonseca MJ (2005) Assays of physical stability and antioxidant activity of a topical formulation added with different plant extracts. J Pharm Biomed Anal 37:287–295
57. Di Mattia CD, Sacchetti G, Mastrocola D et al (2010) Surface properties of phenolic compounds and their influence on the dispersion degree and oxidative stability of olive oil O/W emulsions. Food Hydrocoll 24:652–658
58. Domínguez-Villegas V, Clares-Naveros B, García-López ML et al (2014) Development and characterization of two nano-structured systems for topical application of flavanones isolated from *Eysenhardtia platycarpa*. Colloids Surf B Biointerfaces 116:183–192
59. dos Reis AS, Diedrich C, de Moura C et al (2017) Physico-chemical characteristics of microencapsulated propolis co-product extract and its effect on storage stability of burger meat during storage at −15 °C. LWT-Food Sci Technol 76:306–313
60. Draelos ZD (2010) Nutrition and enhancing youthful-appearing skin. Clin Dermatol 28:400–408
61. Duodu K (2014) Effects of processing on phenolic phytochemicals in cereals and legumes. Cereal Food World 59:64–70
62. Echavarria A, Torras C, Pagán J et al (2011) Fruit juice processing and membrane technology application. Food Eng Rev 3:136–158
63. Ellong EN, Billard C, Adenet S et al (2015) Polyphenols, carotenoids, vitamin C content in tropical fruits and vegetables and impact of processing methods. Food Nutr Sci 6:299
64. Evgeny D, Taisa D (2011) Dissociation energies of O–H bonds of phenols and Hydroperoxides. In: Application of thermodynamics to biological and materials science. IntechOpen
65. Feetham HJ, Jeong HS, McKesey J et al (2018) Skin care and cosmeceuticals: attitudes and trends among trainees and educators. J Cosmet Dermatol 17:220–226
66. Galvan d'Alessandro L, Kriaa K, Nikov I et al (2012) Ultrasound assisted extraction of polyphenols from black chokeberry. Sep Purif Technol 93:42–47. https://doi.org/10.1016/j.seppur.2012.03.024
67. Govindaraj M, Masilamani P, Albert VA et al (2017) Role of antioxidant in seed quality—a review. Agr Rev 38
68. Guerrero MS, Torres JS, Nuñez MJ (2008) Extraction of polyphenols from white distilled grape pomace: optimization and modelling. Bioresour Technol 99:1311–1318
69. Guido LF, Moreira MM (2017) Techniques for extraction of Brewer's spent grain polyphenols: a review. Food Bioprocess Tech 10:1192–1209
70. Gupta J, Gupta A, Gupta AK (2017) The flavonoids ameliorates: protective mechanisms in neurodegenerative diseases. Int J Curr Res Chem Pharm Sci 4:1–7
71. Habibi S (2014) An upscaled study of a membrane filtration process in presence of biofilms. Doctoral Dissertation, Ecole Centrale Paris

72. Hamid A, Aiyelaagbe O, Usman L et al (2010) Antioxidants: its medicinal and pharmacological applications. Afri J Pure Appl Chem 4:142–151
73. Hasler CM (2002) Functional foods: benefits, concerns and challenges—a position paper from the American Council on Science and Health. J Nutr 132:3772–3781
74. Heinrich U, Neukam K, Tronnier H et al (2006) Long-term ingestion of high flavanol cocoa provides photoprotection against UV-induced erythema and improves skin condition in women. J Nutr 136:1565–1569
75. Horžić D, Jambrak AR, Belščak-Cvitanović A et al (2012) Comparison of conventional and ultrasound assisted extraction techniques of yellow tea and bioactive composition of obtained extracts. Food Bioprocess Tech 5:2858–2870
76. Hossain M, Barry-Ryan C, Martin-Diana AB et al (2011) Optimisation of accelerated solvent extraction of antioxidant compounds from rosemary (*Rosmarinus officinalis* L.), marjoram (*Origanum majorana* L.) and oregano (*Origanum vulgare* L.) using response surface methodology. Food Chem 126:339–346
77. Hossain MA, Al-Toubi WA, Weli AM et al (2013) Identification and characterization of chemical compounds in different crude extracts from leaves of Omani neem. J Taibah Univ Sci 7:181–188
78. Hu-Zhe Z, Lee H-R, Lee S-H et al (2008) Pectinase assisted extraction of polyphenols from apple pomace. Chinese J Anal Chem 36:306–310
79. Ibrahim HM, Abou-Arab AA, Salem FMA (2010) Effects on lamb patties quality. J Food Technol 8:134–142
80. Ibrahima NA, Zaini MAA (2017) Solvent selection in microwave assisted extraction of castor oil. Chem Eng 56
81. Jin U-H, Lee J-Y, Kang S-K et al (2005) A phenolic compound, 5-caffeoylquinic acid (chlorogenic acid), is a new type and strong matrix metalloproteinase-9 inhibitor: isolation and identification from methanol extract of *Euonymus alatus*. Life Sci 77:2760–2769
82. Jovanović A, Petrovic P, Đordjevic V et al (2017a) Polyphenols extraction from plant sources. Lekovite Sirovine. https://doi.org/10.5937/leksir1737045j
83. Jovanović AA, Đorđević VB, Zdunić GM et al (2017) Optimization of the extraction process of polyphenols from *Thymus serpyllum* L. herb using maceration, heat-and ultrasound-assisted techniques. Sep Purif Technol 179:369–380
84. Junior MRM, Leite AV, Dragano NRV (2010) Supercritical fluid extraction and stabilization of phenolic compounds from natural sources—review (supercritical extraction and stabilization of phenolic compounds). Open Chem Eng J 4
85. Kapłan M, Najda A (2014) Antioxidant activity of vine fruits depending on their colouring. Chemija 25:51–55
86. Karaaslan M, Ozden M, Vardin H et al (2011) Phenolic fortification of yogurt using grape and callus extracts. LWT-Food Sci Technol 44:1065–1072
87. Karrar EMA (2014) A review on: antioxidant and its impact during the bread making process. Int J Nutr Food Sci 3:592–596
88. Katiyar SK (2011) Green tea prevents non-melanoma skin cancer by enhancing DNA repair. Arch Biochem Biophys 508:152–158
89. Katiyar SK, Elmets CA (2001) Green tea polyphenolic antioxidants and skin photoprotection. Int J Oncol 18:1307–1313
90. Kaufmann B, Christen P (2002) Recent extraction techniques for natural products: microwave-assisted extraction and pressurised solvent extraction. Phytochem Anal 13:105–113
91. Keser S, Celik S, Turkoglu S (2013) Total phenolic contents and free-radical scavenging activities of grape (*Vitis vinifera* L.) and grape products. Int J Food Sci Nutr 64:210–216
92. Khan BA, Akhtar N, Khan H et al (2013) Development, characterization and antioxidant activity of polysorbate based O/W emulsion containing polyphenols derived from *Hippophae rhamnoides* and *Cassia fistula*. Braz J Pharm Sci 49:763–773
93. Khan MK, Abert-Vian M, Fabiano-Tixier A-S et al (2010) Ultrasound-assisted extraction of polyphenols (flavanone glycosides) from orange (*Citrus sinensis* L.) peel. Food Chem 119:851–858

94. Khaw K-Y, Parat M-O, Shaw PN et al (2017) Solvent supercritical fluid technologies to extract bioactive compounds from natural sources: a review. Molecules 22:1186
95. Khoddami A, Wilkes MA, Roberts TH (2013) Techniques for analysis of plant phenolic compounds. Molecules 18:2328–2375
96. Kitagawa S, Tanaka Y, Tanaka M et al (2009) Enhanced skin delivery of quercetin by microemulsion. J Pharm Pharmacol 61:855–860
97. Kitagawa S, Yoshii K, S-y Morita et al (2011) Efficient topical delivery of chlorogenic acid by an oil-in-water microemulsion to protect skin against UV-induced damage. Chem Pharm Bull 59:793–796
98. Koca I, Tekguler B, Yilmaz VA et al (2018) The use of grape, pomegranate and rosehip seed flours in Turkish noodle (erişte) production. J Food Process Pres 42:e13343
99. Kumar GP, Anilakumar K, Naveen S (2015a) Phytochemicals having neuroprotective properties from dietary sources and medicinal herbs. Pharmacogn J 7
100. Kumar H, Choudhary N, Varsha K et al (2014) Phenolic compounds and their health benefits: a review. J Food Res Technol 2:46–59
101. Kumar Y, Yadav DN, Ahmad T et al (2015) Recent trends in the use of natural antioxidants for meat and meat products. Compr Rev Food Sci Food Saf 14:796–812
102. Le Bourvellec C, Guyot S, Renard C (2009) Interactions between apple (*Malus x domestica* Borkh.) polyphenols and cell walls modulate the extractability of polysaccharides. Carbohydr Polym 75:251–261
103. Le Bourvellec C, Renard C (2012) Interactions between polyphenols and macromolecules: quantification methods and mechanisms. Crit Rev Food Sci Nutr 52:213–248
104. Lee J, Koo N, Min DB (2004) Reactive oxygen species, aging, and antioxidative nutraceuticals. Compr Rev Food Sci Food Saf 3:21–33
105. Lee L-S, Lee N, Kim YH et al (2013) Optimization of ultrasonic extraction of phenolic antioxidants from green tea using response surface methodology. Molecules 18:13530–13545
106. Li F, Li S, Li H-B et al (2013) Antiproliferative activity of peels, pulps and seeds of 61 fruits. J Funct Foods 5:1298–1309
107. Li T, Li J, Hu W et al (2012) Shelf-life extension of crucian carp (*Carassius auratus*) using natural preservatives during chilled storage. Food Chem 135:140–145
108. Li T, Shen P, Liu W et al (2014) Major polyphenolics in pineapple peels and their antioxidant interactions. Int J Food Prop 17:1805–1817
109. Lima GPP, Vianello F, Corrêa CR et al (2014) Polyphenols in fruits and vegetables and its effect on human health. Food Nutr Sci 1065–1082
110. Löf D, Schillén K, Nilsson L (2011) Flavonoids: precipitation kinetics and interaction with surfactant micelles. J Food Sci 76
111. Loginov M, Boussetta N, Lebovka N et al (2013) Separation of polyphenols and proteins from flaxseed hull extracts by coagulation and ultrafiltration. J Membrane Sci 442:177–186
112. Loprinzi P, Mahoney S (2014) Association between flavonoid-rich fruit and vegetable consumption and total serum bilirubin. Angiology 66(3):286–290. https://doi.org/10.1177/0003319714537111
113. Loprinzi PD, Mahoney SE (2015) Association between flavonoid-rich fruit and vegetable consumption and total serum bilirubin. Angiology 66:286–290
114. Łozowicka B, Jankowska M, Rutkowska E et al (2012) Comparison of extraction techniques by matrix solid phase dispersion and liquid-liquid for screening 150 pesticides from soil, and determination by gas chromatography. Pol J Environ Stud 21(4):973–992
115. Luo Z, Murray BS, Ross A-L et al (2012) Effects of pH on the ability of flavonoids to act as Pickering emulsion stabilizers. Colloids Surf B Biointerfaces 92:84–90
116. Luo Z, Murray BS, Yusoff A et al (2011) Particle-stabilizing effects of flavonoids at the oil–water interface. J Agric Food Chem 59:2636–2645
117. Machado BAS, de Abreu Barreto G, Costa AS et al (2015) Determination of parameters for the supercritical extraction of antioxidant compounds from green propolis using carbon dioxide and ethanol as co-solvent. PLoS ONE 10:e0134489

118. Maestri D, Nepote V, Lamarque A et al (2006) Natural products as antioxidants. Phytochem Adv Res 37:105–135
119. Mahoney JR, Graf E (1986) Role of alpha-tocopherol, ascorbic acid, citric acid and EDTA as oxidants in model systems. J Food Sci 51:1293–1296
120. Manohar C (2017) Antioxidant and antiproliferative activity of flavonoids from Ontario grown onions by pressurized low polarity water technology. Doctoral Dissertation
121. Maqsood S, Benjakul S (2010) Comparative studies of four different phenolic compounds on in vitro antioxidative activity and the preventive effect on lipid oxidation of fish oil emulsion and fish mince. Food Chem 119:123–132
122. Martorana M, Arcoraci T, Rizza L et al (2013) In vitro antioxidant and in vivo photoprotective effect of pistachio (*Pistacia vera* L., variety Bronte) seed and skin extracts. Fitoterapia 85:41–48
123. Materska M, Perucka I (2005) Antioxidant activity of the main phenolic compounds isolated from hot pepper fruit (*Capsicum annuum* L.). J Agric Food Chem 53:1750–1756
124. Matsui MS, Hsia A, Miller JD et al (2009) Non-sunscreen photoprotection: antioxidants add value to a sunscreen. J Invest Dermatol Symp Proc 1:56–59
125. Mattiello T, Trifirò E, Jotti GS et al (2009) Effects of pomegranate juice and extract polyphenols on platelet function. J Med Food 12:334–339
126. Medina-Torres N, Ayora-Talavera T, Espinosa-Andrews H et al (2017) Ultrasound assisted extraction for the recovery of phenolic compounds from vegetable sources. Agron J 7:47
127. Medvidović-Kosanović M, Šeruga M, Jakobek L et al (2010) Electrochemical and antioxidant properties of (+)-catechin, quercetin and rutin. Croat Chem Acta 83:197–207
128. Mennen LI, Walker R, Bennetau-Pelissero C et al (2005) Risks and safety of polyphenol consumption. Am J Clin Nutr 81:326S–329S
129. Milošević SG, Lepojević ŽD, Zeković ZP et al (2011) Determination of extraction conditions of *Ginkgo biloba* L. leaves by supercritical CO_2 using response surface methodology. Hem Ind 65:147
130. Milutinović M, Radovanović N, Ćorović M et al (2015) Optimisation of microwave-assisted extraction parameters for antioxidants from waste *Achillea millefolium* dust. Ind Crops Prod 77:333–341
131. Miron T, Plaza M, Bahrim G et al (2011) Chemical composition of bioactive pressurized extracts of Romanian aromatic plants. J Chromatogr A 1218:4918–4927
132. Misra K, Dhillon GS, Brar SK et al (2014) Antioxidants. In: Biotransformation of waste biomass into high value biochemicals. Springer, pp 117–138
133. Mordi R (1993) Mechanism of beta-carotene degradation. Biochem J 292:310
134. Morton LW, Caccetta RAA, Puddey IB et al (2000) Chemistry and biological effects of dietary phenolic compounds: relevance to cardiovascular disease. Clin Exp Pharmacol Physiol 27:152–159
135. Mustafa A, Turner C (2011) Pressurized liquid extraction as a green approach in food and herbal plants extraction: a review. Anal Chim Acta 703:8–18
136. Nagore D (2018) Development and quality control of natural products, vol 1. Evincepub Publishing
137. Nahrstedt A, Schmidt M, Jäggi R et al (2007) Willow bark extract: the contribution of polyphenols to the overall effect. Wien Med Wochenschr 157:348–351
138. Naude Y, De Beer W, Jooste S et al (1998) Comparison of supercritical fluid extraction and Soxhlet extraction for the determination of DDT, DDD and DDE in sediment. WATER SA-PRETORIA- 24:205–214
139. Nawaz H, Shi J, Mittal GS et al (2006) Extraction of polyphenols from grape seeds and concentration by ultrafiltration. Sep Purif Technol 48:176–181. https://doi.org/10.1016/j.seppur.2005.07.006
140. Ni Z, Mu Y, Gulati O (2002) Treatment of melasma with Pycnogenol®. Phytother Res 16:567–571
141. Nikitina V, Kuz'mina LY, Melent'ev A (2007) Antibacterial activity of polyphenolic compounds isolated from plants of Geraniaceae and Rosaceae families. Appl Biochem Micro+ 43:629–634

142. O'connell J, Fox P (2001) Significance and applications of phenolic compounds in the production and quality of milk and dairy products: a review. Int Dairy J 11:103–120
143. Ojagh SM, Rezaei M, Razavi SH et al (2010) Effect of chitosan coatings enriched with cinnamon oil on the quality of refrigerated rainbow trout. Food Chem 120:193–198
144. Oksana S, Marian B, Mahendra R et al (2012) Plant phenolic compounds for food, pharmaceutical and cosmetics production. J Med Plants Res 6:2526–2539
145. Oliveira LL, Carvalho MV, Melo L (2014) Health promoting and sensory properties of phenolic compounds in food. Rev Ceres 61:764–779
146. Oniszczuk A, Podgórski R (2015) Influence of different extraction methods on the quantification of selected flavonoids and phenolic acids from *Tilia cordata* inflorescence. Ind Crops Prod 76:509–514
147. Osakabe N, Sanbongi C, Yamagishi M et al (1998) Effects of polyphenol substances derived from Theobroma cacao on gastric mucosal lesion induced by ethanol. Biosci Biotechnol Biochem 62:1535–1538
148. Otero-Pareja MJ, Casas L, Fernández-Ponce MT et al (2015) Green extraction of antioxidants from different varieties of red grape pomace. Molecules 20:9686–9702
149. Ozcan T, Akpinar-Bayizit A, Yilmaz-Ersan L et al (2014) Phenolics in human health. Int J Chem Eng Appl 5:393
150. Pal M, Misra K, Dhillon G et al (2014) Antioxidants. https://doi.org/10.1007/978-1-4614-8005-1_6
151. Park SY, Chin KB (2010) Evaluation of pre-heating and extraction solvents in antioxidant and antimicrobial activities of garlic, and their application in fresh pork patties. Int J Food Sci Technol 45:365–373
152. Pasqualone A, Punzi R, Trani A et al (2017) Enrichment of fresh pasta with antioxidant extracts obtained from artichoke canning by-products by ultrasound-assisted technology and quality characterisation of the end product. Int J Food Sci Technol 52:2078–2087
153. Paszkiewicz M, Budzyńska A, Różalska B et al (2012) Immunomodulacyjna rola polifenoli roślinnych The immunomodulatory role of plant polyphenols. Postepy Hig Med Dosw (Online) 66:637–646
154. Pawlaczyk I, Czerchawski L, Kuliczkowski W et al (2011) Anticoagulant and antiplatelet activity of polyphenolic-polysaccharide preparation isolated from the medicinal plant *Erigeron canadensis* L. Thromb Res 127:328–340
155. Pereira D, Pinheiro RS, Heldt LFS et al (2017) Rosemary as natural antioxidant to prevent oxidation in chicken burgers. Food Sci Technol (Campinas)
156. Pisoschi AM, Pop A, Cimpeanu C et al (2016) Antioxidant capacity determination in plants and plant-derived products: a review. Oxid Med Cell Longev 2016
157. Prakash D, Sharma G (2014) Phytochemicals of nutraceutical importance. CABI
158. Prodanov M, Garrido I, Vacas V et al (2008) Ultrafiltration as alternative purification procedure for the characterization of low and high molecular-mass phenolics from almond skins. Anal Chim Acta 609:241–251
159. Puri M, Sharma D, Barrow CJ (2012) Enzyme-assisted extraction of bioactives from plants. Trends Biotechnol 30:37–44
160. Rahmalia W, Fabre J-F, Mouloungui Z (2015) Effects of cyclohexane/acetone ratio on bixin extraction yield by accelerated solvent extraction method. Procedia Chem 14:455–464
161. Rajbhar K, Dawda H, Mukundan U (2015) Polyphenols: methods of extraction. Sci Revs Chem Commun 5:1–6
162. Ranawana V, Raikos V, Campbell F et al (2016) Breads fortified with freeze-dried vegetables: quality and nutritional attributes. Part 1: breads containing oil as an ingredient. Foods 5:19
163. Rather SA, Masoodi F, Akhter R et al (2016) Advances in use of natural antioxidants as food additives for improving the oxidative stability of meat products. Madridge J Food Technol 1:10–17
164. Ribnicky DM, Roopchand DE, Oren A et al (2014) Effects of a high fat meal matrix and protein complexation on the bioaccessibility of blueberry anthocyanins using the TNO gastrointestinal model (TIM-1). Food Chem 142:349–357

165. Richards MP, Chaiyasit W, McClements DJ et al (2002) Ability of surfactant micelles to alter the partitioning of phenolic antioxidants in oil-in-water emulsions. J Agric Food Chem 50:1254–1259
166. Rodrigo R, Miranda A, Vergara L (2011) Modulation of endogenous antioxidant system by wine polyphenols in human disease. Clin Chim Acta 412:410–424
167. Ruela HS, Sabino K, Leal I et al (2013) Hypoglycemic effect of *Bumelia sartorum* polyphenolic rich extracts. Nat Prod Commun 8:207–210
168. Rusconi M, Conti A (2010) *Theobroma cacao* L., the food of the gods: a scientific approach beyond myths and claims. Pharmacol Res 61:5–13
169. Ruto MC, Ngugi CM, Kareru PG et al (2018) Antioxidant activity and antimicrobial properties of *Entada leptostachya* and *Prosopis juliflora* extracts. J Med Plant Econ Devel 2:1–8
170. Saifuddin N, Saltanat A, Refal H (2014) Enhancing the removal of phenolic compounds from palm oil mill effluent by enzymatic pre-treatment and microwave-assisted extraction. Chem Sci Trans 3:1083–1093
171. Salina HF, Shima AR, Masniza M et al (2013) Enzyme assisted aqueous extraction and phenolic antioxidants of onion oil. Int J Sci Environ Technol 2:949–955
172. Saliou C, Rimbach G, Moini H et al (2001) Solar ultraviolet-induced erythema in human skin and nuclear factor-kappa-B–dependent gene expression in keratinocytes are modulated by a French maritime pine bark extract. Free Radic Biol Med 30:154–160
173. Schillaci C, Nepravishta R, Bellomaria A (2014) Antioxidants in food and pharmaceutical research. AJPhSci 1:9–15
174. Schmidt C, Geweke I, Struck S et al (2018) Blackcurrant pomace from juice processing as partial flour substitute in savoury crackers: dough characteristics and product properties. Int J Food Sci Technol 53:237–245
175. Scognamiglio I, De Stefano D, Campani V et al (2013) Nanocarriers for topical administration of resveratrol: a comparative study. Int J Pharm 440:179–187
176. Selvamuthukumaran M, Shi J (2017) Recent advances in extraction of antioxidants from plant by-products processing industries. Food Qual Saf 1:61–81
177. Shahidi F, Janitha P, Wanasundara P (1992) Phenolic antioxidants. Crit Rev Food Sci Nutr 32:67–103
178. Shamloo M, Babawale EA, Furtado A et al (2017) Effects of genotype and temperature on accumulation of plant secondary metabolites in Canadian and Australian wheat grown under controlled environments. Sci Rep 7:9133
179. Shebis Y, Iluz D, Kinel-Tahan Y et al (2013) Natural antioxidants: function and sources. Food Nutr Sci 4:643
180. Shori AB, Baba AS (2014) Comparative antioxidant activity, proteolysis and in vitro α-amylase and α-glucosidase inhibition of *Allium sativum*-yogurts made from cow and camel milk. J Saudi Chem Soc 18:456–463
181. Simić VM, Rajković KM, Stojičević SS et al (2016) Optimization of microwave-assisted extraction of total polyphenolic compounds from chokeberries by response surface methodology and artificial neural network. Sep Purif Technol 160:89–97
182. Singh R, Kumar R, Venkateshappa R et al (2013) Studies on physicochemical and antioxidant properties of strawberry polyphenol extract-fortified stirred dahi. Int J Dairy Technol 66:103–108
183. Spinelli S (2016) Study of microencapsulated bioactive compounds in food products
184. Stalikas CD (2007) Extraction, separation, and detection methods for phenolic acids and flavonoids. J Sep Sci 30:3268–3295
185. Sulaiman IAZ (2016) Effect of temperature and sonication on the extraction of gallic acid from *Labisia pumila* (Kacip Fatimah), vol 11
186. Tan L, Ji T, Jiang G et al (2014) Simultaneous identification and quantification of five flavonoids in the seeds of *Rheum palmatum* L. by using accelerated solvent extraction and HPLC–PDA–ESI/MSn. Arab J Chem
187. Teresa M-M, Magdalena W, Andrzej K (2011) A systematic review of the effect of vitamin C infusion and vitamin E-coated membrane on hemodialysis-induced oxidative stress. In: Hemodialysis-different aspects. InTech

188. Teruel MR, Garrido MD, Espinosa MC et al (2015) Effect of different format-solvent rosemary extracts (*Rosmarinus officinalis*) on frozen chicken nuggets quality. Food Chem 172:40–46
189. Tomaz I, Maslov L, Stupić D et al (2016) Solid-liquid extraction of phenolics from red grape skins. Acta Chim Slov 63:287–297
190. Tomović V, Jokanović M, Šojić B et al (2017) Plants as natural antioxidants for meat products. In: IOP conference series: earth and environmental science, vol 1. IOP Publishing, p 012030
191. Trindade R, Lima A, Andrade-Wartha E et al (2009) Consumer's evaluation of the effects of gamma irradiation and natural antioxidants on general acceptance of frozen beef burger. Radiat Phys Chem 78:293–300
192. Uma D, Ho C, Aida WW (2010) Optimization of extraction parameters of total phenolic compounds from henna (*Lawsonia inermis*) leaves. Sains Malays 39:119–128
193. Vajić U-J, Grujić-Milanović J, Živković J et al (2015) Optimization of extraction of stinging nettle leaf phenolic compounds using response surface methodology. Ind Crops Prod 74:912–917
194. Vauzour D, Vafeiadou K, Rodriguez-Mateos A et al (2008) The neuroprotective potential of flavonoids: a multiplicity of effects. Genes Nutr 3:115
195. Venkatesh R (2011) Review of the physiological implications of antioxidants in food
196. Venkatesh R, Sood D (2011) Review of the physiological implications of antioxidants in food. Bachelor of Science Interactive Qualifying Project Worcester Polytechnic Institute, Worcester, Massachusetts, US
197. Vergara-Salinas JR, Pérez-Jiménez J, Torres JL et al (2012) Effects of temperature and time on polyphenolic content and antioxidant activity in the pressurized hot water extraction of deodorized thyme (*Thymus vulgaris*). J Agric Food Chem 60:10920–10929
198. Vetal S, Bodhankar SL, Mohan V et al (2013) Anti-inflammatory and anti-arthritic activity of type-A procyanidine polyphenols from bark of *Cinnamomum zeylanicum* in rats. Food Sci Human Wellness 2:59–67
199. Vidhushi C, Gunasekaran S, Sathiavelu M et al (2016) Importance of waste food products and its antimicrobial activity against food spoilage. Res J Pharm Biol Chem Sci 7:1560–1564
200. Vladimir-Knežević S, Blažeković B, Štefan MB et al (2012) Plant polyphenols as antioxidants influencing the human health. In: Phytochemicals as nutraceuticals-global approaches to their role in nutrition and health. InTech
201. Vollmannova A, Margitanova E, Tóth T et al (2013) Cultivar influence on total polyphenol and rutin contents and total antioxidant capacity in buckwheat, Amaranth, and Quinoa Seeds. Czech J Food Sci 31
202. Wallace TC (2010) Analysis of procyanidins and anthocyanins in food products using chromatographic and spectroscopic techniques. The Ohio State University
203. Wallace TC, Giusti MM (2008) Determination of color, pigment, and phenolic stability in yogurt systems colored with nonacylated anthocyanins from *Berberis boliviana* L. as compared to other natural/synthetic colorants. J Food Sci 73
204. Wang L, Bohn T (2012) Health-promoting food ingredients and functional food processing. In: Nutrition, well-being and health. InTech
205. Wang L, Weller CL (2006) Recent advances in extraction of nutraceuticals from plants. Trends Food Sci Technol 17:300–312
206. Wang S, Melnyk JP, Tsao R et al (2011) How natural dietary antioxidants in fruits, vegetables and legumes promote vascular health. Food Res Int 44:14–22
207. Watanabe Y, Nakanishi H, Goto N et al (2010) Antioxidative properties of ascorbic acid and acyl ascorbates in ML/W emulsion. J Am Oil Chem Soc 87:1475–1480
208. Williamson G, Day A, Plumb G et al (2000) Human metabolic pathways of dietary flavonoids and cinnamates. Portland Press Limited
209. Wong Paz JE, Muñiz Márquez DB, Martínez Ávila GCG et al (2015) Ultrasound-assisted extraction of polyphenols from native plants in the Mexican desert. Ultrason Sonochem 22:474–481. https://doi.org/10.1016/j.ultsonch.2014.06.001
210. Yamakoshi J, Sano A, Tokutake S et al (2004) Oral intake of proanthocyanidin-rich extract from grape seeds improves chloasma. Phytother Res 18:895–899

211. Yang Z-F, Bai L-P, W-b Huang et al (2014) Comparison of in vitro antiviral activity of tea polyphenols against influenza A and B viruses and structure–activity relationship analysis. Fitoterapia 93:47–53
212. Yordi EG, Pérez EM, Matos MJ et al (2012) Antioxidant and pro-oxidant effects of polyphenolic compounds and structure-activity relationship evidence. In: Nutrition, well-being and health. InTech
213. Yoshimura M, Amakura Y, Yoshida T (2011) Polyphenolic compounds in clove and pimento and their antioxidative activities. Biosci Biotechnol Biochem 75:2207–2212
214. Yu J, Vasanthan T, Temelli F (2001) Analysis of phenolic acids in barley by high-performance liquid chromatography. J Agric Food Chem 49:4352–4358
215. Yutani R, S-y Morita, Teraoka R et al (2012) Distribution of polyphenols and a surfactant component in skin during aerosol OT microemulsion-enhanced intradermal delivery. Chem Pharm Bull 60:989–994
216. Zillich O, Schweiggert-Weisz U, Eisner P et al (2015) Polyphenols as active ingredients for cosmetic products. Int J Cosmetic Sci 37:455–464
217. Zillich O, Schweiggert-Weisz U, Hasenkopf K et al (2013) Release and in vitro skin permeation of polyphenols from cosmetic emulsions. Int J Cosmetic Sci 35:491–501

Chapter 9
Innovative Techniques and Alternative Solvents for Green Extraction of Proteins from Pulses and Oleaginous Meals as Industrial Sources for Food and Feed

Anne-Gaëlle Sicaire, Frédéric Fine, Alain Quinsac, Meriem Boukroufa, Njara Rakotomanomana and Farid Chemat

Abstract Considering the global population growth driven by emerging countries and the increase in nutritional needs of populations, the valorization of protein fractions of plant resources, particularly oilseed crops, appears to be a major issue to answer the demand. Oilseeds are indeed rich in both oil and protein. Historically, oil being the main compound valued, the current crushing process is optimized to maximize its extraction and the proteins remain in the cake (valued in animal feed). The conditions applied during the process and, in particular, during the step of desolventization of the cake after the extraction of the oil with hexane, cause degradation of the proteins. The valorization of these proteins uses traditional extraction methods as well as innovative techniques in order to preserve the nutritional and functional quality of proteins. This chapter gives a picture of current knowledge on extraction techniques, procedures, and solvents for proteins from pulses and oleaginous meals. The modern innovative and intensified extraction techniques, alternative solvents, and original procedures (ultrasound, microwaves, bio-based solvents, mechanical extraction, enzyme-assisted extraction, natural deep eutectic solvents, pulsed electric fields) are summarized in terms of their principles, processes, applications, benefits, and disadvantages.

Keywords Protein · Intensification · Extraction · Innovative techniques · Original procedures · Alternative solvents

---

A.-G. Sicaire (✉) · F. Fine · A. Quinsac
Terres Inovia, 11 rue Monge, 33600 Pessac, France
e-mail: ag.sicaire@terresinovia.fr

M. Boukroufa · N. Rakotomanomana · F. Chemat
Université d'Avignon et des Pays de Vaucluse, INRA, UMR408, GREEN Extraction Team, 84000 Avignon, France

Y. Li and F. Chemat (eds.), *Plant Based "Green Chemistry 2.0"*,
Green Chemistry and Sustainable Technology,
https://doi.org/10.1007/978-981-13-3810-6_9

9.1 Introduction

Proteins are one of three macronutrients families with carbohydrates and lipids. They consist in a sequence of 20 amino acids, 8 of which are considered essential for human nutrition because they are not biosynthesized and must be provided by food. The amino acids (AA) chain, of variable length, defines the primary structure of the protein. It can fold, take a specific form, to associate with other proteins and thereby determine the structures in space, very complex and vary in size. These structures are called secondary, tertiary, and quaternary, and they define the biological properties of proteins [1]. Protein has different roles in biology. Indeed, two types of proteins can be distinguished: structural proteins (collagen, keratin) that enter into the constitution of tissues (bones, hair, skin) and proteins with biological activity such as enzymes, hormones, muscles, transport proteins between cells and through cell membranes (membrane receptor, hemoglobin), proteins involved in host defense (immune globulin, fibrinogen), and reserve protein. Proteins are essential compounds for human nutrition. The daily requirements of protein established by ANSES are of 0.83 g kg^{-1} d^{-1} which represent an average of 10–15% of total daily energy intake [2]. This protein intake has to be balanced between animal and plant sources on a 1:1 ratio. However, the elderly often eat less than younger, because of physiological changes, such as difficulty chewing or swallowing, leading to avoid chewy meat; a reduced sense of smell and taste with age often causes deleterious food monotony. In fact, approximately 40% of subjects aged over 70 have a protein intake located below this threshold, and 16% use less than 75% of the recommended intake [3]. Furthermore, as a result of aging, the body undergoes inevitable metabolic changes: it secretes less hydrochloric acid needed to digest food and intestinal absorption decreases, resulting in a lower bioavailability of the macronutrients (vitamins, trace elements) [4]. This decline in consumption may reduce the fibrillar protein synthesis or cause a loss of lean body mass [5]. It is therefore important to rebalance the contributions between animal and plant proteins—or other alternatives—in sustainable food systems allowing access to quality protein intake to the entire world population. In order to balance protein diet, several studies aim to find an alternative to animal proteins; first, because the production of animal protein is more expensive regarding water and energy resources than plant-based protein and then, in order to help the elderly to overcome the lack of protein intake. Therefore, pulse proteins or proteins from oleaginous seeds can be good alternatives in sustainable food systems allowing access to quality protein intake due to their balanced amino acid profile and potential functional propriety such as foaming, emulsifying, or gelling abilities.

In order to isolate intracellular proteins, cells must be disrupted and an efficient protocol for cell disruption must be developed to release the protein from its intracellular compartment in a soluble form. The disruption protocol should be as gentle as possible to the protein, as the extraction step is the starting point for all subsequent procedures. This non-exhaustive short review presents perspectives regarding the extraction of proteins from oleaginous seeds and meals. It gives new directions for research and industry by presenting examples of conventional extraction techniques and intensified green extraction processes (ultrasound, microwaves, bio-based solvents, mechanical extraction, enzyme-assisted extraction, pulse electric fields).

9.2 Conventional Extraction

9.2.1 Principle, Process, and Procedure

Protein extract from proteaginous seeds usually contains up to 200 non-protein phytochemicals, including residual fatty acids, phenolic compounds, phytic acid, glucosinolates, isoflavones, and minerals [6]. Although some of these components can offer health benefits, others are not desirable. For example, phytate-protein complexes formed from phytic acid are products which limit the digestibility.

Two different ways are used to extract proteins (protein isolate and protein concentrate). Alkaline extraction with sodium hydroxide is the most widely used method to recover protein isolates from various matrices. This procedure is generally followed by a precipitation step by the addition of a diluted acid. The solid–liquid extraction step is performed by maceration with continuous stirring. The soluble fraction of proteins is obtained by modulating several operating parameters such as pH, temperature, and solid-to-liquid ratio. The mixture is then centrifuged in order to separate the meal residue from the extracted protein and the supernatant. Afterward, pH is adjusted to isoelectric point in order to induce the precipitation of the soluble protein fraction. This step is then followed by ultrafiltration and diafiltration to recover protein isolate.

Proteins concentrates are obtained by a selective extraction of the unwanted water-soluble compounds without solubilizing proteins. The product obtained is thus enriched in proteins (60–90%). Alcohol leaching is commonly used in protein concentrate preparation, because it produces light-colored and bland-flavored products. It is also able to eliminate part of the anti-nutritional components such as oligosaccharides, as well as some soy allergens, including Kunitz trypsin inhibitor [7].

9.2.2 Applications

Protein extraction from pulses requires a mechanical treatment (drooling, dehulling, grinding) prior to solid–liquid extraction. However, the purity of the obtained protein fractions is relatively low (38–65%) [8]. Therefore, it is necessary to use purification methods such as isoelectric precipitation or ultrafiltration and diafiltration to obtain a purity higher than 70% [9]. Boye et al. [10] compared these two processes for the purification of protein extracts from three kinds of pulses (Table 9.1) as well as their effects on physicochemical properties of the obtained proteins. They observed

Table 9.1 Conventional extraction of proteins from pulses

Crop	Solvent	Experimental conditions					References
		Temp. (°C)	Time (min)	Sol/liq ratio	pH	Yield (%)	
Yellow Pea	Water, NaOH	35	60	1:15	9.5	55	[10]
Kabuli Chickpea	Water, NaOH	35	60	1:15	9.5	69	[10]
Lentil	Water, NaOH	25	60	1:10	9	62	[10]
Rapeseed meal	Water, Ethanol, Na_2SO_4 (0.4%), NaCl (0.1 M)	25	60	1:5	/	30	[11]

that all concentrates exhibited good functional properties. However, these functional properties varied according to the type of pulse and manufacturing process. Indeed, for a pH ranging from 1 to 3 and from 7 to 10, red and green lentil concentrates were the most soluble (70–77%); their ultrafiltration concentrates were more soluble at all pH values studied compared to those obtained by isoelectric precipitation. Furthermore, water-holding capacity was the highest for isoelectric precipitation of yellow pea and was the lowest for the ultrafiltration of chickpeas. Moreover, emulsifying properties and foam expansion were generally higher for the chickpea concentrates, but they had less foam stability. Finally, protein extracts from green lentils appeared to have the best gelling properties.

9.3 Enzyme-Assisted Extraction

9.3.1 Principle, Process, and Procedure

Enzyme-assisted extraction of proteins is an alternative approach to the conventional method. It was developed to intensify the extraction of proteins from various plant materials by increasing extraction yields as well as allowing the co-extraction of oil and proteins from many oilseeds and oil-fruits [12–14]. In fact, a strong improvement of conventional method can be obtained as enzymes cause the degradation of the cell-wall structure; these hydrolyzing enzymes can cleave structural molecules such as polysaccharides (pectins for example) and effectively enhance the liberation of more intracellular proteins. Enzymes can also directly hydrolyze proteins to peptides, increasing their solubility and thus their extractability [15]. But it depends on the plant material; if the extract is phytic-acid-free, it may be directly used as food supply (a high-value product, against disorders of digestion). Otherwise, purification would be more tedious (moreover, some di- or tri-peptides give "umami" or bitter flavors to the extract) than conventional extraction where proteins can be easily separated by crystallization at a pH close to pHi of the protein, or by size exclusion chromatography (SEC). Aqueous enzymatic protein extraction is recognized as an environmentally friendly and safe process, by complete avoidance of organic solvents. It also limits damages to proteins, improving their nutritional and functional properties. Nevertheless, the application of this method is limited by time-consuming and high-cost processes because of the use of enzymes and their reprocessing. The manufacture of high-value products such as bioactive peptides (opioids, antithrombotics, anti-ACE (converting enzyme angiotensin), cholesterol-lowering agents, antimicrobials) released during the enzymatic hydrolysis is constituted of very different mixtures and can positively take part in many human physiological processes [16].

9.3.2 Applications

The aim of using enzymes in the extraction of proteins from oilseeds is to help to disrupt the components of cell walls in order to facilitate mass transfers and increase protein yield. Zhang et al. [17] found that the use of aqueous enzymatic extraction of rapeseed protein hydrolysates was enhanced from 47.04 to 73.51%. Jung [18] compared the effects of extrusion pretreatment and protease addition during aqueous extraction processing of soybean and lupine flakes. They found that extrusion improved the enzymatic action, which decreased extraction yield but increased protein extractability from soybean and lupine by 47 and 26%, respectively (Table 9.2).

Table 9.2 Enzyme-assisted extraction of proteins from pulses

Crop	Enzyme	Experimental conditions						References
		Temp. (°C)	Time (min)	Sol/liq ratio	pH	Enzyme	Yield (%)	
Rapeseed	Alcalase 2.4L	60	180	1:5	10	Alcalase 2.4L 1.5%	83.3	[17]
Lentil	Glucoamylase	50	300	1:10	7	Glucoamylase	50.3	[19]
Soybean	Protex 6L	50	60	1:10	9	Protex 6L 0.5%	84.6	[18]
Lupine	Protex 6L	50	60	1:10	9	Protex 6L 0.5%	77.2	[18]

9.4 Ultrasound-Assisted Extraction

9.4.1 Principle, Process, and Procedure

Ultrasound is mechanical waves that can spread over an elastic medium. They are characterized by their frequency range comprised between 20 and 10 MHz, and they can be separated into two categories: (i) diagnostic ultrasound (ranging from 2 to 10 MHz) mostly used in medical field for imaging, (ii) power ultrasound (comprised between 20 and 100 kHz) principally used in sonochemistry and in various industrial fields. The major effect of ultrasound in a liquid medium is attributed to cavitation phenomena, by the formation of microbubbles, thanks to compression and rarefaction cycles generated by the acoustic wave. These bubbles expanse at every cycle, and they collapse when a critical size is attained. Temperature and pressure at the moment of collapse could reach around 5000 K and 100 MPa [20]. If bubbles collapse at the surface of a solid material, microjets and shock waves are generated directly toward the solid surface [21]. If the solid is a plant to extract, the microjet generated by the collapsing of the bubble can destroy cell walls of the matrix and its content will be released into the medium (or solvent) as can be seen in Fig. 9.1. Ultrasound can either be emitted with bath and horn systems (or probe). Probes are often attached to a horn tip known as a sonotrode. These ultrasound systems are developed for both laboratory and industrial scales. With a bath, ultrasound is transmitted in the solvent that plays the role of medium of transmission and the irradiation is made in an indirect way. With a probe, acoustic power generated by the system is much higher than with the bath as it allows a direct irradiation of the medium with less dispersion of ultrasonic energy. Ultrasound can also be operated continuously in systems allowing continuous alimentation in fresh solvent through the reactor submitted to ultrasound. Ranges of devices, Fig. 9.1, have been developed by the company REUS (France) from laboratory (0.5–3 L) to industrial scale (1000 L) or by the company Hielscher (Germany).

9.4.2 Applications

Unlike conventional extraction techniques, ultrasound-assisted extraction causes a particular physical phenomenon, cavitation, which can enhance mass transfer by degrading cell walls of the matrix; its content will then be easily released into the medium. Therefore, the extraction may be performed at low temperature and consequently, thermal damage may be reduced. Moulton and Wang [22] reported that applying continuous high-intensity ultrasound in combination with aqueous or alkaline extraction of protein from soybean meal resulted in the extraction of, respectively, 54 and 23% more protein compared to conventional extraction. Dong et al. [23] also reported an increase of protein extract from rapeseed of 35.43% by using an ultrasonic bath with a temperature of 35 °C. Zhu et al. [24] studied the quantitative effect

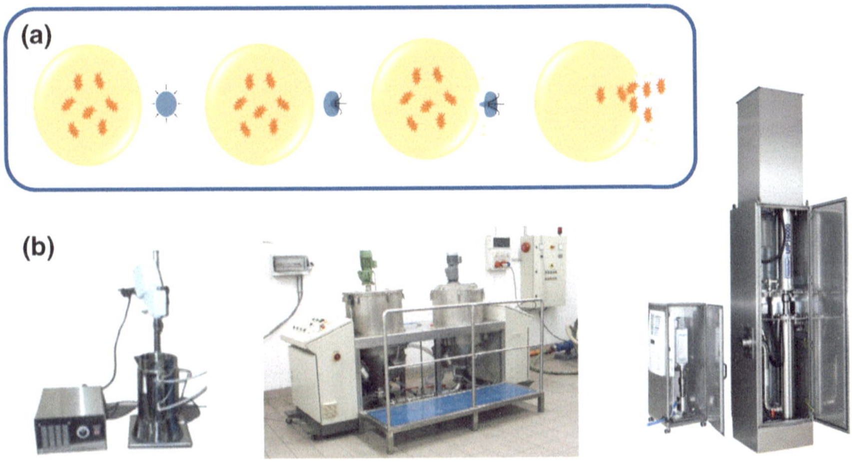

Fig. 9.1 Ultrasound-assisted extraction. **a** Extraction mechanism: collapse of cavitation bubble at the surface of the cell, **b** Ultrasound apparatus REUS and Hielscher (from laboratory to industrial scale)

of ultrasonic power, liquid-to-solid ratio, and extraction time for the extraction of protein from perilla seed meal and found that the optimal extraction parameters to maximize extraction yield (10.77%) were an extraction time of 12 min, a power of 61 W and an extraction temperature of 40 °C. Examples of ultrasound-assisted extraction of proteins from pulses are presented in Table 9.3.

9.5 Subcritical Water Extraction

9.5.1 Principle, Process, and Procedure

Water is often recognized as a green solvent; in addition to the fact that it is an inexpensive and environmental friendly solvent, water has the advantage of being also non-flammable, nontoxic, providing opportunities for clean processing and pollution prevention [25]. Water possesses many thermodynamic properties greatly influenced by temperature and pressure. Depending on conditions, its physical state (solid, liquid, or gas), thermal behavior, density, or viscosity can be modified. By increasing temperature and pressure over a point known as critical point (defined at 221 bar and 374 °C), water can attain the supercritical state. At subcritical conditions, corresponding to temperatures between 100 °C and critical temperature (374 °C) and pressure between 1 bar and critical pressure (221 bar) in order to avoid vaporization, water polarity decreases; it makes water a better solvent for the extraction of various organic bioactive compounds. Indeed, several examples of compounds (essential oils,

Table 9.3 Ultrasound-assisted extraction of proteins from pulses

Crop	Solvent	Experimental conditions							References
		Temp. (°C)	pH	Power (W)	Time (min)	Sol/liq ratio	Ultrasonic device	Yield (%)	
Soybean	Alkaline buffer solution	35	8	451	84	1:24	Ultrasonic bath	88.2	[23]
Perilla seed meal	Water, NaOH	40	/	61	12	1:40	Ultrasonic horn	10.77	[24]
Rapeseed meal	Phosphate buffer	35	11.5	450	84	1:24	Ultrasonic bath	71.33	[23]
Soybean	Water, NaOH	60	8.5	61	2	1:10	Ultrasonic horn	71.30	[22]

phenolic compounds, antioxidants…) extracted by subcritical water can be found in the literature [26–29]. The implementation of the process can be done either (i) in static mode: in closed extraction vessel (batch), with a contact between matrix and solvent during a limited time, (ii) in dynamic mode: in semi-continuous extraction devices where the water percolates through the substrate [30].

9.5.2 *Applications*

Subcritical water extraction is a new approach for the extraction of bioactive compounds from plant material offering a better purity of target compounds as well as reduced extraction times [31]. Subcritical water extraction demonstrates the ability to extract different types of compounds, depending on the temperature used. The most polar compounds are extracted at lower temperature and less polar compound are extracted at higher temperatures [32]. Watchararuji et al. [33] studied the application of subcritical water to the production of value-added products from raw and de-oiled soybean meal by subcritical water hydrolysis (Table 9.4). The optimal conditions were an extraction time of 30 min and a hydrolysis temperature of 210 and 200 °C for raw and de-oiled soybean meal, respectively. Wiboonsirikul et al. [34] obtained a better extraction yield of protein from soybean pulp by using an extraction temperature of 240 °C and an extraction time of 5 min. Both studies [33, 34] showed that the protein content in the extract increased with the increasing temperature treatment and decreased if the experiment takes long time. However, the carbohydrate content decreased at higher temperature (from 170 to 260 °C) treatment. Table 9.4 gives some examples of subcritical water extraction of proteins from pulses.

Table 9.4 Subcritical water extraction from pulses

Crop	Solvent	Experimental conditions					References
		Pressure	Temp. (°C)	Time (min)	Sol/liq ratio	Yield	
Soybean cake	Water	30 bar	200	30	1:5	46.98%	[33]
Soybean	Water	50 bar	250	14	1:10	Amino acid yield 63%	[35]
Rapeseed cake	Water	30 bar	215	36	1:10	Amino acid yield 31.72%	[36]
Soybean	Water	/	210	30	1:5	48.92%	[33]

9.6 Microwave-Assisted Extraction

9.6.1 Principle, Process, and Procedure

Microwaves are electromagnetic waves resulting from the application of an electric field combined with a magnetic field that is self-propagating in space. The mechanism of microwave heating is based on dielectric heating of polar molecules, such as water, by the phenomenon of polar rotation occurring thanks to their dipole moment. The first microwave-assisted extraction has been reported by Ganzler in 1986 [37]. He studied the isolation of target compounds such as fats or pesticides from different samples by comparing Soxhlet extraction to extractions performed in a domestic microwave oven and he demonstrated equivalent extraction performances comparing to the conventional method. These results rapidly generated other applications and microwave extraction technique has been approved for its effectiveness in the extraction of several compounds such as oils, fats, organic compounds, essential oils, and antioxidants in terms of extraction yield [38, 39]. Furthermore, one of the major improvements generated by microwave heating is the reduction of extraction time resulting in a decrease of energy consumed during the extraction process. In conventional solvent extraction, mass transfer takes place from the interior of the matrix to the outside while the temperature gradient occurs from the outside to the inside (Fig. 9.2a). For solvent extraction assisted by microwave, these two processes are exercised in the same direction, from the heart of the matrix to the solvent. Process intensification using microwave-heating may be due to a synergistic combination of two transport phenomena: mass and heat gradients working in the same direction (Fig. 9.2a). Devices from laboratory to pilot scale have been developed by the company Milestone (Italy) as can be seen in Fig. 9.2b.

9.6.2 Applications

Microwave-assisted extraction has been proposed as an alternative technique to conventional methods for the extraction of proteins from pulses thanks to various advantages such as: reduction of extraction time, improvement of efficiency and selectivity with less degradation of the extracts. The use of microwaves has already been successfully applied to the extraction of essential oils [40, 41], polysaccharides [42, 43], phenolic compounds [44], or proteins [45]. Indeed, Choi et al. [46] compared microwave-assisted process to conventional extraction of proteins from soybean meals; they observed that microwave treatment resulted in significantly higher concentrations of soluble protein than conventional treatment under optimum conditions. In these processes, microwave affect cell walls of the matrix which help the diffusivity of soluble protein outside the cells. Some other studies using microwaves on protein extraction are presented in Table 9.5.

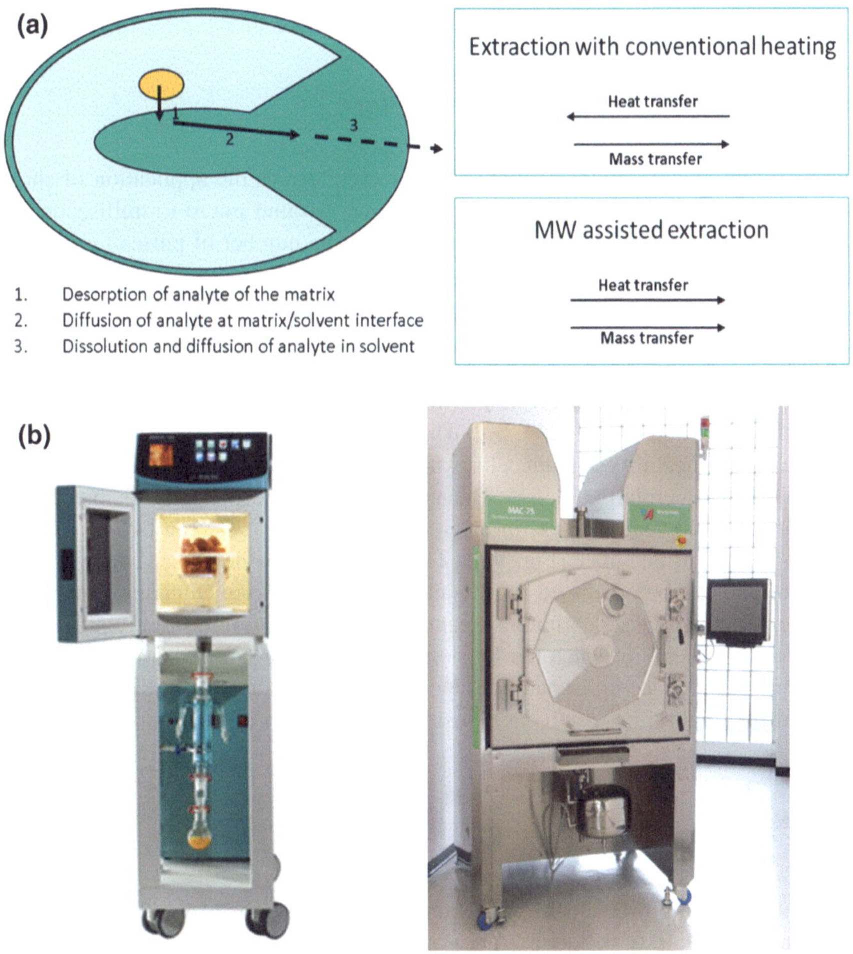

Fig. 9.2 Microwave-assisted extraction. **a** Extraction mechanism: reversed mass and heat transfers compared to conventional extraction, **b** microwave reactors

Table 9.5 Microwave-assisted extraction of proteins from pulses

Crop	Solvent	Experimental conditions						References
		Temp.	pH	Time (min)	Power	Sol/liq ratio	Yield (%)	
Soybean	Water NaOH	60.1 °C	/	30	/	1:12.6	24.27	[46]
Peanut protein	Water NaOH	/	9.73	8	72 W	1:28	88.73	[47]
Rapeseed meal	Water NaOH	35 °C	11.5	24	73 W	1:28	38.76	[48]

9.7 Extraction with Pulsed Electric Fields

9.7.1 Principle, Process, and Procedure

The principle of pulsed electric field (PEF) technology is the application of short pulses of high electric fields during a short time (around micro-to-milliseconds). The processing time is calculated by multiplying the number of pulses times with effective pulse duration. The duration and number of pulses that are generally applied are used to limit the temperature rise during treatment (generally less than 5 °C) [30]. The process is based on the potential formation of pores inside the cell membranes, due to their exposure to low-moderate external electric fields of adequate strength (<10 kV/cm) during a very short time. This phenomenon is called "electroporation" or "electropermeabilization" [49, 50]. The electroporation mechanism is likely to increase the permeability of plant cells, which can enhance the diffusion of the solute through the cell membrane [3, 51]. The application of pulsed electric field consists in treating a biological product, in batch or continuously, between two electrodes inside a treatment chamber (Fig. 9.3). This chamber can be static, where only a given volume can be processed at the same time (for experimental application) or dynamic which allows a continuous treatment, according to industrial requirement. The electric pulses are generated by condensers discharge and pass through the sample inside the processing chamber. One of the major advantages of PEF extraction is a short treatment time which has a positive impact on energy consumption.

9.7.2 Applications

The application of pulsed electric field on plant material leads to temporary or complete electroporation of cell membranes increasing the membrane permeability and

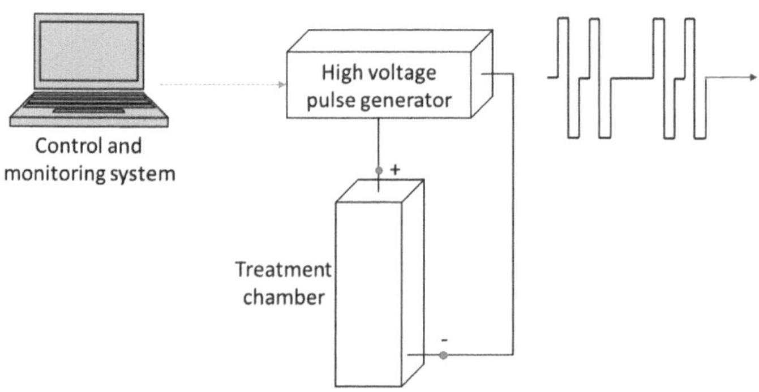

Fig. 9.3 Schematic representation of simplified PEF installation

Table 9.6 Pulsed electric field extraction of proteins from pulses

Crop	Solvent	Experimental conditions					References
		E	Temp. (°C)	Ratio	Time	Yield (%)	
Rapeseed	Water, NaOH	8 kV/cm	20	1:2	2 ms	64.1	[57]
Rapeseed	Water, NaOH	20 kV/cm	20	1:4	20 ms	60	[55]
Rapeseed	Water, NaOH	240 kJ/kg	20	1:20	120 min	23.52	[56]
Rapeseed cake	Water, NaOH	240 kJ/kg	20	1:20	120 min	39.4	[56]

facilitating the passage of intracellular compounds to the surrounding solution [52]. A number of biological tissues and biosuspensions have already been tested for extractive recovery of various compounds of interest such as proteins [53] or phenolic compounds [54]. Yu et al. [55] applied this process to extract proteins from rapeseed, and they compared it to conventional extraction. The conclusion was that optimal conditions ($E = 8$ kV/cm, $t\ PEF = 2$ ms, $P = 10$ bar) offer significant increases in total protein content (0.14 vs. 0.07 g BSA/100 g DM). Barba et al. [56] studied the applicability of high-voltage electrical discharge technology to recover proteins from rapeseeds. An optimal treatment energy input of 240 kJ/kg was observed; beyond this value, the recovery of proteins decreases (Table 9.6).

9.8 Extraction with Deep Eutectic Solvents

9.8.1 Principle, Process, and Procedure

Deep eutectic solvents (DES) are known as a new class of ionic liquids. They are usually obtained by the complexation of a quaternary ammonium salt with a metal salt or hydrogen bond donor [58]. DES are systems characterized by their ability to form a eutectic mixture, meaning that their melting point is much lower than those of the individual components (Fig. 9.4a), with a very large depression of the melting point in the region of 200 °C (Fig. 9.4b). DES can be described by the general formula:

$$\mathbf{Cat^{+}X^{-}zY}$$

where Cat^{+} is the ammonium (or phosphonium, or sulfonium) cation, and X^{-} is generally a halide anion. The complex anionic species are formed between X^{-} and

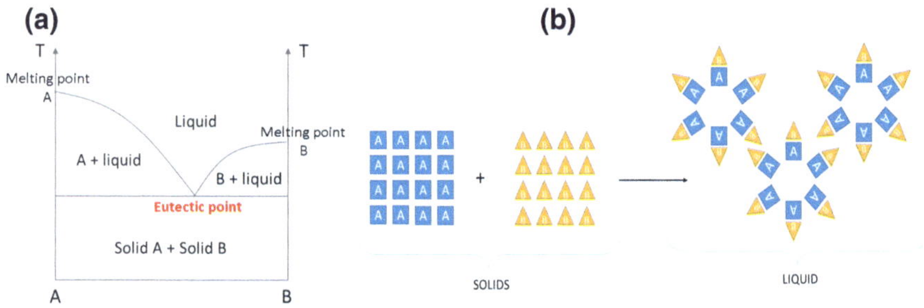

Fig. 9.4 Phase diagram (**a**) and representation (**b**) of a eutectic point for a two-component mixture

either a Lewis or Brönsted acid Y (where z is the number of Y molecules that interact with the anion). One of the most used ammonium salt is choline chloride.

DES have already been applied in many fields of chemistry such as electrochemistry or organic synthesis [59–61] and some recent studies show their potential in the extraction of bioactive compounds like flavonoids [62, 63], caffeic and chlorogenic acids [64], or terpenoïds [65].

9.8.2 *Applications*

Deep eutectic solvents (DESs) have emerged as a new type of promising ionic solvents with a broad range of potential applications. DESs are generally regarded as "green" because they are composed of ammonium salts and H-bond donors (HBDs) which are considered to be eco-friendly. DESs have already been used for the extraction of proteins with an aqueous two-phase system. Xu et al. [66] investigated the use of four types of choline-chloride-(ChCl)-based DESs to extract bovine serum albumin (BSA); in this study, ChCl-glycerol mixture was selected as the most suitable solvent for the extraction of BSA. In fact, experimental results showed that 98.16% of the BSA could be recovered by a single-step extraction under optimized conditions corresponding to pH = 4–9, Temperature = 30 °C, Time = 16 min and Sol/liq ratio = 1:2. On the other hand, under the same conditions, the recovery of another protein, trypsin, attained 94.36%. However, Li et al. [67] observed a better yield recovery (99.82%) of BSA in only 12 min by using Betaine–urea as extraction solvent under the same extraction conditions. For now, this kind of extraction is only applied to a BSA recovery. However, it can also be extrapolated to the extraction of proteins from pulses. Table 9.7 presents some examples of the extraction of proteins using deep eutectic solvents.

Table 9.7 Extraction of proteins using deep eutectic solvents

Protein	Solvent	Experimental conditions					References
		pH	Temp. (°C)	Time (min)	Sol/liq ratio	Yield (%)	
BSA	ChCl-glycerol	4–9	30	16	1:2	98.16	[66]
Trypsin	ChCl-glycerol	4–9	30	16	1:2	94.36	[66]
BSA	Betaine–urea	4–9.9	30	12	1:2	99.82	[67]
BSA	ChCl–urea	/	50	8	1:2	99.94	[68]

9.9 Future Prospects

Advantages and drawbacks of using innovative techniques and alternative solvents for green extraction of proteins have been compared: conventional maceration, subcritical water, enzymatic extraction, DES-assisted extraction, microwave extraction, PEF-assisted extraction, and ultrasound-assisted extraction (Table 9.8). The use of alternative extraction techniques such as ultrasound, microwave, pulsed electric fields, bio-based solvent, and enzyme-assisted extraction allows total recovery of proteins from pulses and oleaginous meals and also reduces extraction time, energy, and solvent consumptions. Conventional techniques are limited by the diffusion of the solvent (mostly water) into vegetal tissues, because of the structure of cell walls; ultrasound or pulsed electric fields can intensify disruption of plant cells in order to accelerate mass transfer, thus recovery of proteins in a short time. Extraction and valorization of proteins is an increasingly important research and industrial sector that directly transfers knowledge into technology for commercial development. Future challenges will be to develop systems that will allow the extraction of proteins with advantages in terms of yield, selectivity, operating time, energy input, and preservation of native proteins. These challenges will be possible by investigating innovative processes or solvents which physical and chemical phenomena are fundamentally different from those applied in conventional extraction techniques. These new systems could be hybrid techniques such as a combination of conventional and innovative techniques or a combination of innovative techniques and alternative solvents.

Table 9.8 Characteristics, main disadvantages and advantages of various processes for protein extraction from pulses and oleaginous meals

Name of extraction process	Conventional	Enzyme assisted	Ultrasound assisted	Subcritical water	Microwave assisted	Pulse electric fields
Pretreatment	Mechanical	– Extrusion – Enzyme treatment	Ultrasound			
Description system	– Alkaline extraction with NaOH – Precipitation by acidification–Centrifugation	Enzyme treatment	Ultrasound system 20–100 kHz	Water at 30–50 bars	Microwave 2.45 GHz	Pulsed electric fields
Usual pH	9–11	7–11	8–11	Slightly acidic (5.6–6.1)	9.73	Alkaline
Temperature (°C)	25–35	50–60	35–60	100–200		20
Extraction time (min)	60	60–300	2–84	5–30	8–30	2–120
Extraction yield for instance rapeseed meal	20.3% [10]	16% [16]	47% [22]	13.6% [35]	13.2% [47]	15% [55]
Investment	Low	High	Low	High	Medium	High
Ease of operation	High	Medium	Medium	Medium	Medium	Difficult ease of operation
Scale	>1000 L	>1000 L	600 L	Static to dynamic mode	150 L	Continuous
Main disadvantages		Cost of Enzymes		Hydrolysis or Maillard reaction	Hot spots	Difficult ease of operation
Main advantages	Large scale	Precise conditions	High cell disruption	Enhance mass transfer	Cell disruption	Electroporation of wall cells

References

1. Gargaud M, Amils R, Quintanilla JC et al (2011) Encyclopedia of astrobiology. Springer-Verlag, Berlin
2. AFSSA (2007) Apport en protéines : consommation, qualité, besoins et recommandations
3. Lebovka N, Vorobiev E, Chemat F (2011) Enhancing extraction processes in the food industry. CRC Press
4. Whitbourne SK (1985) The aging body: physiological changes and psychological consequences. Springer-Verlag, New York
5. Backx EMP, Tieland M, Borgonjen-van den Berg KJ et al (2016) Protein intake and lean body mass preservation during energy intake restriction in overweight older adults. Int J Obes (Lond) 40:299–304. https://doi.org/10.1038/ijo.2015.182

6. Fang N, Yu S, Badger TM (2004) Comprehensive phytochemical profile of soy protein isolate. J Agric Food Chem 52:4012–4020. https://doi.org/10.1021/jf049842y
7. Laroque D, Inisan C, Berger C et al (2008) Kinetic study on the Maillard reaction. Consideration of sugar reactivity. Food Chem 111:1032–1042. https://doi.org/10.1016/j.foodchem.2008.05.033
8. Boye J, Zare F, Pletch A (2010) Pulse proteins: processing, characterization, functional properties and applications in food and feed. Food Res Int 43:414–431. https://doi.org/10.1016/j.foodres.2009.09.003
9. Han X-Z, Hamaker BR (2002) Partial leaching of granule-associated proteins from rice starch during alkaline extraction and subsequent gelatinization. Starch—Stärke 54:454–460. https://doi.org/10.1002/1521-379X(200210)54:10%3c454:AID-STAR454%3e3.0.CO;2-M
10. Boye JI, Aksay S, Roufik S et al (2010) Comparison of the functional properties of pea, chickpea and lentil protein concentrates processed using ultrafiltration and isoelectric precipitation techniques. Food Res Int 43:537–546
11. Chabanon G, Chevalot I, Framboisier X et al (2007) Hydrolysis of rapeseed protein isolates: kinetics, characterization and functional properties of hydrolysates. Process Biochem 42:1419–1428. https://doi.org/10.1016/j.procbio.2007.07.009
12. Domínguez H, Núñez MJ, Lema JM (1994) Enzymatic pretreatment to enhance oil extraction from fruits and oilseeds: a review. Food Chem 49:271–286. https://doi.org/10.1016/0308-8146(94)90172-4
13. Rosenthal A, Pyle DL, Niranjan K (1996) Aqueous and enzymatic processes for edible oil extraction. Enzyme Microb Technol 19:402–420. https://doi.org/10.1016/S0141-0229(96)80004-F
14. Zhang SB, Wang Z, Xu SY (2007) Downstream processes for aqueous enzymatic extraction of rapeseed oil and protein hydrolysates. J Amer Oil Chem Soc 84:693–700. https://doi.org/10.1007/s11746-007-1080-2
15. Martínez-Maqueda D, Hernández-Ledesma B, Amigo L et al (2013) Extraction/fractionation techniques for proteins and peptides and protein digestion. In: Toldrá F, Nollet LML (eds) Proteomics in foods. Springer, US, pp 21–50
16. Meisel H (1997) Biochemical properties of regulatory peptides derived from milk proteins. Biopolymers 43:119–128. https://doi.org/10.1002/(SICI)1097-0282(1997)43:2%3c119:AID-BIP4%3e3.0.CO;2-Y
17. Zhang SB, Wang Z, Xu SY (2007) Optimization of the aqueous enzymatic extraction of rapeseed oil and protein hydrolysates. J Am Oil Chem Soc 84:97–105
18. Jung S (2009) Aqueous extraction of oil and protein from soybean and lupin: a comparative study. J Food Process Preserv 33:547–559. https://doi.org/10.1111/j.1745-4549.2009.00400.x
19. Bildstein M, Lohmann M, Hennigs C et al (2008) An enzyme-based extraction process for the purification and enrichment of vegetable proteins to be applied in bakery products. Eur Food Res Technol 228:177
20. Mason TJ, Lorimer JP (2002) Applied sonochemistry: the uses of power ultrasound in chemistry and processing. Wiley VCH, Weinheim
21. Vinatoru M (2001) An overview of the ultrasonically assisted extraction of bioactive principles from herbs. Ultrason Sonochem 8:303–313
22. Moulton KJ, Wang LC (1982) A pilot-plant study of continuous ultrasonic extraction of soybean protein. J Food Sci 47:1127–1129. https://doi.org/10.1111/j.1365-2621.1982.tb07632.x
23. Dong X-Y, Guo L-L, Wei F et al (2011) Some characteristics and functional properties of rapeseed protein prepared by ultrasonication, ultrafiltration and isoelectric precipitation. J Sci Food Agric 91:1488–1498. https://doi.org/10.1002/jsfa.4339
24. Zhu J, Fu Q (2012) Optimization of ultrasound-assisted extraction process of perilla seed meal proteins. Food Sci Biotechnol 21:1701–1706. https://doi.org/10.1007/s10068-012-0226-7
25. Petigny L, Özel MZ, Périno S et al (2015) Water as green solvent for extraction of natural products. In: Chemat F, Strube J (eds) Green extraction of natural products. Wiley-VCH Verlag GmbH & Co, KGaA, pp 237–264

26. Ozel MZ, Gogus F, Lewis AC (2003) Subcritical water extraction of essential oils from *Thymbra spicata*. Food Chem 82:381–386. https://doi.org/10.1016/S0308-8146(02)00558-7
27. Ibañez E, Kubátová A, Señoráns FJ et al (2003) Subcritical water extraction of antioxidant compounds from *rosemary* plants. J Agric Food Chem 51:375–382. https://doi.org/10.1021/jf025878j
28. Ju Z, Howard LR (2005) Subcritical water and sulfured water extraction of anthocyanins and other phenolics from dried red grape skin. J Food Sci 70:S270–S276
29. Kumar MSY, Dutta R, Prasad D, Misra K (2011) Subcritical water extraction of antioxidant compounds from *Seabuckthorn* (*Hippophae rhamnoides*) leaves for the comparative evaluation of antioxidant activity. Food Chem 127:1309–1316. https://doi.org/10.1016/j.foodchem.2011.01.088
30. Chemat F (2014) Eco-extraction du végétal - Procédés innovants et solvants alternatifs. Dunod, Paris
31. Asl AH, Khajenoori M (2013) Subcritical water extraction. Mass transfer—advances in sustainable energy and environment oriented numerical modeling. https://doi.org/10.5772/54993
32. Smith RM (2006) Superheated water: the ultimate green solvent for separation science. Anal Bioanal Chem 385:419–421. https://doi.org/10.1007/s00216-006-0437-y
33. Watchararuji K, Goto M, Sasaki M, Shotipruk A (2008) Value-added subcritical water hydrolysate from rice bran and soybean meal. Bioresour Technol 99:6207–6213. https://doi.org/10.1016/j.biortech.2007.12.021
34. Wiboonsirikul J, Mori M, Khuwijitjaru P, Adachi S (2013) Properties of extract from Okara by its subcritical water treatment. Int J Food Prop 16:974–982. https://doi.org/10.1080/10942912.2011.573119
35. Pińkowska H, Oliveros E (2014) Application of the Doehlert matrix for the determination of the optimal conditions of hydrolysis of soybean protein in subcritical water. Ind Eng Chem Res 53:1320–1326. https://doi.org/10.1021/ie403451b
36. Pińkowska H, Wolak P, Oliveros E (2014) Hydrothermolysis of rapeseed cake in subcritical water. Effect of reaction temperature and holding time on product composition. Biomass Bioenerg 64:50–61
37. Ganzler K, Salgó A, Valkó K (1986) Microwave extraction. A novel sample preparation method for chromatography. J Chromatogr 371:299–306
38. Craveiro AA, Matos FJA, Alencar JW, Plumel MM (1989) Microwave oven extraction of an essential oil. Flavour Fragr J 4:43–44. https://doi.org/10.1002/ffj.2730040110
39. Vian MA, Fernandez X, Visinoni F, Chemat F (2008) Microwave hydrodiffusion and gravity, a new technique for extraction of essential oils. J Chromatogr A 1190:14–17. https://doi.org/10.1016/j.chroma.2008.02.086
40. Franco-Vega A, Ramírez-Corona N, Palou E, López-Malo A (2016) Estimation of mass transfer coefficients of the extraction process of essential oil from orange peel using microwave assisted extraction. J Food Eng 170:136–143. https://doi.org/10.1016/j.jfoodeng.2015.09.025
41. Benmoussa H, Farhat A, Romdhane M, Bouajila J (2016) Enhanced solvent-free microwave extraction of Foeniculum vulgare Mill. Essential oil seeds using double walled reactor. Arab J Chem. https://doi.org/10.1016/j.arabjc.2016.02.010
42. Chen Z, Zhang W, Tang X et al (2016) Extraction and characterization of polysaccharides from Semen Cassiae by microwave-assisted aqueous two-phase extraction coupled with spectroscopy and HPLC. Carbohydr Polym 144:263–270. https://doi.org/10.1016/j.carbpol.2016.02.063
43. Chen R, Jin C, Tong Z et al (2016) Optimization extraction, characterization and antioxidant activities of pectic polysaccharide from tangerine peels. Carbohydr Polym 136:187–197. https://doi.org/10.1016/j.carbpol.2015.09.036
44. Rodríguez-Pérez C, Gilbert-López B, Mendiola JA et al (2016) Optimization of microwave-assisted extraction and pressurized liquid extraction of phenolic compounds from Moringa oleifera leaves by multiresponse surface methodology. Electrophoresis 37:1938–1946. https://doi.org/10.1002/elps.201600071

45. Kale A, Varadan R, Davis SC (2016) Methods for extracting and purifying non-denatured proteins
46. Choi I, Choi SJ, Chun JK, Moon TW (2006) Extraction yield of soluble protein and microstructure of soybean affected by microwave heating. J Food Process Preserv 30:407–419. https://doi.org/10.1111/j.1745-4549.2006.00075.x
47. Zhou L, Xu Z, Zhang M et al (2011) Research of ultrasonic-microwave-assisted extraction technology of peanut protein. Guangdong Agric Sci 4:039
48. Jin J, Xu Z, Wei Z, Liu J (2009) Ultrasonic and microwave assisted extraction of *rapeseed* protein. Mod Food Sci Technol
49. Weaver JC, Chizmadzhev YA (1996) Theory of electroporation: a review. Bioelectrochem Bioenerg 41:135–160. https://doi.org/10.1016/S0302-4598(96)05062-3
50. Pakhomov AG, Miklavcic D, Markov MS (2010) Advanced electroporation techniques in biology and medicine, 1st edn. CRC Press, Boca Raton
51. Lebovka NI, Bazhal MI, Vorobiev E (2001) Pulsed electric field breakage of cellular tissues: visualisation of percolative properties. Innovative Food Sci Emerg Technol 2:113–125. https://doi.org/10.1016/S1466-8564(01)00024-8
52. Vorobiev E, Lebovka N (2008) Electrotechnologies for extraction from food plants and biomaterials. Springer-Verlag, New York
53. Polikovsky M, Fernand F, Sack M et al (2016) Towards marine biorefineries: selective proteins extractions from marine macroalgae *Ulva* with pulsed electric fields. Innovative Food Sci Emerg Technol 37:194–200. https://doi.org/10.1016/j.ifset.2016.03.013
54. Bouras M, Grimi N, Bals O, Vorobiev E (2016) Impact of pulsed electric fields on polyphenols extraction from Norway *spruce bark*. Ind Crops Prod 80:50–58
55. Yu X, Bals O, Grimi N, Vorobiev E (2015) A new way for the oil plant biomass valorization: polyphenols and proteins extraction from rapeseed stems and leaves assisted by pulsed electric fields. Ind Crops Prod C:309–318. https://doi.org/10.1016/j.indcrop.2015.03.045
56. Barba FJ, Boussetta N, Vorobiev E (2015) Emerging technologies for the recovery of isothiocyanates, protein and phenolic compounds from rapeseed and rapeseed press-cake: effect of high voltage electrical discharges. Innovative Food Sci Emerg Technol Complete:67–72. https://doi.org/10.1016/j.ifset.2015.06.008
57. Yu X, Gouyo T, Grimi N et al (2016) Pulsed electric field pretreatment of rapeseed green biomass (stems) to enhance pressing and extractives recovery. Bioresour Technol 199:194–201. https://doi.org/10.1016/j.biortech.2015.08.073
58. Smith EL, Abbott AP, Ryder KS (2014) Deep eutectic solvents (DESs) and their applications. Chem Rev 114:11060–11082. https://doi.org/10.1021/cr300162p
59. Figueiredo M, Gomes C, Costa R et al (2009) Differential capacity of a deep eutectic solvent based on choline chloride and glycerol on solid electrodes. Electrochim Acta 54:2630–2634. https://doi.org/10.1016/j.electacta.2008.10.074
60. Jhong H-R, Wong DS-H, Wan C-C et al (2009) A novel deep eutectic solvent-based ionic liquid used as electrolyte for dye-sensitized solar cells. Electrochem Commun 11:209–211. https://doi.org/10.1016/j.elecom.2008.11.001
61. Gore S, Baskaran S, Koenig B (2011) Efficient synthesis of 3,4-dihydropyrimidin-2-ones in low melting tartaric acid–urea mixtures. Green Chem 13:1009. https://doi.org/10.1039/c1gc00009h
62. Nam MW, Zhao J, Lee MS et al (2015) Enhanced extraction of bioactive natural products using tailor-made deep eutectic solvents: application to flavonoid extraction from Flos sophorae. Green Chem 17:1718–1727. https://doi.org/10.1039/C4GC01556H
63. Bi W, Tian M, Row KH (2013) Evaluation of alcohol-based deep eutectic solvent in extraction and determination of flavonoids with response surface methodology optimization. J Chromatogr A 1285:22–30. https://doi.org/10.1016/j.chroma.2013.02.041
64. Park HE, Tang B, Row KH (2014) Application of deep eutectic solvents as additives in ultrasonic extraction of two phenolic acids from herba *artemisiae scopariae*. Anal Lett 47:1476–1484. https://doi.org/10.1080/00032719.2013.874016
65. Tang B, Bi W, Zhang H, Row KH (2013) Deep eutectic solvent-based HS-SME coupled with GC for the Analysis of Bioactive Terpenoids in *Chamaecyparis obtusa* Leaves. Chromatographia 77:373–377. https://doi.org/10.1007/s10337-013-2607-3

66. Xu K, Wang Y, Huang Y et al (2015) A green deep eutectic solvent-based aqueous two-phase system for protein extracting. Anal Chim Acta 864:9–20. https://doi.org/10.1016/j.aca.2015.01.026
67. Li N, Wang Y, Xu K et al (2016) Development of green betaine-based deep eutectic solvent aqueous two-phase system for the extraction of protein. Talanta 152:23–32. https://doi.org/10.1016/j.talanta.2016.01.042
68. Zeng Q, Wang Y, Huang Y et al (2014) Deep eutectic solvents as novel extraction media for protein partitioning. Analyst 139:2565–2573. https://doi.org/10.1039/C3AN02235H

Chapter 10
Carob as Source for Sustainable Ingredients and Products

Ikram Boublenza, Ibtissam Boublenza, Abdelhak Boublenza, Sara Madji, Anne-Sylvie Fabiano-Tixier and Farid Chemat

Abstract The carob is a dome-shaped evergreen tree that is a member of the pea family (Fabaceae). Its fruit has been used for human consumption and as animal feed in the Mediterranean countries for centuries and was also widely used as a herbal remedy in traditional folk's medicine. In this paper, an in-depth literature review was conducted about this species discussing its history, origins, main current uses, components (both nutrient and bioactive), biological activities, and potential applications as the source of sustainable ingredients and products. Among the many bioactive compounds that were found in carob, the most well studied are polyphenols. The levels of these compounds were found to vary depending on many factors, such as the variety, plant part, geographic location, roasting and extraction conditions, and even analysis methods. In the light of these researches, it was recognized that carob offers several biological activities including antioxidant, anticancer, and anti-diabetic due to the presence of a high quantity of pinitol and antibacterial. Rich in sugars, proteins, and minerals, the carob has a vast range of industrial applications such as cacao substitute, sugars substitute, free from gluten making it cereal-derived foods for celiac people.

Keywords Carob · Nutritional components · Bioactive molecules · Biological activities

I. Boublenza (✉) · I. Boublenza · A. Boublenza
Sarl Boublenza Agroalimentaire et Produits Agricoles, Rue Bouabderrahmane, Fg Agadir, 13000 Tlemcen, Algeria
e-mail: sarl.boublenza.bik@gmail.com

S. Madji · A.-S. Fabiano-Tixier · F. Chemat
Université d'Avignon et des Pays de Vaucluse, INRA, UMR408, GREEN Extraction Team, 84000 Avignon, France

Y. Li and F. Chemat (eds.), *Plant Based "Green Chemistry 2.0"*,
Green Chemistry and Sustainable Technology,
https://doi.org/10.1007/978-981-13-3810-6_10

10.1 Introduction

Ceratonia siliqua L., commonly known as carob tree, is a Mediterranean evergreen tree that belongs to the family *Fabaceae* (subfamily of *Caesalpinioideae*) [1]. The scientific name derives from the Greek *keras* meaning horn and the Latin *Siliqua* alluding to the hardness of the pod [2]. It is a tree that is distinguished by its great longevity, several times centenary [1] This tree is a dioecious species with some hermaphroditic forms [2]. Known for its hardness, it requires little maintenance and adapts to different types of soils except hydromorphic and salty soils [1]. Carob has been used for thousands of years by the habitants of the Mediterranean countries as feed and food for its sugary pulp (48–56% of sugars) [2]. The carob pod is the fruit of the carob tree (*Ceratonia siliqua* L.), which is a flat, curved brown pod measuring 10–30 cm long and 1.5–3.5 cm wide [2]. It has two main products: carob kernels or seeds which represent 10–20% of the fruit weight, and carob kibbles or the remaining pulp with 80–90% of the fruit weight obtained after deseeding(Fig. 10.1) [3].

The seeds are covered by a brown and very hard hull which accounts for 30–33%. They contain a white and translucent endosperm which weigh 42–46% called carob gum or locust bean gum (LBG). The seeds also contain the embryo called carob germ, which accounts for 23–25% of the seed weight [4]. The germ is recovered as the by-product of the seed processing [2]. Locust bean gum is widely used in the food industry as a natural food additive (E 410) to function as a stabilizing and thickening agent in food preparations because of its ability to form a viscous solution at relatively low concentration. It is also used in the cosmetic, pharmaceutical, petroleum, paper, textile, paint, oil drilling, and construction [5]. The flour of carob germ is used as dietetic human food or as a potential ingredient in cereal-derived foods for celiac people [6]. Carob kibbles can be used raw mainly for animal feeding or roasted as a

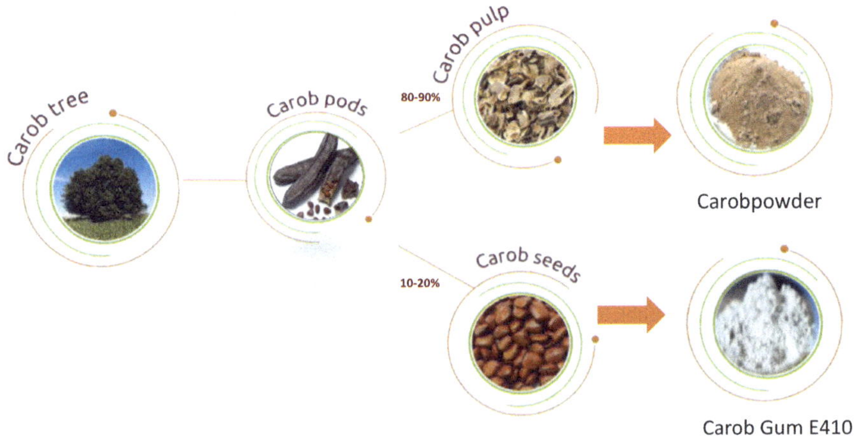

Fig. 10.1 Carob tree and carob pod (containing pulp and seeds)

cocoa substitute. They are also used to make beverage, compote, liqueur and syrup [7], or by fermentation for ethanol and citric acid production [8].

The biological activities of the fruit of carob tree are mainly related to the sugars, polyphenols, and dietary fiber present in the kibbles, galactomannan in the seed endosperm and protein content in the germ of the kernel of this fruit. Many reports state that carob has shown good effects on prevention of colon cancer, reducing the risk of type 2 diabetes, and lowering cholesterol [5]. In regard to the potential uses of carob for food products considering its health benefits, this paper overviews the nutritional, medicinal, and biological profile of this fruit. The main industrial applications are also reviewed and discussed.

10.2 Brief History of Carob

The carob tree has a very ancient origin and its appearance dates back to prehistory. Many names are attributed to this tree. In Spain, it is known by the name "El garrobo"; in Algeria, it is known as "elkherroub," "karrûba," "taslighoua," and "tikharoubt" [9]; and it is also known as St. John's bread in the English-speaking countries in reference to its presumed use by St. John, the Baptist [9]. The scientific name of carob tree (*Ceratonia siliqua* L.) derives from the Greek *keras*, "horn," and Latin *siliqua*, indicating to the hardness of the pod. The origin of the word carob is the Hebrew word *kharuv* [2]. In the past, Arabs used its uniform seeds as unit of weight (0.2 g), they called it qirat or carat; it then became the unit of weight for gold and precious stones. Furthermore, the ancient Egyptians are reputed to have used the gum as an adhesive in mummy binding [10]. Carob pods can be stored and transported long distances; it was eaten by children as snacks or sweet fruit during the famine period in Spain [9].

10.3 Botanical Description, Distribution, and Cultivation

Ceratonia siliqua L. belongs to the order of *Rosales*, *Leguminosae* (syn Fabaceae) family, in the tribe *Cassieae* of the subfamily *Caesalpinioideae*. The origin of *C. siliqua* is not clear. De Candolle and Vavilov stated that it originates from Eastern Mediterranean region, more precisely from Syria and Turkey [11, 12]. It was propagated by the Arabs on North Africa and Spain then dispersed to the south of France and south of Portugal. The Greeks introduced it in some European countries, like Greece and Italy.

The carob tree is an evergreen shrub that can reach 8–15 m height [13]. Its leaves are 10–20 cm long, pinnate, dark green, and shiny above. The carob is known for its great resistance to dryness. However, it fears frost which limits its spread. The flowering is conducted in winter and the pollination is mainly provided by insects

Fig. 10.2 Carob flowers (**a** male/**b** female)

and wind (Fig. 10.2). Carob flowers are small and numerous, and they are 6–12 mm long and distinguished by a greenish-red color [1, 2].

The tree has been cultivated since ancient times throughout the Mediterranean basin in mild and dry places. It was also introduced to the United States of America (California), the Philippines, Iran, South Africa, and Australia [2].

10.4 Culinary and Folk Medicine Uses

Many plants have been used as important ingredients in therapeutic practices and traditional cuisine in various civilizations. *Ceratonia siliqua* L. is well known in most parts of the Mediterranean and Arabic countries, used both for food and "folk pharmacopeia." Considering its high amounts of sugars, the carob kibbles are traditionally used to make a traditional concentrate syrup, named "pekmez" in Turkey, which is mainly consumed in Arab countries during the month of Ramadan [14, 15]. In North Africa and the Middle East, carob powder and kibbles were usually used for medicinal purposes against digestive disorders, as purgative, anti-diarrheal, mouth inflammation, and anti-ulcer [16]. In Turkey, about 8.4% of the high cholesterol sufferers use the carob in order to lower their cholesterol. They usually mash the carob kibbles and consume it either with yogurt or raw [17]. Moreover, reports state that in Kingdom of Jordan, the carob kibble is known as stomach strengthener and phlegm clearer; the seed is usually used as purgative and toothache calmative, and the carob jam is also used for tongue sores and stomachache treatments [18] (Table 10.1).

Table 10.1 Uses of carob pulp

Uses	Advantages
Animal feed	– Antidiarrheal [19] – Improving body weight and carcass weight [20] – Good source of energy [21] – Increasing the polyunsaturated fatty acids in muscle and reducing of saturated fatty acids [22] – Improving meat lighter [23]
Human food	– Extender of cocoa after roasting and milling (Fewer calories, neither caffeine/theobromine) [24] – As alternative for people with cocoa allergy [25] – Low glycemic index [26] – Natural sweetener [27] – Color and flavor similar to chocolate [28] – Suitable for the feeding of diabetics and celiac [29] – Anti-diarrheic product [30] – Improving digestion [31] – High antioxidative potential [31] – Lipid utilization [31] – Positive effect on cholesterol metabolism [31] – Potential regulation effect on blood glucose levels [31] – No presence of oxalic acid (good absorption of minerals) [5] – Anxiolytic and sedative effect [32] – Much lower fat and sodium content [33]

10.5 Nutritional Composition

The proximate compounds, vitamins, minerals, fatty acids, and amino acids are found in different parts of the fruit of carob tree. The composition of carob on macronutrient and micronutrient differs according to the differences in species, varieties, growth conditions, harvesting times, soil properties, climate, origin, and geographic parameters [3]. Many factors affect the chemical composition of pods; it can be influenced by cultivar, temperature, dryness, irrigation, fertilization, and salinity [34]. Carob fruit is high in sugars and low in fat. It generally contains 50–60% of sugars, 0.2–0.8% of lipids, and contains 1–5% of proteins as well as crude fiber of about 11% in addition to significant amounts of minerals (1–6%). It is found that the mineral content of carob like calcium, potassium, magnesium, iron, and phosphorus is numerous [35]. Carob does not contain oxalic acid present in cocoa, allowing a better absorption of minerals. All in all, this fruit is considered as healthy natural food (Fig. 10.3).

Carob powder either roasted or not contains more sugar than cocoa powder and about 75% or more of those sugars as sucrose. Thus, carob and its products could be considered as natural sweeteners for formulating food products. Regarding the composition and nutritional function of carob kibble, it can be an excellent substitute of cocoa [36]. Unlike cocoa powder, carob is free from the two anti-nutrients caffeine and theobromine which is valuable advantage [15]. Analyzing the carob seed germ, Dakia et al. [37] reported the following composition: moisture 8.3%, ash 6.5%,

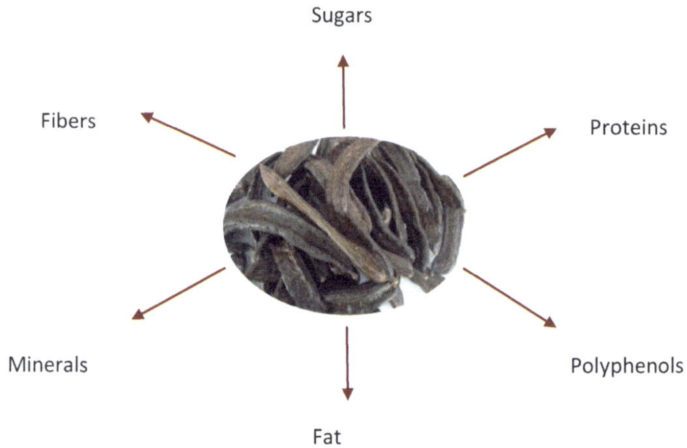

Fig. 10.3 Composition of carob pulp

lipids (neutral and polar) 6.6% containing 21% of polar lipids, crude proteins 54.7%, and energy value 17.5 kJ/g. They reported Oleic acid (34.4%) and linoleic acid (44.5%) as the major fatty acids, while palmitic acid (16.2%) and stearic acid (3.4%) were recognized as the main saturated ones. The study also showed that the carob germ protein fraction was extremely heterogeneous, making this seed component interesting in different diets and particularly for celiac diet.

10.6 Composition on Bioactive Compounds

The main bioactive compounds of carob are shown in Table 10.2. The last years, very researches have focused on the valuable molecules with the health promoting effects [38]. The composition of the carob in polyphenols may be influenced by various parameters such as the development of fruit, the transformation that the fruit undergoes, roasting method, the nature of the solvent used for the extraction, and its duration.

Polyphenols are reactive secondary metabolites abundant in plant-derived foods particularly fruits, seeds, leaves, and stems. They are unequally distributed in plants from a quantitative and qualitative point of view. They are non-nutritional molecules but influence sensory quality such as color and astringency and that have well-noted antioxidant properties and exert preventive activity against chronic diseases which continue to be recognized in all domains [39]. In the aim of obtaining phenolic-rich extracts with biological activities, Roseiro et al. [40] conducted supercritical fluid extraction (SFE) on Carob kibbles. They also conducted an ultrasound assisted (UAE) and conventional solid–liquid extractions with two different solvent systems (100% H_2O and 70% acetone). A diversity of phenolic compounds was recognized.

Table 10.2 Composition on bioactive compounds of carob pulp

Compound	Biological effects
Polyphenols and tannins	– Strong antioxidant capacity [42]
	– Natural antioxidant in food [46]
	– Anti-diarrheal effect [16, 30]
	– Anti-ulcer [16]
Fibers	– Cholesterol lowering effect [47]
	– Increasing excretion of bile acid [47]
	– Prevention of diabetes mellitus, cardiac infarctus, and colon carcinoma [47]
	– Inhibition of proliferation of both adenoma and adenoma–carcinoma cells colon [48]
Pinitol	– Regulation of blood glucose [45]
	– Protection of sensible organs as liver and kidney [45]
	– Hypolipidemia [45]
	– Anti-inflammatory [49]

However, conventional and ultrasound extracts contained mainly gallic acid. The highest phenolic concentration and antioxidant capacity were also found in the SFE extract.

Cepo et al. [24] studied the antioxidant activity and gastrointestinal solubility of polyphenols during thermal processing of carob powder in different roasting time–temperature combinations. They showed that the antioxidant activity of carob powder can be significantly increased by roasting due to improved polyphenols solubility. In another study done by Roseiro et al. [41], it was observed that temperature had the most significant effect on phenolic profile and antioxidant property of carob kibbles. They obtained the maximum of total polyphenols and tannins extraction in 98.5 °C for 17 min. Under these conditions, inhibition of DPPH and ABTS radicals were 85 and 90%, respectively, showing the promising use of this biomass as an antioxidant source, which could be used, for example, in functional beverages. Makris and Kefalas [42] have used various solvent systems, in order to evaluate and optimize the conditions for the recovery of polyphenols from carob kibbles. The 80% acetone seemed to be the best solvent for the maximum quantities of polyphenolic components and flavonol content. Papagiannopoulos's et al. [43] researches showed the extraction of carob fibers and roasting influence quantitatively and qualitatively on polyphenols. Results showed that 448 mg/kg extractable polyphenols contain gallic acid, hydrolyzable and condensed tannins, flavonol-glycosides, and traces of isoflavonoids were recognized in carob pods.

The fibers contain a wide variety of phenolic compounds. About twenty-four structures were identified by different chromatographic techniques and showed that gallic acid is the majority compound as well as Gallotannins and methyl gallate [44]. Recently, it has evaluated a good amount of D-pinitol, and a cyclitol which is highly soluble in water [45].

10.7 Processing Effects

Carob pods are collected manually during the harvest period, and then transported to processing rooms where they will undergo various technological processes. The different stages of transformation are schematized in Fig. 10.4.

Different parts of carob are usually marketed in the form of powder or flour resulting from the fruit milling. Carob kibbles (Grounded) are very often sold raw for animal feeding which gives the same results as a barley diet [50]. As part of a human diet, it is more often roasted for the food industry [5]. It is also marketed in other forms: whole pods, crushed, and syrup. The advantages of concentrated liquid extract over other forms include lower storage costs and increased concentration and stability of the active substances [5]. Raw carob which has not undergone any technological treatment has a characteristic smell that some consider undesirable. This smell is due to the presence of isobutyric acid. The roasting process reduces this smell considerably while releasing different aromas, and this brings a higher sensory quality which is a factor to consider when establishing the roasting time [51]. Other factors such as the functional properties of the powder as well as the energy cost must be taken into account as they are also influenced by roasting time and temperature [51]. It was shown by Iipumbu [34] that roasting improved the sensory acceptability of carob while an increase in roasting time significantly reduced the sugar and protein and had no significant effect on fat content. Roasting of carob kibbles has been done under different conditions (roasting time and temperature combinations) in order to evaluate the physical characteristics, chemical composition, and biological activities of the resulting product. Cepo et al. [24] showed that roasting carob at 130 °C for just 30 min could be proposed as the procedure of choice to obtain carob powder

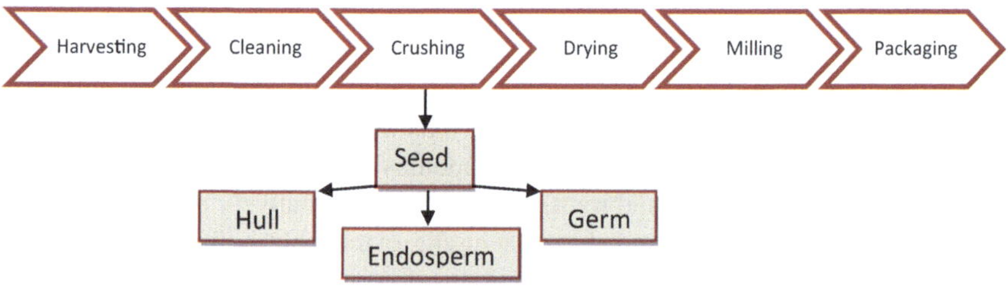

Fig. 10.4 Different steps of carob processing

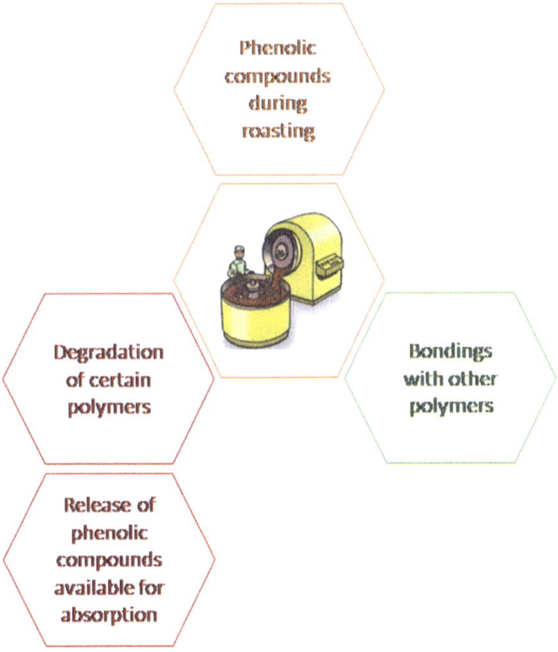

Fig. 10.5 Diagram illustrating the degradation of phenolic compounds during roasting

with high antioxidant activity. Figure 10.5 summarizes the evolution of polyphenols during the roasting process.

It was demonstrated that the process of grinding and roasting carob has a significant influence on the antioxidant activity. In fact, the increase in antioxidant activity is due to the improvement of the solubility of polyphenols [24]. By studying the physicochemical characteristics of roasted carob powder, Yousif and Alghzawi [15]concluded that the best organoleptic quality of the product results from roasting at 150 °C for 60 min [15].The volatile fraction of carob bean pulp has been also analyzed during a roasting process by Cantalejo [52], and 137 components were identified. 91.4% of the identified compounds in raw carob contained acids, alcohols, and aldehydes. During the roasting process, the amount of these compounds decreased sharply, showing 51.2% of the total as a result of the increasing amounts of other volatile compounds in roasted carob, whose smell is more pleasant than that of raw one [52]. Although, roasting is the main process done on carob, extraction of phenolic compounds and D-pinitol compound have been studied through different methods and conditions. The highest phenolic compounds and antioxidant capacity were found in the SFE extract. Thus, the potential association of two environmentally clean processes which are UAE and SFE used to extract polyphenols from carob kibbles will permit bioactive compounds within a sustainable and integrated recovery process [40]. These residues have been subject to water extraction at mild operating conditions by Almanasrah et al. [53]. This study showed that this offers a sustainable, selective process for the recovery of sugars and phenolic compounds from carob residues. In fact, it has been shown that the one-step extraction recovered only about 20% of the phenolic compounds, with an extraction yield of 0.6 g Gallic acid equivalents (GAE)/100 g dry mass of carob kibbles, which contained significant amount of carbohydrates (110 g/L).

In contrast, the developed two-step extraction enabled higher compound selectivity with an increase in the yield of the phenolic compounds to about 70%, corresponding to 1.9 g GAE/100 g carob dry matter. Furthermore, Almanasrah et al. [53] state that the two-step extraction is an effective method to obtain significantly separated carbohydrates and polyphenol-rich streams that can be further processed, e.g., in biorefineries or food industries, respectively [53].

Using the membrane technology, Almanasrah et al. [54] tried to obtain two distinct extracts from carob kibbles, one extract enriched in catechin and its derivatives with a reduced concentration of gallic acid and monomeric sugars and a second extract concentrated in sugars. The high concentration of sugars in carob kibbles compounds is taken into account, and it is clear that the proposed separation of phenolic was challenging. This molecular mass is similar for all compounds under study. Thus, using water as a biocompatible extraction solvent was proposed in the above process which fulfills the zero discharge principle. In the obtained extract enriched on phenolic compounds, all the target compounds were fully recovered, the sugar extract was free of catechin and its derivatives and had a minor gallic acid contamination which can be eliminated [54]. Cháfer and Berna [49] studied the influence of operating conditions on the SFE of D-pinitol processed from carob pods. They showed that the D-pinitol yield increases if the temperature or the pressure increases; the choice of these parameters will be determined for economic studies. The optimization of the conditions of D-pinitol compound extraction from carob pods by using ultrasonication was studied by Tetik and Yuksel [45]. They obtained the highest D-pinitol concentration of 11.98 g/L with ultrasonic power, a dilution rate of 1:4, and an extraction time and temperature of 120 min and 50 °C, respectively. The writers suggested the ultrasonic assisted extraction as an alternative to conventional hot water extraction to recover D-pinitol from the carob pods. It is an advantageous alternative because it reduces the extraction time and thus the energy consumption [45].

The heat treatment can lead to two processes resulting in different compounds that are involved in the organoleptic characteristics of a product. From the roasting, two chemical reactions are triggered which are the Maillard reaction and the caramelization. In the Maillard reaction, the flavors and colors of caramels can appear during an intermediate heating at around 100 °C. While roasted notes so much appreciated by the consumer take over at a higher temperature. The intensity of the color increases with the time of exposure to a certain temperature [55]. Maillard reaction or non-enzymatic browning is a reaction process that is triggered by a reducing sugar and an amino acid. It can therefore have two impacts on foodstuffs, either negatively during storage periods, causing the synthesis of molecules that are sometimes undesirable organoleptic characteristics, since it can take place at room temperature. Or, a positive impact that makes it possible to enhance the organoleptic properties of a food caused by heat treatment [56]. Caramelization is a heat treatment of sugars above

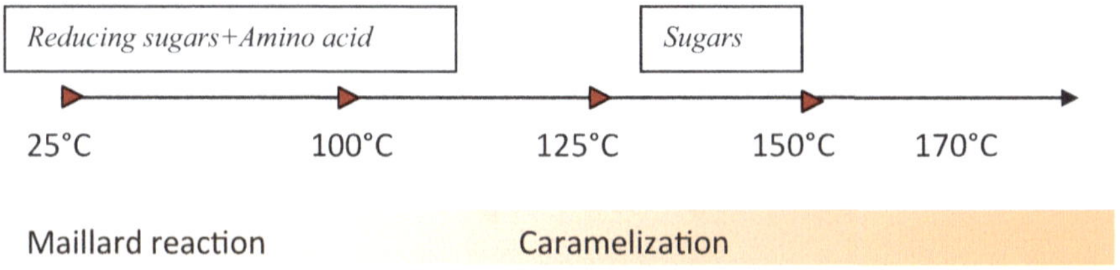

Fig. 10.6 Maillard reaction and caramelization

their melting point and starts at a high temperature. The breakdown of sugars by heat, catalyzed by acids and bases, leads to the formation of volatile compounds (caramel flavor) and brown compounds (caramel color) that depend on the nature of sugars [57]. The different of the two reactions are presented in Fig. 10.6.

10.8 Biological Activities

Biological activities of carob are often attributed to their major constituents. However, some minor compounds already exist and participate synergistically with other compounds (Fig. 10.7). Carob is a fruit known since antiquity for its content on bioactive compounds that have many virtues on human health. Several compounds of different classes are responsible for the biological activities of the plant: inositols such as pinitol, polyphenols, fat, sugars, fibers, and tannins. New analytical techniques have highlighted the identification and quantification of these compounds as well as the study of their effects. The influence of dietary polyphenols on human health and inositols on the quality of food products is widely recognized. They are known to have a positive influence on various chronic diseases. Today, several biological activities are attributed to carob. The chemical compounds listed in Table 10.2 are responsible for the properties of this plant.

10.8.1 Antioxidant Activity

The antioxidant activity of the carob may be attributed to the polyphenols contents of the carob kibbles. Gallic acid and polymers of flavan-3-ol and flavan-3, 4-diol subunits are the major polyphenols present in carob. Carob kibble also comprises tannins which are complex polyphenolic compounds and can be classified into two hydrolysable and condensed groups [48]. Joslyn et al. [58] reported that many kinds of hydrolyzable tannins, such as galloyl glucose compounds, were present in carob pods in large amounts. Carob kibble is rich in condensed tannins, containing flavonoid polymers, formed by groups of flavon-3-ol and their gallic esters, gallic acid, catechins, epicatechingallate, epigalocatechingallate, and quercitin glycosides [59].

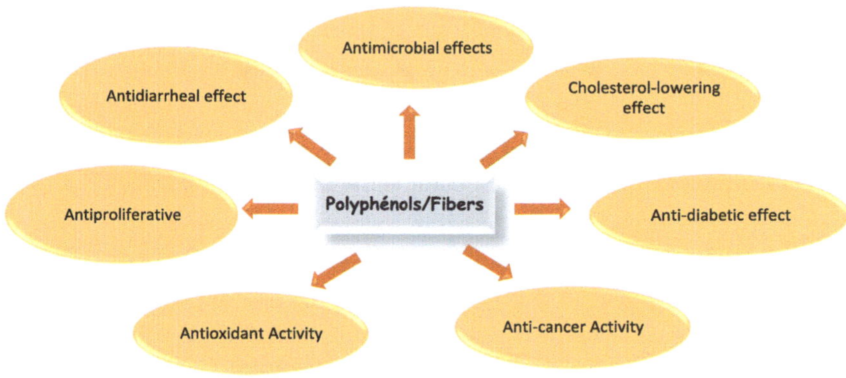

Fig. 10.7 Biological activity of carob pulp

Polyphenols reduce oxidative stress by scavenging free radicals, and thus delaying the formation of free radicals, and decreasing the rate of oxidation. Condensed tannins are the main polyphenols constituents of carob; they contain the flavan nucleus which is insoluble in usual organic solvents. Inositols also contribute to the antioxidant potential of carob [5]. In recent years, much research has focused on plants and their by-products to extract natural and low-cost antioxidants that can replace synthetic additives such as butylated hydroxyanisole [60] and hydroxytoluene, which might be liver-damaging and carcinogenic [61].

10.8.2 Anticancer Activity

Klenow and Glei [48] found that carob fiber extracts have protective effects against oxidative stress in adenoma cells, which can stop or slow down the carcinogenic process in colorectal cancer [62]. Recognized gallic acid, (-) epigallocatechin-3-gallate and (-) epicatechin-3-gallate in pod and leaf extracts as the main antiproliferative agents of carob. Consisting of polyphenols, carob fiber inhibits proliferation of both adenoma and adenocarcinoma cells of the human colon [48].

10.8.3 Anti-diabetic Effect

Carob is a low-glycemic index food because it is very rich in dietary fiber. Its high fiber content makes it a slowly digestible food which leads to a drop in blood sugar and contributes in the protection against diabetes and hyperlipidemia [63]. Also, the work of Banuls et al. [64] focused on forty euglycemic subjects that showed the positive effect of chronic consumption of an inositol-enriched carob extract on insulin sensitivity and postprandial glucose levels. In addition to that, their data suggest that the improvement in insulin sensitivity is due to an increase in the size of low density lipoprotein (LDL) particles. These findings endorse the efficacy of this dietary intervention to avoid hyperglycemia and other conditions caused by insulin resistance and reducing the risk of cardiovascular disease [64]. Administration of crude polyphenols extract of carob pods to male rats showed a significant reduction in blood glucose after 30, 60, and 120 min when compared with the control group [65]. Both animal experiments and human clinical trials showed significant anti-diabetic effects of inositol [66].

10.8.4 Cholesterol Lowering Effect

The consumption of polyphenol-rich carob fiber has shown beneficial effects on the human blood lipid profile [67]. Zunft et al. [47] found that the cholesterol-lowering effect of carob fibers is due to increased cholesterol 7 α-hydroxylase activity. The consumption of 15 g of carob supplement per day, as an ingredient, reduces after seven weeks 7.1% of the total cholesterol concentration and 10.6% of low-density lipoprotein (LDL). Besides this, inositols increase high-density lipoprotein cholesterol (HDL) levels in the blood (good cholesterol), lower-serum LDL cholesterol and triglycerides, and reduce the risk of cardiovascular disease [47].

10.8.5 Antimicrobial Effects

Some studies have shown that carob kibble shows antimicrobial activities against some microorganisms including *Listeria monocytogenes*, *Salmonella enteritidis, and Brochothrixthermosphacta* [68]. Carob fiber binds almost three times its weight in water which can limit microbial growth and thus increases the shelf-life of the final product [5]. Carob leaf extracts have acted on the bacterium *pectobacterium atrosepticum*, which causes potato rot [69]. This can introduce carob as a natural ingredient for food preservation.

10.8.6 Other Effects

Lee et al. [70] observed that using lupin kernel flour, which is a novel food rich in protein and fiber as a food ingredient results in more satiety, and because of being slowly digestible, it decreases the energy intake and helps losing weight. Carob can have the same effect on energy intake and satiety because it is very rich in fiber. Carob bean juice is also effective in curing infantile diarrheal disturbance and bacterial dysentery [71]. Carob fruits were also used traditionally against cough and warts [72].

10.9 Potential Industrial Applications and Future Trends

Recently, healthy foods and lifestyle became a tendency. Subsequently, food industry followed up and carob is now considered as a commercial healthy ingredient in different kinds of industries. Carob kibbles are used in many ways, ranging from using them in a raw state for animal feeding to roasting them for consumption as a cocoa substitute. Other uses include extracting sugars for making syrup or bioethanol. Food industries use the locust bean gum as a thickening and stabilizing agent in food preparations considering its ability to form a viscous solution at relatively low concentration. However, it is also used in the cosmetic, pharmaceutical, textile, paper, petroleum, paint, oil drilling, and construction industries. In addition, carob germ flour is recognized as dietetic human food or as a potential ingredient in cereal-derived foods for celiac people. In the following, the potential and current industrial application of carob are presented:

Tsatsaragkou et al. [72] investigated the rheological properties of gluten-free doughs from rice flour containing different amounts of carob flour. The water content was proved to decrease the resistance of dough to deformation by the creep data evaluation. In contrast, and due to its high fiber contents, carob flour increased the elastic character and structural strength of the dough. Thus, by increasing carob flour amounts, the bread dough elastic character increases [72]. Carob fiber has been successfully added to different baked goods, including bread, rolls, cakes, and cookies. Haber [73] showed that adding carob fiber to wheat, rye, and multigrain breads and rolls, without changing the formulations, increases water absorption and reduces the stickiness of dough. Mis et al. [74] found that the dough made from the additions of carob fiber were rheologically more stable during mixing compared to the common one.

Carob fiber can also be milled into a fine powder which is suitable for using in a variety of food products, such as health bars, baked goods, extruded products, dairy drinks, and dietary supplements [75]. Sanchez-Muniz [76] showed that carob fiber can be successfully used as an additive to edible oils to increase shelf-life and decrease the potential toxicity of the heated oil; upon introducing 50 mg/kg of carob fiber to sunflower oil, triacylglycerol polymerization, tocopherol loss, and oxidation were reduced during frying [77]. Known for its high content of sugar, the carob is used for the preparation of syrup as an alternative to refined sugar in various food products such as the flavoring of yogurt [78]. Carob juice can be a good alternative to grape juice because of its sensory properties and its higher levels of total polyphenols [79]. In recent years, carob has attracted considerable attention because of its high carbohydrate and mineral content [80]. Using carob via fermentation process produced many high-value-added products such as lactic acid [81], mannitol [82], citric acid [83], and pullulan [84]. Ethanol has also been produced using carob pod extract. Mannitol extracted by fermentation of syrup is used in chemistry and in the pharmaceutical field [85]. For example, Turhan et al. [86] performed ethanol production from carob pod extract by using *Saccharomyces cerevisiae.*

Carob powder has nutritional advantages over cocoa powder by having lower contents of fat (0.6%) and higher dietary fiber (40%) as compared to cocoa powder which contains 23% fat and 5% fiber. It has great potential to replace cocoa powder [28]. In addition to that, carob powder does not contain theobromine, caffeine, and oxalic acid [15]. It also contains a higher content of phenolic compounds and antioxidant capacity [87]. Many studies have reported a cocoa-like an aroma in roasted carob [88].

References

1. Rejeb NM (1995) Le caroubier en Tunisie: situation et perspectives d'amélioration. Quel avenir pour l'amélioration des plantes? Ed. AUPELF-UREF. John Libbey Eurotext, pp 79–85
2. Battle I, Tous J (1997) Carob tree. *Ceratonia siliqua* L., promoting the conservation and use of under-utilised and neglected crops. Institute of Plant Genetics and Crop Plant Research and Gatersleben/International Plant Genetic Resource Institute
3. Oziyci HR, Tetik N, Turhan I, Yatmaz E, Ucgun K, Akgul H, Gubbuk H, Karhan M (2014) Mineral composition of pods and seeds of wild and grafted carob (*Ceratonia siliqua* L.) fruits. Sci Hortic 167:149–152
4. Durazzo A, Turfani V, Narducci V, Azzini E, Maiani G, Carcea M (2014) Nutritional characterisation and bioactive components of commercial carobs flours. Food Chem 153:109–113
5. Nasar-Abbas SM, Huma Z, Vu TH, Khan MK, Esbenshade H, Jayasena V (2016) Carob kibble: a bioactive-rich food ingredient. Compr Rev Food Sci Food Saf: 63–72
6. Bengoechea C, Romero A, Villanueva A, Moreno G, Alaiz M, Millan F, Guerrero A, Puppo MC (2008) Composition and structure of carob (*Ceratonia siliqua* L.) germ proteins. Food Chem 107:675–683
7. Youssef MKE, El-Manfaloty MM, Ali MH (2013) Assessment of proximate chemical composition, nutritional status, fatty acid composition and phenolic compounds of carob (*Ceratonia siliqua* L.). Food Public Health 3(6):304–308
8. Bernardo-Gil MG, Roque R, Roseiro LB, Duarte LC, Girio F, Esteves P (2011) Supercritical extraction of carob kibbles (*Ceratonia siliqua* L.). J Supercrit Fluids 59:36–42
9. Ali-Delille L (2010). Les plantes médicinales d'Algérie, vol 82. Berti Editions
10. Rol F (1973) Locust bean gum. In: Whistler RL (ed) Industrial gums: polysaccharides and their derivatives, 2nd edn. Academic Press, New York, pp 323–337
11. De Candolle A (1883) L'origine des plantes cultivées. Balière, París
12. Vavilov NI (1951) The origin, variation, immunity, and breeding of cultivated plants [translated from the Russian by K.S. Chester]. The Ronald Press Co., New York
13. Shepperd VD (2008) Fabaceae-pea family *Ceratonia siliqua* L. woody plant seed manual, pp 371–373
14. Tetik N, Turfan I, Karhan M, Oziyci HR (2010) Characterization of, and 5-Hydroxymethylfurfural concentration in carob pekmez. Gida 35(4):1–6
15. Yousif AK, Alghzawi HM (2000) Processing and characterization of carob powder. Food Chem 69:283–287
16. Jaradat NA (2005) Medical plants utilized in Palestinian folk medicine for treatment of diabetes mellitus and cardiac diseases. Al-Aqsa University
17. Cakılcıoğlu U, Turkoğlu I (2007) Plants and fruits used for cholesterol treatment by the folk in Elaziğ. Phytologia Balcanica 13:239–245

18. Lev E, Amar Z (2002) Ethnopharmacological survey of traditional drugs sold in the Kingdom of Jordan. J Ethnopharmacol 82:131–145
19. Gutzwiller A (2016) Aliments diététiques contenant de la caroube contre la diarhhée du porcelet sevré. Recherche agronomique suisse 7(6):290–295
20. Kotrotsios N, Christaki E, Bonos E, Florou-Paneri P (2012) Dietary carob on growth performance and meat quality of fattening pigs. Asian-Aust J Anim Sci 25:880–885
21. Marakis S (1996) Carob bean in food and feed: current status and future potentials—a critical appraisal. J Food Sci Technol-Mysore 33(5):365–383
22. Gravador RS, Luciano G, Jongberg S, Bognanno M, Scerra M, Andersen ML, Lund MN, Priolo A (2015) Fatty acids and oxidative stability of meat from lambs fed carob-containing diets. Food Chem 182:27–34
23. Priolo Q, Vasta V (2007) Effects of tannin-containing diets on small ruminant meat quality. Ital J Anim Sci 6(SUPPL. 1):527–530
24. Cepo DV, Mornar A, Nigovic B, Kremer D, Radanovic D, Dragojevic IV (2014) Optimization of roasting conditions as an useful approach for increasing antioxidant activity of carob powder. LWT-Food Sci Technol 58:578–586
25. Barroso LS, de Oliveira VR, Garcia AV, Doneda D, Ouriques LA, Vieira MM (2015) Physico-chemical and sensory evaluation of sandwich cookies made with carob. Adv J Food Sci Technol 9(4):290–295
26. Dos Santos LM, Tulio LT, Campos LF, Domeles MR, Hecke Krüger CC (2015) Glycemic response to carob (*Ceratonia siliqua* L.) in healthy subjects and with the *in vitro* hydrolysis index. Nutricion hospitaliria 31(1):482–487
27. Baston O (2016) Production and analysis of *Ceratonia siliqua* L. Powders Ann Food Sci Technol 17:50–54
28. Rosa CS, Tessele K, Prestes RC, Silveira M, Franco F (2015) Effect of substituting of cocoa powder for carob flour in cakes made with soy and banana flours. Int Food Res J 22:2111–2118
29. Makris DP, Kefalas P (2004) Carob pods (*Ceratonia siliqua* L.) as a source of polyphenolic antioxidants. Food Technol Biotechnol 42:105–108
30. Santos M, Rodrigues A, Teixeira JA (2005) Production of dextran and fructose from carob pod extract and cheese whey by *Leuconostoc mesenteroides*, NRRL B512 (f). Biochem Eng J 25:1–6
31. Bernardo-Gil MG, Roque R, Roseiro BL, Duarte LC, Girio F, Esteves P (2011) Supercritical extraction of carob kibbles (*Ceratonia siliqua* L.). J Supercrit Fluids 59:36–42
32. Nasar-abbas SM, Huma Z, Vu T-H, Khan MK, Esbenshade H, Jayasena V (2016) Carob kibble: a bioactive-rich food ingredient. Compr Rev Food Sci Food Saf 15:63–72
33. Avallone R, Cosenza F, Farina F, Baraldi C, Baraldi M (2002) Extraction and purification from *Ceratonia siliqua* of compounds acting on central and peripheral benzodiazepine receptors. Fitoterapia 73:390–396
34. Iipumbu L (2008) Compositional analysis of locally cultivated carob (*Ceratonia siliqua*) cultivars and development of nutritional food products for a range of market sectors. University of Stellenbosch, Stellenbosch
35. Correiaa PJ, Saavedraa T, Gamaa F, Miguela MG, Varennesb A, Pestana M (2018) Biologically active compounds available in *Ceratonia siliqua* L. grown in contrasting soils under Mediterranean climate. Sci Hortic 235:228–234
36. Ozcan MM, Arslan D, Gokcalik H (2007) Some compositional properties and mineral contents of carob (*Ceratonia siliqua*) fruit, flour and syrup. Int J Food Sci Nutr 58:652–658
37. Youssef MKE, El-Manfaloty MM, Ali HM (2013) Assessment of proximate chemical composition, nutritional status, fatty acid composition and phenolic compounds of carob (*Ceratonia siliqua* L.). Food Public Health:304–308
38. Dakia PA, Wathelet B, Paquot M (2007) Isolation and chemical evaluation of carob (*Ceratonia siliqua* L.) seed germ. Food Chem 102:1368–1374
39. Goulas V, Stylos E, Chatziathanasiadou MV, Mavromoustakous T, Tzakos AG (2016) Review functional components of carob fruit: linking the chemical and biological space. Int J Mol Sci 17:1875. https://doi.org/10.3390/ijms17111875

40. Sarni-manchado P, Cheynier V (2006) Les polyphénols en agroalimentaire. TEC&DOC, Lavoisier
41. Roseiro LB, Duarte LC, Oliveira DL, Roque R, Bernardo-Gil MG, Martins AI, Sepúlveda C, Almeida J, Meireles M, Gírio FM, Rauter AP (2013a) Supercritical, ultrasound and conventional extracts from carob (*Ceratonia siliqua* L.) biomass: effect on the phenolic profile and antiproliferative activity. Ind Crops Prod 47:132–138
42. Roseiro LB, Tavares CS, Roseiro JC, Rauter AP (2013) Antioxidants from aqueous decoction of carob pods biomass (*Ceratonia siliqua* L.): optimisation using response surface methodology and phenolic profile by capillary electrophoresis. Ind Crops Prod 44:119–126
43. Papagiannopoulos M, Wollseifen HR, Mellenthin A, Haber B, Galensa R (2004) Identification and quantification of polyphenols in carob fruits (*Ceratonia siliqua* L.) and derived products by HPLC-UV-ESI/MSn. J Agric Food Chem 52:3784–3791
44. Owen RW, Haubner R, Hull WE, Erben G, Spiegelhalder B, Bartsch H, Haber B (2003) Isolation and structure elucidation of the major individual polyphenols in carob fibre. Food Chem Toxicol 41:1727–1738
45. Tetik N, Yüksel E (2014) Ultasound-assisted extraction of D-pinitol from carob using response surface methodology. Ultrason Sonochem 21:860–865
46. Kumazawa S, Taniguchi M, Suzuki Y, Shimura M, Kwon M, Nakayama T (2002) Antioxidant activity of polyphenols in carob pods. J Agric Food Chem 50:373–377
47. Zunft HJF, Lüder W, Harde A, Haber B, Graubaum HJ, Koebnick C, Grünwald J (2003) Carob pulp preparation rich in insoluble fibre lowers total and LDL cholesterol in hypercholesterolemic patients. Eur J Nutr 42:235–242
48. Klenow S, Glei M (2009) New insight into the influence of carob extract and gallic acid on hemin induced modulation of HT29 cell growth parameters. Toxicol In Vitro 23:1055–1061
49. Cháfer A, Berna A (2014). Study of kinetics of the D-pinitol extraction from carob pods using supercritical CO_2. J Supercrit Fluids:212–215
50. Priolo A, Lanza M, Biondi L, Pappalardo P, Young OA (1998) Effect of partially replacing dietary barley with 20% carob pulp on post-weaning growth, and carcass and meat characteristics of Comisana lambs. Meat Sci 50(3):355–363
51. Berna, A, Perez-Gago MB, Guardiola VG, Salazar D, Mulet A (1997) Effect of temperature on isobutyric acid loss during roasting of carob kibble. J Agric Food Chem 45:4084–4087
52. Cantalejo MJ (1997) Effects of roasting temperature on the aroma components of carob (*Ceratonia siliqua* L.). Agric Food Chem 45:1345–1350
53. Almanasrah M, Roseiro LB, Bogel-Lukasik R, Carvalheiro F, Brazinha C, Crespo J, Kallioinen M, Mänttäri M, Duarte LC (2015) Selective recovery of phenolic compounds and carbohydrates from carob kibbles using water-based extraction. Ind Crops Prod 70:443–450
54. Almanasrah M, Brazinha C, Kallioinen M, Duarte LC, Roseiro LB, Bogel-Lukasik R, Carvalheiro F, Mänttäri M, Crespo JG (2015) Nanofiltration and reverse osmosis as a platform for production of natural botanic extracts: the case study of carob by-products. Sep Purif Technol 149:389–397
55. Jeantet R, Croguennec T, Schuck P, Brule G (2006) Sciences des aliments. Biochimie. Microbiologie. Procédés. Produits. I. Stabilisation biologique et physico-chimique. Techniques & Documentation. 122
56. Amarowicz R (2009) Antioxidant activity of Maillard reaction products. Eur J Lipid Sci Technol 111:109–111
57. Kroh LW (1994) Caramelisation in food and beverages. Food Chem 51:373–379
58. Joslyn MA, Nishira H, Ito S (1968) Leucoanthocyanins and related phenolic compounds of carob pods (*Ceratonia siliqua*). J Sci Food Agric 19:543–550
59. Saura-Calixto F, Perez-Jimenez J, Tourino S, Serrano J, Fuguet E, Torres JL, Goni I (2010) Proanthocyanidin metabolites associated with dietary fiber from in vitro colonic fermentation and proanthocyanidin metabolites in human plasma. Mol Nutr Food Res 54:939–946
60. Babovic N, Djilas S, Jadranin M, Vajs V, Ivanovic J, Petrovic S, Zizovic I (2010) Supercritical carbon dioxide extraction of antioxidant fractions from selected Lamiaceae herbs and their antioxidant capacity. Innovative Food Sci Emerg Technol 11:98–107

61. Ak T, Gulcin I (2008) Antioxidant and radical scavenging properties of curcumin. Chem Biol Interact 174:27–37
62. Corsi L, Avallone R, Cosenza F, Farina F, Baraldi C, Baraldia M (2002) Antiproliferative effects of *Ceratonia siliqua* L. on mouse hepatocellular carcinoma cell line. Fitoterapia 73:674–684
63. Wolever TMS (1990) The glycemic index. In: Borne GH (ed) Aspects of some vitamins, minerals and enzymes in health and disease. Karger, Basel, Switzerland, pp 120–185
64. Banuls C, Rovira-Llopis S, Falcon R, Veses S, Monzo N, Víctor MV, Rocha M, Hermandez-Mijares A (2016) A chronic consumption of an inositol-enriched carob extract improves post-prandial glycaemia and insulin sensitivity in healthy subjects: a randomized controlled trial. Clin Nutr 35:600–607
65. Mohamed DA, Hamed IM, Al-Okbi SY (2008) *Ceratonia siliqua* pods as a cheap source of functional food components. DtschLebensmittRundsch 104:25–29
66. Croze ML, Soulage CO (2013) Potential role and therapeutic interests of myo-inositol in metabolic diseases. Biochimie 95:1811–1827
67. Ruiz-Roso B, Quintela JC, Fuente E, Haya J, Pérez-Olleros L (2010) Insoluble carob fiber rich in polyphenols lowers total and LDL cholesterol in hypercholesterolemicsujects. Plant Foods Hum Nutr 65:50–56
68. Tassou CC, Drosinos EH, Nychas GJE (1997) Weak antimicrobial effect of carob (*Ceratonia siliqua*) extract against food-related bacteria in culture media and model food systems. World J MicrobBiot 13:479–481
69. Meziani S, Oomah BD, Zaidi F, Simon-Lever TA, Bertrand C, Zaidi-Yahiaoui R (2014) Antibacterial activity of carob (*Ceratonia siliqua* L.) extracts against phytopathogenic bacteria *Pectobacterium atrosepticum*. Microb Pathog 78:95–102
70. Lee YP, Mori TA, Sipsas S, Barden A, Puddey IB, Burke V, Hall RS, Hodgson JM (2006) Lupin-enriched bread increases satiety and reduces energy intake acutely. Am J Clin Nutr 84(5):975–980
71. Aksit S, Caglayan S, Cukan R, Yaprak I (1998) Carob bean juice: a powerful adjunct to oral rehydration solution treatment in diarrhoea. Paediatr Perinat Epidemiol 12(2):176–181
72. Tsatsaragkou K, Yiannopoulos S, Kontogiorgi A, Poulli E, Krokida M, Mandala I (2014) Effect of carob flour addition on the rheological properties of gluten-free breads. Food Bioprocess Technol 7:868–876
73. Haber B (2002) Carob fiber benefits and applications. Cereal Food World 47(8):365–9
74. Mis A, Grundas S, Dziki D, Laskowski J (2012) Use of farinograph measurements for predicting extensograph traits of bread dough enriched with carob fiber and oat wholemeal. J Food Engr 108(1):1–12
75. Çag Lar A, Eroll N, Elgün MS (2012) Effect of carob flour substitution on chemical and functional properties of Tarhana. J Food Process Preserv:1745–4549
76. Sanchez-Muniz FJ, Botega DZ, Lorenzo L, Marmesat S, Bastida S, Perez-Olleros L, Ruiz-Roso B (2007) A non-extractable condensed-tannins fiber reduces thermal oxidation in oils at frying temperature. Eur J Lipid Sci Tech 109(12):1218–1225
77. Botega ZD, Bastida S, Marmesat S, Perez-Olleros L, Ruiz-Roso B, Sanchez-Muniz FJ (2009) Carob fruit polyphenols reduce tocopherol loss, triacylglycerol polymerization and oxidation in heated sunflower oil. J Am Oil Chem Soc 86(5):419–425
78. Atasoy AF (2009) The effects of carob juice concentrates on the properties of Yoghurt. Intl J Dairy Technol 62(2):228–233
79. Rababah TM, Al-u'datt M, Ereifej K, Almajwal A, Al-Mahasneh M, Brewer S, Alsheyab F, Yang W (2013) Chemical, functional and sensory properties of carob juice. J Food Qual 36(4):238–44
80. Vaheed H, Shojaosadati SA, Galip H (2011) Evaluation and optimization of ethanol production from carob pod extract by Zymomonasmobilis using response surface methodology. J Ind Microbiol Biotechnol 38:101–111
81. Turhan I, Bialka KL, Demirci A, Karhan M (2010) Enhanced lactic acid production from carob extract by Lactobacillus casei using Invertase pretreatment. Food Biotechnol 24:364–374

82. Carvalheiro F, Moniz P, Duarte LC, Esteves MP, Gírio FM (2011) Mannitol production by lactic acid bacteria grown in supplemented carob syrup. J Ind Microbiol Biotechnol 38:221–227
83. Pramod T, Lingappa K (2012) Immobilization of Aspergillus niger in Hen Egg White for the production of Citric acid using carob pod extract. J Microbiol Biotechnol Res 2
84. Roukas T, Biliaderis CG (1995) Evaluation of carob pod as a substrate for pullulan production by Aureobasidium pullulans. Appl Biochem Biotechnol 55:27–44
85. Andrade CT, Azero EG, Luciano L, Goncalves MP (1999) Solution properties of the galactomannans extracted from the seeds of *Caesalpiniapulcherrima* and *Cassia javanica*: comparison with locust bean gum. Intl J Biol Macromol 26(2–3):181–185
86. Turhan I, Bialka KL, Demirci A, Karhan M (2010) Ethanol production from carob extract by using Saccharomyces cerevisiae. BioresourTechnol 101:5290–5296
87. Sahin H, Topuz A, Pischetsrieder M, Ozdemir F (2009) Effect of roasting process on phenolic, antioxidant and browning properties of carob powder. Eur Food Res Technol 230(1):155–161
88. Arrighi WJ, Hartman TG, Ho C-T (1997) Carob bean aroma dependence on roasting conditions. Perfumer Flavorist 22(1):31–41

Chapter 11
Green Extracts from Chinese Medicinal Plants as Bio-source of Drugs

Xinguang Qin, Wanrong Li, Yi He, Zhenzhou Zhu, Gang Liu, Jie Cai, Shuyi Li and Qi Wang

Abstract Traditional Chinese medicine (TCM), which is an important part of Chinese culture, presents several challenges in terms of preparation, such as high extraction temperature, long extraction time, and large amounts of impurities. In recent years, TCM-related engineering has made great progress in many new extraction technologies for Chinese herbs, thereby improving yield and purity of effective components. This chapter describes the extraction assisted by supercritical and subcritical fluids, pulsed electric field, and ultrasonic intensification techniques, especially for the application of Ultrasound-enhanced supercritical CO_2 and subcritical water extraction. These highly efficient approaches have facilitated the development of environmentally friendly technologies and provided remarkable social, economic, and environmental benefits to human beings.

Keywords Extraction · Chinese medicinal plants · Supercritical · Subcritical · Ultrasonic intensification techniques · Pulsed electric field

11.1 Introduction

Chinese medicinal plants exhibit complex chemical composition systems. Usually, a sample contains dozens or even hundreds of components. The content of the active component, as well as ineffective and toxic, vary greatly. The extraction and separation of active ingredients have always been very complicated tasks.

The active ingredients extracted from traditional Chinese medicines (TCMs) refer to compounds with unique functions and biological activities, many of which are material bases for disease prevention and physical fitness. Extraction is the first step in the analysis. Only after extracting the active ingredients of the drug can the follow-up analysis be continued. The active ingredients, such as alkalis, glycosides, ketones, and sugars, found in TCMs mostly belong to the cell contents and are surrounded

X. Qin · W. Li · Y. He · Z. Zhu · G. Liu (✉) · J. Cai · S. Li · Q. Wang
College of Food Science and Engineering, Wuhan Polytechnic University, Wuhan 430023, China
e-mail: lg820823@163.com

Y. Li and F. Chemat (eds.), *Plant Based "Green Chemistry 2.0"*,
Green Chemistry and Sustainable Technology,
https://doi.org/10.1007/978-981-13-3810-6_11

by hard cell walls. The extraction of Chinese herbal ingredients involves sending the solvent into the interior of the plant cells to promote the dissolution of various internal components into the solvent to diffuse out the cells, that is, the transfer of the active components from the solid phase to the liquid phase. An appropriate extraction method can extract the active ingredients and remove the ineffective ingredients. Therefore, TCM extraction methods are particularly important for improving the intrinsic quality and clinical efficacy of TCM.

At present, in the extraction method of active ingredients of TCM, commonly used traditional extraction methods include boiling, dipping, Soxhlet extraction, and so on. However, the traditional extraction method has the following deficiencies: time-consuming; requires a large amount of organic solvents; easily produces solvent residues; exhibits low extraction efficiency; the presence of a large quantity of impurities; and complicated subsequent process [1]. In order to reduce the use of organic solvents and increase extraction efficiency, some new extraction methods have emerged in recent years. This chapter reviews the principles, characteristics, and application status of several new extraction and separation technologies for modern Chinese medicine and discusses its existing problems and future research directions in order to provide a reference to the study of Chinese medicinal plants.

11.2 Techniques

11.2.1 Supercritical Fluid Technology

Supercritical fluid, which refers to the state in which a substance is located at or above its critical temperature, exhibits numerous properties that differ from traditional solvents. A supercritical fluid not only displays low gas viscosity and high diffusion coefficient but also possesses high liquid density and good dissolving ability; moreover, it possesses the properties of a fluid near its critical point (density, viscosity, diffusion coefficient, dielectric constant, and interfacial tension). It demonstrates mutation and adjustability, and it can easily control phase equilibrium characteristic, transfer characteristic, and reaction characteristic of a system by adjusting the temperature and pressure to render chemical processes, such as separation and reaction, more controllable.

Being a "green" enhancement method, supercritical fluid technology not only can greatly reduce environmental pollution caused by chemical processes but also can display a considerably larger diffusion coefficient than that of common solvents, thus significantly improve mass transfer effect and improve the efficiency of separation, reaction, and other chemical processes [2].

Supercritical fluid extraction technology is the most studied type of extraction technology. Studies have focused mainly on theoretical aspects, including determination and correlation of density and viscosity of supercritical fluid, determination of phase equilibrium data in supercritical fluid, and establishment of a thermodynamic model. The mass transfer kinetics is significant in supercritical fluid research. In recent years, researchers have studied microscopically the molecular interactions in supercritical state, aiming to explain the mechanism of selective extraction at the molecular level. In applications, the supercritical fluid extraction technology is mainly used in extraction of active ingredients in natural products and in removal of trace components, such as metal ions and pesticides. Since 1978, when a company in Germany set up the first plant to remove caffeine by using supercritical fluid extraction, this technology has gone through a rapid industrialization. Production plants were established in the USA, Japan, Canada, Italy, and China. Although research on supercritical fluid extraction technology for the extraction of hop essences, natural flavors, pigments, and greases is increasingly becoming mature, the weak solubility of polar substances in CO_2 remains the bottleneck restricting the development of this technology. Therefore, supercritical CO_2 has been employed to strengthen the methods greatly and expand the scope of the application of supercritical fluid extraction [3].

11.2.1.1 Principle and Characteristics

The supercritical fluids, such as supercritical CO_2, are based on the principle of the method to influence the solubility of supercritical fluid by pressure and temperature [4], so as to selectively extract the components of polarity, boiling point, and molecular weight. Separation of the extracted components is achieved by means of decompression, wherein the supercritical fluid is heated until it turns into a gas. A light color stability test has shown that a pigment extracted by the above method is more stable than that obtained through ethanol solvent extraction. The solvent is the temperature and pressure of the CO_2-controlled separation vessel, which can improve the separation effect. Compared with the solvent method, it has the advantages of environmental protection, no toxicity, and good controllability, and can replace the organic solvent which has serious pollution to the environment. In the supercritical state, the reaction rate and selectivity of a target product can be greatly improved, which can reduce or avoid by-product formation, reduce reactant circulation, reduce or eliminate subsequent separation units, promote effective utilization of resources and energy, and reduce emissions and other benefits of great significance[5].

Supercritical fluid extraction, which uses supercritical fluid as a solvent and the characteristics of high permeability and high solubility in this state, is a new technology for extraction and separation of mixed substances. When the fluid is in the supercritical state, the density of the fluid is very close to that of the liquid, and the fluid changes significantly with the change in the fluid pressure and temperature. However, the solubility of the solute in a supercritical fluid increases with the increase in density. Under high pressure, supercritical fluid extraction exploits this property to dissolve solutes in the fluid, and solutes are then precipitated by lowering or raising the pressure or temperature of the fluid solution to achieve specific solute extraction [6].

Supercritical fluid as a reaction medium possesses the following characteristics [7]:

(1) High solubility
By simply changing the pressure, one can already control a reaction phase. We can either make a reaction homogeneous or heterogeneous. A supercritical fluid dissolves most solid organic compounds, allowing a reaction to proceed in a homogeneous phase. Increasing the concentration especially of hydrogen and other gasses with high solubility is conducive to speed up a reaction rate.
(2) High diffusion coefficient
Generally, a solid catalyst is a porous material, and during a liquid–solid reaction, diffusion of liquid into the catalyst occurs very difficultly; thus, the reaction can only be carried out on the surface of a solid catalyst. However, in the supercritical state, due to the large diffusion coefficient of the components of the supercritical fluid and the high solubility of the gas, some reactions controlled by the diffusion can significantly increase the reaction rate.
(3) Effective control of reaction activity and selectivity
Fluids with continuous variation of physical properties (density, viscosity, and polarity) can be influenced by the molecular forces between the solute and the solvent, and the local cohesion directly controls reaction activity and selectivity.
(4) Non-toxic and nonflammable
A supercritical fluid (such as CO_2 and H_2O) is non-toxic and non-combustible and is suitable for safe production; moreover, the use of supercritical fluid is favorable for cost reduction because of its abundance and low cost.

Chemical reaction under supercritical conditions displays the following characteristics [7]:

(1) The homogeneous reaction rate, which is controlled by the diffusion rate, is increased because the diffusion coefficient in the supercritical phase state is larger than that in the liquid phase.
(2) Overcomes interface resistance, increasing the solubility of reactants.
(3) The solubility of solute in a supercritical fluid with a certain molecular mass, temperature, and pressure can be considerably changed to achieve coupling reaction and separation; one can use this property in a timely manner to remove the reaction products from a reaction system to achieve larger conversion rate.
(4) Extends the life of a solid catalyst to prolong its activity.
(5) Pressure in the supercritical medium increases the reaction rate constant.

Features of supercritical extraction technology [8]:

(1) A low operating temperature can better preserve the active ingredients in an extract; production of secondary compounds is not induced; reactions can be completed at near normal extraction temperature; heat-sensitive foods and food flavors do not change; and especially for heat-sensitive, highly oxidative oxidation decomposition substances.

(2) Selective extraction. Separation of natural substances in a high pressure, closed, and inert environment is selective; under optimum process conditions, the extraction rate can be as close to 100%, greatly improving product yield and resource utilization.
(3) Extraction process is relatively simple, efficient, and pollution-free; raw materials and the supercritical fluid are placed together in an extraction kettle. The components of the raw material are selectively dissolved in the supercritical fluid. Following dissolution of the extract in the supercritical fluid, a constant temperature or pressure or a constant temperature heating process is applied in the separation kettle; the extract and the supercritical fluid separate in the separation kettle. The supercritical fluid that separated from the extract can be recycled after refining.

11.2.1.2 Application

As early as 1822, Cagniard discovered the existence of a critical phenomenon. In 1869, Andrews measured the critical parameters of CO_2 [9]. In 1879, Hanny and Hogarth found that supercritical fluid has the ability to dissolve solids, which provides a basis for the application of supercritical fluid technology [10]. Although it has been more than a hundred years since the discovery of the critical phenomenon, its rapid development has only lasted for nearly 30 years. With the deepening of theoretical and applied research in recent years, supercritical fluid has been widely used in the extraction, reaction, granulation, chromatography, cleaning, and other technical processes, and in the chemical, pharmaceutical, food, environmental protection, materials and other fields this technical shows a broad application prospects [11].

Carbon dioxide supercritical extraction has good solvent properties and is easy to operate. While extracting of flavonoids in *Artemisia annua* by carbon dioxide supercritical extraction technology, the extraction rate is 11–59% higher than traditional industrial solvent methods (gasoline and dilute ethanol). At an extraction pressure of 12 Mpa, an extraction temperature of 36 °C, and a time of 3 h, components such as octadecanol can be easily extracted from the plant named *A. annua*. The extraction of tanshinone from *Salvia miltiorrhiza* by supercritical CO_2 extraction is greatly improved compared to the traditional solvent method. Extraction of sea buckthorn oil from sea buckthorn seed and its skin with CO_2 supercritical fluid extraction yields high-quality, residue-free sea buckthorn oil, which successfully overcomes the shortcomings of existing extraction methods. However, supercritical fluid extraction technology belongs to high-pressure technology, and the technology of the process is high. The equipment investment is very large, and the cost is higher than the traditional method, which is not conducive to large-scale production.

The following will describe in detail the active ingredients in supercritical fluid extraction of traditional Chinese medicines plants.

(1) Extraction and separation of terpenoids and volatile oils

Terpenoids are an important class of natural pharmaceutical chemical components with a large number of skeletons, a large variety, a large number, an ever-changing structure, and a wide range of biological activities. From the chemical structure, they are the polymerization of isoprene. In the body and its derivatives, the skeleton is generally based on five carbons as the basic unit, a few exceptions. However, a large number of experimental studies have demonstrated that mevalonic acid (MVA) is the most critical precursor to the pathway of terpenoids, not isoprene. Therefore, derivatives derived from dimethylglutaric acid and having a formula $(C_5H_8)_n$ are referred to as quinone compounds.

Terpenoids are widely distributed in nature and have a wide variety of species. In addition to their major distribution in plants, a large number of terpenoids have recently been discovered from marine organisms. According to incomplete statistics, more than 22,000 terpenoids have been identified. In the study of the chemical constituents of natural medicines, the study of anthraquinones has been a relatively active field and an important source for finding and discovering biologically active ingredients of natural medicines.

Volatile oils contain more complex components, but they are usually rich in terpenoids. The past oil production processes are mostly mechanical pressing, high-temperature steam distillation, or organic solvent (e.g., ethanol) extraction. The purity of the mechanical press oil is extremely low, which greatly reduces its value. The latter two have disadvantages such as low oil yield, poor thermal stability, easy decomposition, complex processes, difficulty in ensuring product quality, and residual organic solvents.

Although the chemical constituents of volatile oils are quite inconsistent due to their different sources, they have a relatively low boiling point, a small molecular weight, and a small polarity. They have good solubility in SC-CO_2, and most of them can be directly extracted with pure CO_2. The required operating temperature is generally low, which avoids the destruction and decomposition of the active ingredients therein, so that not only the product quality is good but also the yield is much higher than the conventional method, and it is the most suitable component for extracting with SC-CO_2. Table 11.1 lists some research examples.

Artemisinin is a sesquiterpene lactone compound from *A. annua* [12]. It is the only anti-malarial drug in China that is internationally recognized. The traditional gasoline method has low yield, high cost, and the danger of inflammability and explosion. With CO_2 supercritical extraction method, from 0.1, 5 L equipment small-scale test to 25, 50 L pilot scale, until the 200 L industrial production proves that the supercritical CO_2 extraction process is superior to the traditional method (gasoline method), the product yield increased by 1.9 times, the production cycle is shortened to 100 h, the cost is reduced, the risk of flammability and explosion is avoided, and the process is greatly simplified.

Ligusticum chuanxiong is the most commonly used traditional Chinese medicine for activating blood circulation, and it has the effect of invigorating blood, removing wind, and relieving pain, which was used to cure the disease of irregular menses,

Table 11.1 Study on extraction of from volatile components of Chinese herbs

Family and genus	Name	Medicinal parts	The yield and characteristics of the extract	
			SFE method	SD method
Zingiberaceae	Fresh ginger	Rhizome	4.3% (3 h), orange thick oil, rich in gingerol, keeping natural flavor	2.2% (5 h), 120 °C superheated steam, basically no gingerol, flavor and ginger are quite different
	Turmeric	Lateral rhizome	4.00% (2 h), oil is fragrant and pure	1.66% (6 h), lighter aroma
	Zingiber corallinum Hance	Rhizome	8–12% (1–2 h), strong fragrance	3–5% (8–12 h), lighter odor
	Galangal	Rhizome	3.2%, yellow oil, pure aroma	0.46% (9 h), light yellow oil
Elliferae	Curcuma zedoary	Rhizome	1.0% (3 h), it can provide different terpene compounds	
	Radix Angelicae Sinensis	Root	1.5% (3 h), twilight oil, ligustilide content 19.82%	0.32% (5 h), brown oil liquid
	Ligusticum wallichii	Rhizome	Ligustilide content 1.57%, 10.3% (4 h), light yellow oil, pure aroma	1.40%(methanol extraction), 0.18% (6 h), light yellow oil, lighter aroma
	Radix Bupleuri	Root	1.86% (4 h)	0.24% (12 h)
	Saposhnicovia divaricata	Root	4.8% (3 h)	0.15% (10 h)
	Notopterygium root	Root and rhizome	8.9% (3 h)	1.38% (5.5 h)
	Radix Angelicae Pubescentis	Root	3.4% (3 h)	0.22% (5.5 h)

(continued)

Table 11.1 (continued)

Family and genus	Name	Medicinal parts	The yield and characteristics of the extract	
			SFE method	SD method
	Fructus Cnidii	Fruit	10% (3 h), the osthole content is 22.61%; 6.3% (reduced pressure); there is a lot of crystallization	Light yellow oil
	Fennel	Fruit	6.8% (3 h), light yellow translucent oil	1.5% (6 h), light yellow oil
	Juniperus formosana	Leaf	Yellowish oily substance, for the first time 20 components such as tretinoin	0.53% (12 h), to hydroquinone lactone content 11.5% 1.31% (5.5 h)
The composite family	Costustoot	Root	2.52% (2 h), hydroquinone lactone content 37%	
	Rhizoma Atractylodis	Rhizome	5.12% (1 h)	
	Artemisia annua	Herba	3.5%, artemisinin content is 16%; a range of components that cannot be extracted by the SD method is obtained	
Celastraceae	*Catha edulis*	Stem leaf	Four types of xanthanes	
Liliaceae	Garlic	Geophyta bulbosa	Allicin 3.77 g/kg raw material	Allicin 1.48 g/kg raw material
Elaeagnaceae	Sea buckthorn	Fruit	Oil yield up to 90%	
Rue family	Fructus evodiae	Fruit	Direct coupling of SFE-GC, better separation of light components	
	Pepper	Pericarp	6.5%	

(continued)

Table 11.1 (continued)

Family and genus	Name	Medicinal parts	The yield and characteristics of the extract	
			SFE method	SD method
Onagraceae	Evening primrose	Seed	Pure color, γ-linolenic acid 9.5%	Color is not ideal, γ-linolenic acid 7.8%
Magnoliaceae	Flower bud of lily magnolia	Flower	80 min, 4.15%	2.4%
	Illicium verum	Fruit	9.5%	
Burseraceae	Frankincense		Get alcohols and esters	
	Myrrh	Resin	Obtained terpene compounds Obtained furose sesquiterpene	
Orobanchaceae	Cistanche	Stem	Get alkanes and esters	
Cyperaceae	Rhizoma Cyperi	Tuber	2.3% (3 h), yellow oil	0.70% (6 h), light red oil
Lauraceae	Cinnamon	Fur	5.8%, pure aroma	
	The root of three-nerved spicebush	Tuberous root	2.9%, there is a white solid precipitated	
Aristolochiaceae	Asarum	Herba	2.4%	
Verbenaceae	Fructus Viticis	Fruit	1.6%, cream, pure	
The grass family	Semen Coicis	Kernel	5.8%	
Cucurbitaceae	*Trichosanthes kirilowii* Maxim	Fruit	11%	

amenorrhea, and algomenorrhea [13]. Volatile oil is its main active ingredient. The traditional steam distillation method yields 0.45–1%, using CO_2 supercritical extraction method from a 1 L laboratory device to a 300 L industrial device. The production results showed that the yield of SFE method chuanxiong oil was 5–8 times that of steam distillation method. It was separated and identified by GC/MS. The chemical composition included two parts, high volatility and low volatility, and 39 components were obtained by SFE method. The steam distillation yields only about 20 components, and the same components account for about 95% of the products obtained by the two methods, mainly compounds with a relative molecular mass of 180 or more, especially some bioactive lactone compounds, such as butylphthalide, butylidenephthalide, ligustilide, neocinllide, senkyunolide, among which ligustilide is the relative contents of SFE; steam distillation products were 38.12 and 45.09%, respectively, and the total lactone compounds were 80 and 70%, respectively. Although the relative content of ligustilide is lower than that of steam distillation, the extraction efficiency of SFE from Rhizoma Chuanxiong is approximately 3 times that of steam distillation, and the entire extraction process of supercritical fluid extraction is completed in 20 min. Steam distillation requires more than ten hours, so the actual extraction of ligustilide will be much higher. The 20 components are different, mainly compounds with a relative molecular weight of 180 or less, especially compounds with a molecular weight of 136. There are fewer small-molecule compounds contained in the volatile oil obtained by steam distillation. These two small-molecule compounds account for only about 5% of the products obtained by the two extraction methods. The possible reason for these differences is that SFE is extracted under low-temperature and closed conditions. Some small-molecule volatile substances are not easy to lose, they more accurately reflect the chemical components in the medicinal materials, and 17 components have been quantified, mainly by α-Pinene, cinnamene, 3-butylene phthalide, etc., in which ethyl palmitate, 9,12-octadecadienoic acid, ferulic acid, and other components were not found in steam distillation products. The content of components such as 3-butylidenephthalein, neonacrol, cnstralactone, artemisinol, ethyl palmitate, 9, 12 stearyldiene, and ferulic acid is several times that of water vapor. The distillation of the product not only benefits the more efficient use of Chinese medicine resources, but also provides a source of material for the new use of the variety, which fully demonstrates the superiority of the supercritical CO_2 extraction technology in the process. The product has been produced on a large scale.

Zanthoxylum bungeanum is a dried ripe peel of Rutaceae or green pepper [14]. It has the effects of analgesic, insecticidal, and itching. It is used for abdominal cold pain, vomiting and diarrhea, puerperal abdominal pain, and tsutsugamushi disease. It is used for external treatment of eczema. It has strong pharmacological activity on the cardiovascular system, digestive system, immune function, coagulation function, analgesia, sedation, anti-inflammatory, antibacterial, insecticidal, and antitumor. The chemical components are mainly alkaloids, aclignin, coumarin, volatile and fatty acids. Volatile oil has more important pharmacological effects and is mainly produced by steam distillation. Chen et al. used supercritical CO_2 fluids to extract essential oils from *Z. bungeanum* under the following conditions: extraction pressure 32 MPa; temperature 40 °C; extraction time 1 h; 58 chemical components were obtained; and 57 chemical groups were identified [15]. Divided by steam distillation, only 22 chemical components were obtained. It can be seen that the supercritical CO_2 fluid is extracted at low temperature, and it can more truly reflect the chemical composition of medicinal materials. Ma et al. used supercritical fluid to extract *Ganoderma lucidum* triterpenes from the wild red fruitbody. By adding 10% entrainer to methanol, the density and pressure of the extraction were greatly reduced, and the extraction was complete only in 10 min.

(2) Extraction and separation of quinones

Quinonoids are natural pigment compounds with unsaturated cyclic diketone structure in the molecule, are a relatively important active ingredient in natural products, and are mainly divided into four types: benzoquinone, naphthoquinone, phenanthrenequinone, and pyrene. Natural quinones and naphthoquinones are mostly free. Therefore, the solvent extraction method can be used. For those containing phenolic hydroxy groups, alkali soluble acid precipitation method can be used for extraction. For the steroids with small molecular weight and volatility, steam distillation can be used. The hydrazines extracted by the above method can be further improved and purificated by chromatography. Terpenoids often exist in their free state and their glycosides in Chinese herbal medicines. Extraction of such components usually uses ethanol or methanol as a solvent, extracts free terpenes and their glycosides, and concentrates the extracts, for the quinone ingredients further separation and purification.

The majority of hydrazine and its derivatives are more polar, so the use of supercritical CO_2 extraction requires the use of higher extraction pressure, and usually also add an appropriate entrainer. Examples of extracts from Chinese medicinal plants are listed in Table 11.2.

S. miltiorrhiza is the most commonly used traditional Chinese medicine for activating blood circulation. The known active ingredients are the tanshinone-based lipid-soluble substances represented by tanshinone IIA and the water-soluble substances represented by danshensu and protocatechuic aldehydes, and the properties differ greatly from each other. The traditional production process of compound Danshen tablet produced by conventional water extraction and ethanol thermal reflux method cannot achieve both of its active ingredients. In the process of extraction and separation, tanshinone IIA is prone to chemical degradation under hot and humid conditions for a long time. Southwest Chemical Research and Design Institute 3 uses different surfactants and entrainers, and adopts CO_2 supercritical extraction method. First, the first entrainer is used for extraction to obtain tanshinone-based fat-soluble active substances. After the basic extraction of tanshinones is completed, a second entrainer is used to continue the extraction and water-soluble active substances mainly composed of danshensu and protocatechualdehyde are obtained. After 1 L labora-

Table 11.2 Examples of extraction from quinones in Chinese herbs

Family and genus	Name	Medicinal parts	Main target component
Labiatae	The root of red-rooted salvia	Roots and rhizomes	Tanshinone IIA
Boraginaceae	Lithospermum	Root	Naphthoquinone pigment
	Arnebia euchroma	Root	Alkannin
Polygonaceae	Rhubarb	Roots and rhizomes	Rhein and emodin
	Polygonum multiflorum	Tuber	Rhein, emodin, and monomethyl ether

tory research and 300 L plant industrial trials, satisfactory results were obtained. This method avoids the chemical changes caused by the moist heat factor in solvent extraction. The primary content of tanshinone IIA is greater than 20%, the content of water-soluble extract is greater than 35%, the yield is more than twice that of conventional methods, and the process is simple.

Xia et al. used supercritical CO_2 extraction technology to extract naphthoquinone pigment from Xinjiang Lithospermum erythropolis. The whole process was only 2 h. Compared with the traditional solvent method, the extraction rate was high, there was no residual solvent, and there were many unknown components [16]. *Rhubarb* is a world-famous Chinese special medicinal material and has been the main medicinal material for treating diarrhea, sputum, and detoxification. It has been reported in the literature that emodin is used as a quantitative indicator of rhubarb herbs and preparations. Its conventional extraction methods are cumbersome and time-consuming, and the general extraction needs several hours or even days, especially in the extraction accompanied by significant alcoholysis reaction, the loss of active ingredients. The supercritical extraction efficiency is high, and the extract can be directly analyzed by spotting. Yuan et al. used supercritical CO_2 extraction technology to extract rhein, emodin, and rhubarb from the radix polygoni multiflori, compared it with the ultrasonic extraction method, and concluded that the supercritical CO_2 extraction has a short time, high efficiency, and simple post-treatment [17].

(3) Extraction of alkaloids

Alkaloids are very important in the active ingredients of traditional Chinese medicines, and they are also the first group of ingredients that people study in plant herbs. The more precise description of the definition of alkaloids is that alkaloids are cyclic compounds that are present in biological organisms and contain an oxidized nitrogen atom. Alkaloids are generally extracted by the solvent method and the precipitation method. The main drawback is that there are many extraction and separation steps and the amount of pollutants is large, because a large amount of acid or alkaline reagents are used in the extraction and separation process. Examples of extraction of alkaloids from Chinese medicinal plants are shown in Table 11.3.

Table 11.3 Examples of extraction of alkaloids from Chinese medicinal plants

Family and genus	Name	Medicinal parts	Effective constituent
Angiospermae	Cusia	Branches or leaves	Indigo yulan
Polygonaceae	Indigo	Branches or leaves	Indigo yulan
Cruciferae	Woad	Branches or leaves	Indigo yulan
Poppy	Corydalis tuber	Tubers	Tetrahydropalmatine
Solanaceae	*Datura metel*	Flowers	Scopolamine
Strychnaceae	Nux-vomica	Seeds	Strychnine
Liliaceae	Mushroom	Bulbs	Colchicine

Strychnos nux-vomica is rich in resources in Yunnan Province in China, and its efficacy is to relieve pain, swelling, and swollen circulation. It is used to treat rheumatism, bruises, and other diseases, such as rheumaatic ioint nain. Its active ingredient is alkaloids, mainly containing strychnine, its conventional content determination requires extraction with alkaline chloroform, cumbersome and time-consuming, and pollutes the environment. Yu et al. used CO_2-SFE to directly extract strychnine from Strychnos nucifera. HPLC detection showed that the extraction was relatively complete with less impurity components and satisfactory results. At the same time, the effects of extraction temperature, extraction pressure, alkalinization agent, entrainer, static extraction time, and dynamic extraction amount on the extraction yield were systematically investigated. The experiment shows that the CO_2-SFE method is effective against Strychnos nucifera. The extraction yield was good at 101 °C, 34.47 MPa pressure, 0.1 mL ammonia as alkalinizer, 0.5 mL acetone as entrainer, CO_2 static extraction for 5 min, and dynamic extraction 5 mL. HPLC assay showed that the yield was basically the same as the solvent method [18].

Datura flower is a commonly used traditional Chinese medicine, and its source is the dry flower of the white plant of the family Succo. With Dingchuan, hurricane, anesthetic analgesic, and other effects, the active ingredient is alkaloids, mainly containing scopolamine, the content of which is routinely extracted with chloroform to prepare samples, and the entire process is cumbersome and time-consuming, and pollutes the working environment. Yu et al. used CO_2-SFE method to extract scopolamine from datura flower, which was time-saving and simple. It took only about 20 min for an extraction process, and the extraction was complete without residual solvent [19]. Detection by HPLC showed that the extraction was complete with less impurities and satisfactory results.

Since alkaloids are often combined with the acidic components of plants in the presence of salts, the use of pure CO_2 is often difficult to put forward. It must be alkalized with an alkaline reagent such as ammonia prior to extraction to make them all convert to free base. Also use suitable modifiers or entrainers to enhance the fluid's ability to dissolve or increase selectivity. Commonly used entraining agents are a static, ethanol, isopropyl alcohol, ethyl acetate, chloroform, and water. Jiang et al. used supercritical fluids to extract colchicine, an antitumor drug, from light mushrooms and used 76% ethanol as an entrainer [20]. The extraction rate was 1.25 times that of the reflux extraction method, and the extraction time was reduced by 0.45 times. Li et al. used supercritical carbon dioxide fluid to extract piperine from lotus root and added 0.4 mL of methanol as a modifying agent to perform extraction automatically [21]. The whole process can be completed in 20 min, and the extract can be directly injected without filtration. The extraction has a certain degree of selectivity, less impurities extracted, and almost no pollution to the environment, which is unmatched by the traditional solvent extraction.

(4) Extraction of sugar and glycoside

The sugars contained in traditional Chinese medicines include monosaccharides, oligosaccharides, and polysaccharides. Sugar has been abandoned as an ineffective ingredient in the past. In recent years, due to the discovery that certain polysaccharides have a variety of biological activities, it has increasingly attracted people's attention. Substances represented as anti-tumor polysaccharides are obtained from Basidiomycetes (Polyporaceae) of Prunus mosella polysaccharides. Later, people further discovered that polysaccharides have complex and multifaceted functions. For example, the sugar chains of recently discovered polysaccharides play a decisive role in molecular biology. They can control and regulate cell division, differentiation and growth, and aging. These have attracted the attention of scientists at home and abroad. For polysaccharides in traditional Chinese medicine, most of the traditional extraction methods use water and dilute alkali solutions at different temperatures for extraction, and are then separated and purified by distributed precipitation method, salting out method, metal coordination method, and quaternary salt precipitation method. For various kinds of glycosides, due to the different structures of aglycone, the type and number of sugars connected are not the same, and it is difficult to have a uniform extraction method. If the extraction is performed in the order of polarity from small to large with solvents of different polarities, the presence of glycosides may be present in each portion. Due to the large molecular weight of sugars and glycoside compounds, many hydroxyl groups, and large polarity, the extraction yield with pure CO_2 is low, and adding entrainers or increasing the pressure can increase the yield.

Liao found that when SC-CO_2 was used to extract the Tibetan medicinal herbs *G. lucidum*, it was found that without SC-CO_2 extraction of the entrainer, even if the pressure was increased, the total saponins and polysaccharides could not be extracted [22]. Under the same conditions, non-gradients with different entrainers were added. With the increase of the polarity of the entrainer, the polysaccharide yield of the extract increased, and the yield of the crude saponin was gradually reduced. In the gradient SC-CO_2 extraction, the yield of the crude total saponin and polysaccharide was traditionally 189 and 162 times the extraction process.

Ginsenoside is a disaccharide saponin found in the roots, stems, and leaves of ginseng, which has the effect of improving the body's immune function. Zhang used SC-CO_2 extraction technology to study the influencing factors of extracting ginsenoside from ginseng leaves [23]. The content of ginsenosides in the extract can reach 30%. Shi et al. discussed and determined the optimal conditions for the extraction of diosgenin by SC-CO_2 and conducted a pilot scale-up [24]. It was found that the yield of this process is high, and the production cycle can be greatly shortened. Liao et al. extracted total saponins from *G. lucidum* and polysaccharides, compared with the three extraction processes, and found that the crude product and yield of a total saponin increased by 18.9- and 1.62-fold, respectively, after SFE plus entrainer [22].

(5) The extraction of coumarin and lignin

Coumarin, also known as coumarin, can be seen as a lactone derived from the dehydration of cis-hydroxycinnamic acid and is a class of compounds having a benzo[a]pyrone ketone core.

Lignin is a natural compound that is polymerized from bimolecular phenylpropanoids. Most of them are formed by the polymerization of β-carbon atoms in the side chain, so their mother nucleus consists of two C_6–C_3.

In the traditional extraction method, the small-molecule coumarin is volatile and can be extracted by steam distillation. For other coumarins, alkali dissolution, acid precipitation, and system solvent methods can be used for extraction and separation, and chromatographic methods can also be used to separate coumarins with similar structure. Free lignin is a lipophilic component that can be obtained through multiple solvent extractions with higher purity. Lignin is often coexisted with a large amount of resinous substances in plants, so it should be first proposed with hydrophilic solvents such as ethanol and acetone, and the resulting extracts should be extracted by chloroform, ether, and the like. The separation of water quality factors should be based on the structural characteristics of specific compounds, using alkaline extraction, adsorption chromatography, or partition chromatography. Examples of extraction from coumarin and lignin in Chinese medicinal plants are shown in Table 11.4.

Radix Angelicae Pubescentis is dry root of the Umbelliferae family. It has the effect of dehumidification and analgesia during the night. It is a commonly used medicine for the treatment of rheumatism. Modern pharmacological studies have shown that the main components of coumarins contained in Duhuo have the effect of inhibiting platelet aggregation and anti-thrombosis and thus have the effect of promoting blood circulation and collaterals. Cai et al. applied the supercritical CO_2 fluid technology to the extraction of coumarins in singho, started from the three factors of pressure, temperature, and modifier addition amount, and adopted orthogonal experiments with three factors and three levels to screen out. The best extraction conditions have achieved better extraction results [25]. At the same time, the experimental results show that the addition of entrainer has almost no effect on the content of total coumarin in Soho.

Table 11.4 Examples of extraction from coumarin and lignin in Chinese medicinal plants

Family and genus	Name	Medicinal parts	Main target component
Umbelliferae	Celery seeds	Seed	N-Butylphathlide
	Fructus cnidii	Fruit	Furocoumarin
Rue family	Asiatic toddalia root	Heel cover	Aculeatin, toddalolactone
Magnoliaceae	*Magnolia*	Flower	Methoxy and magnolol
	Virginiana L.		
	Magnolia officinalis	Stem bark, root bark, and bark	Magnolol
The composite family	*Artemisia capillaris*	Herba	Capillarisin
Moraceae	White mulberry root-bark	Heel cover	Morusin, sanggenon

Paconol, also known as paeonol, peonol, is chemically named 2-hydroxy-4-methoxyacetophenone. It is an effective chemical ingredient in the whole plant of Tree Peony Bark and Luo Moke plant Xu Changqing. Pharmacological studies have shown that paeonol has anti-inflammatory, analgesic, antipyretic, and immunomodulatory effects. Paeonol is an ideal drug since it can be sublimed slightly and has aroma and good permeability. Paeonol is generally extracted by steam distillation or solvent method. The solvents used are mainly water and ethanol. However, the extracted product must be purified by repeated recrystallization. Paeonol has poor water solubility. Fang et al. selectively extracted paeonol from the paeonol with supercritical CO_2 [26]. The results are as follows.

① Paeonol can be extracted from paeonol with high selectivity using supercritical CO_2.
② The paeonol content in the paeonol was 1.53%.
③ The optimum conditions for extracting paeonol from the paeonol by supercritical CO_2 are: extraction temperature 45 °C, pressure 10 MPa, extraction time 75 min. Under this condition, the average extraction rate of paeonol was 99% (mass fraction), and the product paeonol mass fraction was 96.249%.

(6) Extraction of flavonoids

Flavonoids refer to a class of compounds whose basic nucleus is 2-phenylchromone. Refer to a chemical composition that has two (C_6–C_3–C_6) benzene rings (A ring and B ring) linked to each other through intermediate three carbon chains. Because of the keto group in the molecular structure, its color is often yellow, so it is called flavonoids.

The traditional methods for extracting flavonoids include alcohol extraction, alkaline or alkali alcohol extraction, and hot water extraction. The separation of the crude product is mainly based on its polarity difference, acidity strength, molecular size, and special structure, and adopts appropriate methods, separation by methods such as system solvent method, pH gradient extraction method, boric acid complex method, and lead salt precipitation method, etc. These traditional extraction methods have the disadvantages of a large amount of pollutants discharged, large loss of effective components, low extraction efficiency, and high cost. Examples of extraction of flavonoids from Chinese medicinal plants are shown in Table 11.5.

Research on *Ginkgo biloba leaves* has become one of the hotspots in the development of modern plant drugs in the world. There are many researches on the extraction and separation of active chemical components. The main production process adopted abroad is that 60% acetone is used as the extraction solvent, and the product *G. biloba* extract is obtained through a series of separation and purification steps; ethanol extraction and resin adsorption and extraction are widely used in China. These processes are cumbersome and have high costs, low yields, and poor product intrinsic quality. Moreover, residual organic solvents and heavy metals in solution extraction methods may also cause toxic and side effects. Deng et al. established a set of experimental equipment and experimental methods for small-scale and intermediate-scale testing of SFE leaves [27] (*G. biloba* flavonoids and lipids). By comparison with

Table 11.5 Examples of extraction of flavonoids from Chinese medicinal plants

Family and genus	Name	Medicinal parts	Main target component
Ginkgoaceae	Ginkgo leaf	Leaf	Ginkgo flavone, ginkgolide
Leguminosae	Liquorice	Disporum sessile	Glycyrrhizin, glycyrrhizin A, licorice chalcone B
Zingiberaceae	Galangal	Disporum sessile	Galangin
Theaceae	Tea	Sporophyll	Tea polyphenols
Rosaceae	Saffron	Flower	Ink red pigment
Moraceae	Maclura	Heel cover	Flavanone, xanthone
	White mulberry root-bark	Heel cover	Flavonoid

the solvent extraction method, it can be seen that the SFE process is short and the extraction yield is high, which is 2 times higher than the solvent extraction method (1%); the extraction batch test operation time is much shorter than the solvent extraction; the quality of the effective components of the *G. biloba* leaves is high. In the internationally recognized quality standards, CO_2 is used as the extraction medium and the extraction operation is performed at 35–40 °C. The natural quality of the effective components of *G. biloba* leaves is maintained, and there are no residual organic solvents and heavy metals.

Tea polyphenols have significant antioxidant activity and positive ability to scavenge free radicals. They have broad application prospects in the food and pharmaceutical industries, and they are actively carrying out extraction research at home and abroad. However, most of the extraction of tea polyphenols is currently used the solvent extraction, and then the separation method. This method consumes a large amount of solvent, and the separation process is cumbersome. Li et al. first tried to use supercritical CO_2 to extract tea polyphenols from tea leaves [28]. The relationship between tea polyphenol extraction and CO_2 was determined in a flow-making device. Finally, the extracts were simply separated and the relative purity of tea polyphenols was 95.45%. Tea polyphenols. In addition, it was also found in the experiment that the addition of aqueous ethanol as a carrier enhances the extraction of tea polyphenols by about 10 times.

(7) Extraction of other compounds

In addition to the above-mentioned research system, the extraction of medicinal active ingredients such as some fat-soluble seed oils, natural vitamins, plant sterols, and phenols has also been studied in detail. Table 11.6 lists relevant research examples.

Modern pharmacological studies have demonstrated that fat oil of the *Perilla frutescens* has a lipid-lowering effect, and it has the functions of enhancing intelligence, improving memory, and visual acuity. It is currently the most abundant plant resource of α-linolenic acid, which has important physiological activity and can inhibit the occurrence and metastasis of cancer. Extraction of perilla oil by

Table 11.6 Extraction of other ingredients in Chinese medicine plants

Family and genus	Name	Medicinal parts	Extracting ingredients
Lamiaceae	Perilla	Seeds	Alpha-linolenic acid in fat oil
Rosaceae	Almond	Seeds	Fatty oil
Leguminous	Bitter split peas	Seeds	Fatty acids
Cruciferae	Rapeseed	Seeds	Fatty acids
	Rapeseed	Oil residue	Natural Ve and grease proof
Oleaceae	Fructus Forsythiae	The fruit	Fat-soluble constituent
Rangoon creeper division	Fructus chebulae immaturus	The fruit	Fatty oil
Ginkgo biloba	Ginkgo	Leaf	Phenols and acids
Zingiberaceae	Ginger	Rhizome	Ginger hydroxybenzene
Compositae	*Artemisia annua*	Herba	Octadecanol and beta-glusterol
Ranunculaceae	Cortex moutan	Bark	Demethol (content > 80%, oil yield 2.5%)
Gold millet blanc	Grass coral	Herba	Concrete

SC-CO_2 method can increase the content of α-linolenic acid to 73.46%, which is 3.69–4.25% higher than traditional solvent method (petroleum ether method) and pressing method, and can be extracted. The components that cannot be extracted by the solvent method do not leave any solvent residue, which greatly shortens the extraction time. The specific comparison of the fatty oils obtained by the two methods is shown in Table 11.7.

The famous *Sarcandra glabra* medicine toothpaste and grass coral lozenge are all configured with squid coral all-grain boiled extract. Due to the high boiling temperature and long time, many ingredients will decompose, oxidize, or hydrolyze, and the process of concentrating boiled liquid will make the extract brownish black, which is not conducive to the manufacture and development of the product. In order to overcome the above deficiencies, Li et al. took ethanol as an entrainer and used SC-CO_2 to extract scombroid corals to obtain light green substances with a fresh

Table 11.7 Comparison of two extraction methods for extracting perilla oil

Project	SFE method	Solvent method (petroleum ether method)
Oil rate/%	28.1	18
α-Linolenic acid content	77.3	52
Extraction time/h	4	50
Appearance	Clear, golden, or tan oil with special fragrance	Yellowish or brownish yellow muddy oily liquid with fragrance

aroma [29]. Their main component content index was higher than that required by squid coral extracts.

Almonds are the seeds of Rosaceae plant *Prusmandshurica* and have antitussive and anti-asthmatic effects. The almond oil was extracted with SC-CO_2. Compared with the traditional petroleum ether method, the oil yield increased by 1.54 times while the extraction time was shortened by 40 h. The obtained fatty oil was methylesterified and analyzed by GC-MS to obtain five components such as octadecenoic acid.

11.2.1.3 Prospect

With the rapid development of supercritical fluid technology, the theory and applied research on this technology also gradually deepened. For more than three decades, not only the field of application of supercritical fluid technology did continue to expand; rather, a variety of supercritical fluid technologies have also been industrialized. However, due to being non-ideal of supercritical fluid and the lack of research methods suitable under high pressure, theoretical research works have focused mostly on the macro-level thermodynamics and kinetics, and relatively few studies have dealt on the molecular level. Given that research on supercritical fluid remains considerably scarce, the theoretical basis for supercritical fluid technology research needs to be further strengthened. Since molecular simulation can obtain useful information, such as thermodynamics, transfer properties, and spectroscopic properties without involving high-pressure experiments, molecular simulation will play an increasingly key role in the future research. Although technologies such as supercritical fluid extraction and reaction have been industrialized, the capacity of the supercritical fluid to dissolve is limited, resulting in large equipment volume and high investment; thus, the use of this technology is only applicable to high value-added substances. Although adding a modifier can improve the dissolving ability of supercritical fluid to a certain extent, it also reduces its "greenness" and increases the difficulty of post-treatment. A breakthrough in the scale-up and automation of new media and high-pressure equipment for supercritical fluids will greatly enhance the economics of supercritical fluid technology. In addition, the coupling of supercritical fluid technology with other technologies (such as rectification, adsorption, and membrane separation) can also help improve the efficiency of chemical processes and reduce costs, and such approach will become a new trend in research on supercritical fluid technology [30].

Supercritical fluid extraction technology is a relatively new type of extraction separation technology; it originated in the 1940s, developed until the twentieth century, and successfully used in industry in the 1970s. With the advancement of science and technology and the improvement of living standard, people have obtained a new understanding of health and environment, putting forward higher standards and requirements for physical and mental health-related products, such as cosmetics, food, and medicine and their production processes. Supercritical fluid extraction technology has shown a good application prospect in food processing industry, fine chemical industry, pharmaceutical industry, and environment, and it is a unique, efficient, and clean novel extraction method; moreover, it has become the primary option as an alternative to the traditional chemical extraction methods [31]. Governments worldwide have made considerable progress in focusing on the basic theories of supercritical fluid extraction, extraction equipment, and industrial applications in terms of manpower and resources. This method has developed rapidly in the fields of food, medicine, and chemical industry, and has made great progress, especially in the extraction of active ingredients of biological resources. Additionally, it has become a new hotspot in many industries.

With the growing awareness on environmental protection and the call toward sustainable development, people are increasingly calling the attention of certain industries, such as the chemical, food, and pharmaceutical industries. The pursuit of a high-quality, efficient, clean, and green industrial development route will inevitably become a trend. Supercritical fluid extraction technology is called a "green technology" and offers many benefits, such as environmental protection and recyclability. Therefore, the development of supercritical technology is promising and its future market is huge. This technology will undoubtedly bring a huge social, economic, and environmental benefit.

11.2.2 Subcritical Water Extraction Technology

Subcritical water refers to the state of water at a temperature above the boiling point, below the critical point of the liquid water, under certain pressure conditions. The decrease in polarity of water with increasing temperature has become one of the current research hotspots. This technology has been used to extract active ingredients, such as volatile oil and tannin. The advantages of this technology include environmental protection, controllability, and time-saving; however, subcritical water is a high-temperature and high-pressure water and can only be used to extract substances with good heat resistance [32]. For heat-sensitive components, striking a balance between extraction temperature and extraction rate is difficult. Therefore, investigation on a highly efficient and environment-friendly subcritical water extraction green strengthening technology offers important scientific value and practical significance.

11.2.2.1 Principle

Subcritical water, also known as high-pressure hot water, super-hot water, or hot liquid water, refers to a water with a certain pressure (usually 5–20 MPa) and a temperature above 100 °C; the critical temperature is below 374 °C, hence called high-temperature water, and it maintains its liquid state [33]. Subcritical water extraction involves the use of subcritical water as an extractant in airtight, anaerobic, high-pressure containers, according to the principle of similar compatibility of organic matter through the extraction of materials and extractants in the soaking process of molecular diffusion process to achieve solid materials. In the new extraction and separation technology, fat-soluble components are transferred into the liquid extractant and then concentrated through dehydration of the extraction solvent; target product separation is subsequently performed to obtain the final product [34].

At a low temperature, the distance between water molecules is small. Under subcritical condition, as the temperature increases, the molecular kinetic energy increases, and the intermolecular distance also increases. Moreover, the hydrogen bonding, ionic hydration, ion association of the fluid microstructure, and cluster-like structure change; thus, the physical and chemical properties of subcritical water at room temperature and atmospheric pressure quite differ [35]. Dielectric constant is a measure of a solvent's ability to dissolve. Water polarity is considerably stronger at room temperature and pressure, and the dielectric constant *e* of water decreases with increasing temperature in a subcritical state. Hydrogen bonds in water are turned on or weakened to reduce drastically the polarity of water, thereby allowing water molecules to be brought closer together with less polar solute molecules to increase attractiveness. Such a phenomenon is similar in nature to organic solvents, with which the dissolution of solute molecules is promoted, thus allowing the extraction of different polar target substances. Therefore, by controlling the temperature and pressure of subcritical water, the polarity of water is remarkably changed to realize the continuous extraction of substances, ranging from water-soluble to fat-soluble active substances, in natural products. Subcritical water extraction has become the focus on the field of extraction and separation given that only water is used as solvent with advantages of being non-toxic, safe, efficient, environment-friendly, and easy to operate.

11.2.2.2 Influence Factors

Subcritical water extraction technology is characterized by the following: simple operation, nonuse of organic solvents and other toxic reagents, pollution-free, and high degree of automation. Compared with the traditional extraction method, the innovative technology offers the advantages of short extraction time, high extraction efficiency, low energy consumption, and multi-polarity gradient extraction by controlling the extraction temperature. These advantages render subcritical water technology a research hotspot in many professional fields, such as environment, light industry, chemical industry, and food industry. However, current research on subcritical water extraction technology remains at the laboratory scale for the sample preparation stage. The extraction device offers the following advantages: small scale, simple design, and no unified design standard; these characteristics have not yet been scaled up for industrial application. Therefore, the subcritical water technology remains to be further developed given that it is still in its exploratory stage both locally and abroad.

The main factors influencing the efficiency of subcritical water extraction include extraction temperature, extraction time, use of entrainer (organic solvent or surfactant), pressure, particle size, geometric characteristics of extraction kettle, and the flow of water on the solid substrate as raw material; the critical water extraction efficiency was not significantly affected [36] and was often used as a secondary factor. When determining the extraction protocol, it is often compared with other methods such as Soxhlet extraction and ultrasound enhanced solvent extraction.

Extraction Temperature

Since subcritical water alters polarity by increasing the temperature to achieve the extraction and separation of organic compounds, the temperature is one of the most crucial factors affecting extraction efficiency. Temperature also considerably affects the other extraction methods. The biggest difference between the subcritical water technology and the traditional Soxhlet extraction involves temperature. Subcritical water extraction temperature is often above the boiling temperature; by contrast, Soxhlet extraction temperature cannot exceed the boiling point of the extraction solvent. For organic pollutants in the environment, subcritical water extraction temperature will seriously affect the recovery rate of the target substances; for example, some weakly polar organic pollutants and a temperature control of above 300 °C will help increase the solubility of a target substance [37–39], thereby improving extraction efficiency.

Temperature also considerably affects the extraction of active substances from natural plants or Chinese herbal medicines. Heating at 125–180 °C effectively increases the recovery of volatile components, such as kavalactone, and essential oils in raw plant materials [40–43]. Most polar and nonpolar compounds found in medicinal plants display thermal instability. To effectively increase the extraction rate of nonpolar substances, the temperature must be set between 150 and 250 °C. Continuous heating easily leads to thermal degradation of the target compounds and to other side effects. During subcritical water warming, the recovery rates of some of the natural metabolites found in Chinese herbal medicines, such as berberine, strychnine, aristolochic acid, 5,6,7-trihydroxyflavone, glycyrrhizin, and tanshinone IA and IIA, increase, but these compounds started to degrade after the extreme temperatures were reached [36, 44–47]. Subcritical water technology was used to extract essential oils from natural medicines, and antioxidants were extracted from *Rosmarinus officinalis*. Similar phenomena occurred in extraction of epicatechin and catechins in tea and grape skins [48–52]. In addition, subcritical water extraction was performed on several labeled compounds in *Astragali* and *Codonopsis*, when the temperature was raised from 80 to 160 °C, the subcritical water extraction rate of three labeled substances showed a downward trend, whereas the three labeled substances showed a good stability at 160 °C in methanol solvent [36]. Thus, subcritical water extraction system and its temperature factors may adversely affect the target compounds. Therefore, the temperature factors in subcritical water extraction of the target substances in biomass or plant matrices require a comprehensive review.

Extraction Pressure

The main effect of increasing the pressure from 10 to 80 bar is that it maintains the water in a liquid state and prevents the water from boiling at very high temperatures. Therefore, the pressure should be maintained at a certain level. Pressure exerts only a marginal effect on the recovery of the target substances as seen during the subcritical water extraction of organic contaminants from soil samples in the environment [36]. In addition, pressure had no effect on the recovery of essential oil from plant materials and the recovery of ginseng from American ginseng [40, 41, 53–55]. Studies have reported that pressure was maintained at the lowest level. In addition to safety considerations, another possible reason is to avoid the compression of solid matrix particles caused by high pressure, which would affect mass transfer.

Extraction Time

Subcritical water extraction can take two different modes [34]: static equilibrium mode, such as the use of Dionex ASE200 extraction unit, and dynamic cycle extraction mode, such as a laboratory setup that requires two high-pressure circulating pump systems. Few studies have compared the extraction times of the traditional solid–liquid two-phase extraction and subcritical water extraction. One advantage of the subcritical water dynamic cycling extraction mode is that the solvent water is continuously renewed throughout the extraction process; however, a greater water volume is required than that in the static mode. The problem with the static mode is that complete extraction of a target compound may not be achieved due to the use of a limited amount of solvent water. In the static extraction mode, 10–15 min or less is needed to complete a quantitative recovery. In the dynamic circulation extraction mode, one can achieve the ideal recovery of a target compound in Chinese herbal medicine or environmental solid samples within 5 min to 2 h. Studies have shown that for many active substances in medicinal plants, the ideal recovery rate can be obtained within 40 min using the subcritical water dynamic circulation mode extraction [56–58]. Thus, the time and water volume required in subcritical water extraction should be confirmed and optimized through specific experimental procedures.

Use of Entrainer

Subcritical water extraction process uses two broad categories of entrainer, namely organic solvents and surfactants. Extraction of catechins from tea and grape seeds and of phenolic compounds in grapes demonstrated the lowest recovery of subcritical water as an extraction solvent compared with the recovery of organic solvents, such as methanol, ethanol, and ethyl acetate [49, 57]. Compared with Soxhlet extraction, the use of subcritical water extraction of *Astragalus*, *Codonopsis*'s labeled compounds, and tanshinone IIA in *S. miltiorrhiza* yields lower recovery [36, 59, 60]. Increasing the subcritical water temperature alone did not improve the extraction efficiency of tanshinone IIA, but increasing the amount of ethanol entrainer (0–30) in subcritical water effectively increased the extraction yield. Similar results were obtained from the extraction of berberine from rhizomes [59].

Addition of an organic solvent entrainer, such as ethanol, to subcritical water significantly improves the recovery of the target compounds in medicinal plants [36, 46, 60, 61]. When ethanol is maintained at 0–20% in subcritical water, the solubility of the natural metabolite in a plant feedstock in fluid is effectively increased. Moreover, the consumption of organic solvents is considerably reduced compared with that in the traditional solvent extraction method. Using the same method to analyze benzalkonium chloride in sediment samples and artificially discarded indicator compounds, it was found that the use of acetonitrile/water with a volume ratio of 6:4 or 7:3 or isopropyl alcohol with a volume ratio of 50:50 alcohol/water as a subcritical fluid resulted in a more satisfactory recovery. When used as a subcritical fluid, mixed solvents, such as isopropanol/water, can effectively reduce sample preparation, solvent consumption, and minor background interference in the system [36].

11.2.2.3 Application

Subcritical water extraction technology was first proposed by a research team represented by Hawthorne in the mid-1990s. They found that water polarity considerably changes at a high temperature [62–64]. On the basis of this finding, subcritical water technologies have been successively used in numerous fields of study, including environmental testing [65, 66], extraction of functional components [67], extraction and separation of the target compounds from raw botanical materials and medicinal plants [36], pretreatment of raw food materials [68], and chromatographic analysis [69]. Subcritical water technology does not involve a wide range of fields; in addition, studies on its extraction mechanism and mass transfer process are conducted to promote and further establish the foundation subcritical water technologies. The recently reported application of subcritical water extraction technology is summarized as follows:

Extraction and separation of active compounds are affected by biomass of raw materials, the use of different rhizomes or seeds, the application of subcritical water extraction and separation of compounds with different structures, and varying polarities and activities. Recent literature compared traditional steam distillation, Soxhlet extraction and ultrasonic-enhanced solvent extraction, subcritical water extraction using head lavender, coriander seeds, Damascus rose, chrysanthemum, traditional Chinese herbal medicine, flax, and qingning. Subcritical water extraction offers the advantages of simple operation, the absence of solvent residue, environmental protection, and high efficiency.

Foodstuffs have also been used by some researchers as an alternative to study subcritical water extraction technologies; these researchers include Herrero et al., who extracted compounds with special functional activity or high added value from different vegetable materials [70]. It has also been reported that the use of defatted rice as a raw material and the extraction and preparation of functional proteins or amino acids [71], or citrus or dried tangerine extract gum [72], have yielded good results. In addition, subcritical water technology has been used to extract valuable substances, such as catechins and anthocyanins, from by-products or wastes in breweries [52], and anthocyanins from onion wastes [73]. Applications of this technology in the food sector include preparation of banana and nutraceuticals [74, 75], preparation of barley functional beverages [50], and extraction of wandanone antibacterial compound from milk [76]; the use of this technology in the analysis of methamidophos residue in vegetables [77] opens up new directions for subcritical water applications.

In addition, the use of subcritical water extraction of some of the natural metabolites of medicinal plants involves numerous problems; for instance, the extraction rate is not high or it easily leads to thermal degradation during the heating process. Methods to improve extraction efficiency by forming micelles in subcritical water have been used to remove PAHs from soil and environmental samples. The specific operation is as follows: Sodium dodecyl sulfate (SDS) is added into the subcritical water fluid, and then the sample is subjected to static–dynamic combined extraction mode. The fluid pressure is maintained at 50 bar, and the temperature is controlled at 150–225 °C. To increase the recovery of labeled compounds in the root of *C. pilosula*, the surfactant was combined with subcritical water technology, and the sample was treated at a lower temperature of 95 °C. The effects of different surfactants, such as SDS and trinitrotoluene, were studied. Results showed that quantification of labeled compounds in *Astragalus* and *Codonopsis* by subcritical water with trinitrotoluene is superior to that of Soxhlet extraction [46]. Another research team selected American ginseng as a solid matrix model, and they studied the use of liquid nonionic surfactant in a subcritical water system as an alternative solvent. Apart from ultrasonic organic solvent extraction, the extraction method involving a nonionic trinitrotoluene surfactant, which is commonly used in subcritical water at a concentration above the critical micelle concentration, can be used to obtain pharmacologically active substances from ginseng root [53]. In addition to the extraction temperature and extraction time, the type and amount of surfactant are the parameters that must be considered to optimize the surfactant-assisted subcritical water extraction. However, studies have also shown that the use of surfactants, such as trinitrotoluene, at concentrations greater than the critical micelle concentration at certain temperatures has no significant effect on the amount of extracted labeled compound [46, 53].

The use of trinitrotoluene to increase the solubility of a target compound for surfactant-assisted subcritical water extraction in a dynamic cycle extraction mode and the continuous renewal of the aqueous solvent flowing through the solid substrate ensure that the target component in the sample is completely dispersed into the mobile phase. The natural metabolites in *C. pilosula* and *Panax ginseng* have a strong interaction with the solid matrix. The surfactant, which is added into the subcritical fluid, is intended to interfere with the natural forces in medicinal plants and ultimately improve the extraction efficiency. Moreover, extraction at lower temperatures does not significantly reduce the extraction rate but can effectively avoid the degradation of heat-labile components in medicinal plants. Studies have shown that the addition of surfactant in subcritical water can extract more hydrophobic species from medicinal plants [46, 53].

Compared with the earlier reports on the active ingredients extracted by subcritical water from natural products, the recent reports have a more extensive research scope and research objects. The research objects have become more extensive, involving marine organisms obtained in different regions to marine biomass, and the research contents are comprehensive. Moreover, the influence of extraction conditions on a specific biological activity of the target substance was tackled. The effects of extraction conditions in solvent extraction, supercritical CO_2 extraction, and subcritical water extraction were compared and analyzed to obtain a sound reference to the extraction of active compounds from natural products.

11.2.2.4 Summary and Prospect

Subcritical water technology as a new technology, numerous basic theoretical studies, and comprehensive have not yet been done. Domestic research on this technology has just started. Thus, the related literature is limited, and the application range remains narrow. Only extraction of essential oils is reported. Moreover, research content is monotonous and is focused on the initial stage of process optimization and without further analysis and discussion. At the same time, subcritical water extraction mostly involves solid raw materials. Due to the properties of solid matrices and the unique requirements and limitations of high temperature and high pressure of subcritical water, installation of a stirring device in an extraction kettle is impossible. Most of the materials are stacked together. Low efficiency, heterogeneity of the extraction process, and defects manifest with the increasing amount of material and packing height. In addition, the extraction rate of subcritical water technology requires improvement, especially for heat-sensitive substances whose extraction method required the addition of organic solvents or surfactants (e.g., entrainer method) to achieve a more satisfactory extraction and recovery rates; these heat-sensitive substances greatly limit the application and development of subcritical water extraction technology.

These problems restrict the application of subcritical water extraction technology in varying degrees. The main difficulties in the extraction and separation of active ingredients of natural products include alleviating the harsh temperature conditions and ensuring high extraction efficiency. Therefore, one of the effective ways to benefit environmental protection offered by subcritical water extraction is to further develop this technology.

11.2.3 Ultrasonic Strengthening Technology

Ultrasonic extraction broadly refers to the use of ultrasonic energy to accelerate or enhance the extraction of specific components from plants, animals, minerals, and other substances. It is a new application of power ultrasound in the fields of medicine and chemical extraction. Power ultrasound has long been widely applied in the industrial, agricultural, military, medical fields, environmental protection, and cosmetics, among others. With the gradual development in the medical and food industries, an in-depth study on the extraction and on the role of active ingredients has become important, especially in the current modernization of Chinese medicine. Given its rapid development, the use of power ultrasonic extraction technology has become a research hotspot in the medical field and food industry (e.g., foodstuffs, oilseeds, and spices).

The application of ultrasonic technology in analytical chemistry is constantly developing. Ultrasound is very useful in pretreatment of solid samples. It promotes and accelerates extraction process and prevents damage caused by high temperature to the target components. The advantages of ultrasound include simple operation, low amount of by-products, and easy separation of the target, and ultrasound can achieve better results than conventional extraction [78].

11.2.3.1 Ultrasound-Enhanced Supercritical CO_2 Extraction Technology

Ultrasound-enhanced supercritical CO_2 extraction technology [79] involves supercritical CO_2 extraction while attached to an ultrasound field; this technology can significantly enhance the mass transfer process during extraction and separation, shorten the extraction time, reduce extraction pressure and energy consumption, increase extraction rate, and reduce production costs, thus breaking the current constraints. The cost of supercritical fluid extraction has become a bottleneck in industries; the use of ultrasound as a means to enhance supercritical CO_2 extraction attracts the interest of researchers and thus has become a research hotspot. The experiment flowchart of ultrasound-enhanced supercritical CO_2 extraction is shown in Fig. 11.1.

Principle

① Cavitation Effect

Ultrasound itself cannot directly act on a molecule but through the physical effects of the surrounding environment in order to affect the molecule; thus, the roles of ultrasound and the environment are closely related. The cavitation effect, a unique form of interaction between acoustic energy and matter, is a very effective means of process enhancement. This phenomenon is the basis of the study on the mechanism of the enhanced separation process in ultrasonic fields [80].

The cavitation effect is a dynamic process of vibration, growth, and collapse of nucleus in a liquid under the action of sound intensity. Under the action of pressure waves, the average distance between molecules in liquid changes with the vibration of molecules. When a sufficient negative pressure is applied to a liquid, the intermolecular distance exceeds the critical molecular distance for maintaining the

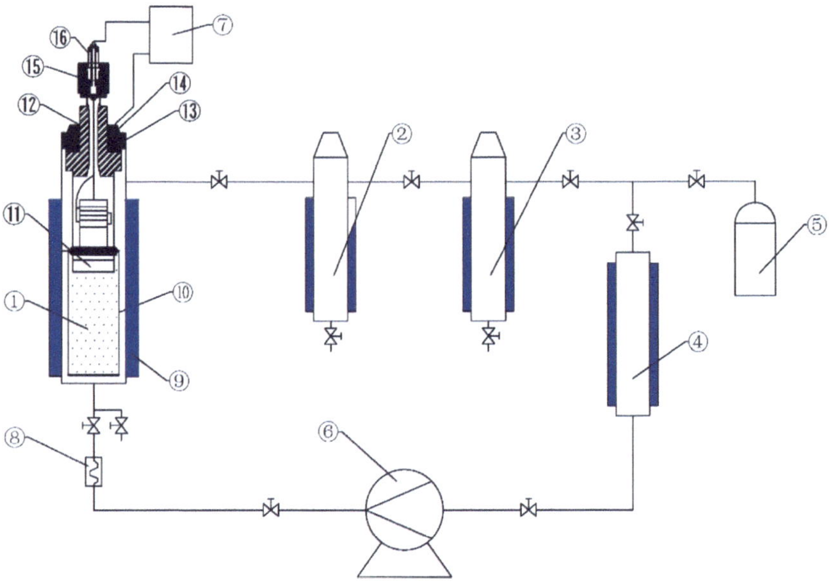

①Extraction of material ②First separating kettle ③Second separation kettle ④Cooling tank ⑤Gas cylinder ⑥High-pressure pump ⑦Ultrasonic generator ⑧Preheater ⑨Circulating water insulation cover ⑩Charging ladle ⑪Ultrasonic transducer ⑫Circlip for shaf ⑬The extraction kettle lid ⑭Big nut ⑮Jam nut ⑯Binding post

Fig. 11.1 Experiment flowchart of ultrasound-enhanced supercritical CO_2 extraction

liquid, and holes are formed. Once a cavity is formed, it will grow until the negative sound pressure reaches a maximum value, but these holes will be compressed again in the successive positive phase of the sound wave. As a result, some cavitation bubbles will enter a continuous oscillation. Others will completely collapse. Cavitation is classified into transient cavitation (cavitation bubbles with a shorter life cycle when the sound intensity is greater than 10 W cm^{-2}) and steady cavitation (cavitation bubbles produced when the sound intensity is low, and the size of the cavitation bubble oscillates around its equilibrium size with a cycle of several cycles). These two kinds of cavitation can be generated simultaneously in a liquid. Under certain conditions, steady-state cavitation can be transformed into transient cavitation, and transient cavitation adiabatic shrinkage can occur. A very short time to the collapse can produce a high temperature of above 5000 K and a high pressure of approximately 5.05×10^7 Pa in a very small space around the cavitation bubble. Temperature change rate reaches up to 109 K s^{-1}, accompanied by strong shock waves and micro-jets with speeds reaching up to 400 km h^{-1}. The jet has a strong impact on the solid surface and forms a jet beam, resulting in hole erosion, causing the outer layer of solid particles to flake off. The cavitation that causes the solid exfoliation produces a strong impact force that can open the strong binding bond (376.8–418.6 kJ mol^{-1}) and promote "water-phase combustion" reaction. The cavitation effect in the liquid is mainly determined by the physical effects of rapid heating and cooling caused by the cavitation bubble implosion process. The characteristics of the implosion temperature and reaction can easily change factors, such as acoustic frequency, acoustic intensity, ambient temperature, static pressure, and the ambient gas used. When various solution cavitation bubbles implode to generate "hotspots," the solution molecules at the "hotspots" may be excited and jump to a high-energy state to speed up the mass transfer process.

Thus far, no direct method that measures the cavitation effect exists, and many observations have been made through indirect methods, such as aluminum foil corrosion. Since the ultrasound-enhanced supercritical extraction process is performed in a high-pressure environment where pressure exceeds 10 MPa, the difficulty of observation is further increased. In the related research of Yang et al. [81], Ultrasound-enhanced supercritical fluid extraction device with built-in transducer can produce 200 W power ultrasonic in a 1 L extraction kettle. Although no corrosion of aluminum foil was observed, the conditions for cavitation in the supercritical state and voids were generated. The relevant characteristics of the bubble were theoretically deduced, and the conditions for the closed sound pressure and closed-in sound intensity required for ultrasonic cavitation were as follows:

Threshold sound pressure:

$$P_{\mathrm{B}} = P_{\mathrm{l}} - P_{\mathrm{v}} + \frac{2}{3}\sqrt{\frac{\left(\frac{2\sigma}{R_{\mathrm{l}}}\right)^{3}}{3\left(P_{\mathrm{l}} - P_{\mathrm{v}} + \frac{2\sigma}{R_{\mathrm{l}}}\right)}}$$

Threshold sound intensity:

$$I_{\mathrm{B}} = \frac{1}{2\rho c}\left(P_{\mathrm{l}} - P_{\mathrm{v}} + \frac{2}{3}\sqrt{\frac{\left(\frac{2\sigma}{R_{\mathrm{l}}}\right)^{3}}{3\left(P_{\mathrm{l}} - P_{\mathrm{v}} + \frac{2\sigma}{R_{\mathrm{l}}}\right)}}\right)^{2}$$

where P_{l} is the supercritical fluid pressure, P_{V} is the cavitation bubble vapor pressure, σ is the cavitation bubble surface tension coefficient, R_{l} is the cavitation bubble initial radius, ρ is the fluid density, and c is the sound velocity in the fluid.

Based on the theoretical calculation, cavitation occurs in supercritical CO_2, its closed-value sound pressure must reach 7.5551×10^{6} Pa, and its ultrasonic output power must reach 4.88×10^{5} W. The test device cannot meet the above power requirements, and thus no cavitation was observed.

According to Yang's formula [81] for calculating the natural vibration frequency of cavitation bubbles, the initial radius of the cavitation bubbles at a frequency of 20 Hz Ultrasound-enhanced supercritical fluid extraction is 1.73 mm, which is larger than the corresponding initial radius of cavitation bubbles (0.139 mm) in water.

$$f_r = \frac{1}{2\pi R_l}\left[\frac{3K}{\rho}\left(P_l + \frac{2\sigma}{R_l}\right) - \frac{2\sigma}{\rho R_l}\right]^{\frac{1}{2}}$$

where R_l is the initial radius of the cavitation bubble, K is the variable index, P_l is the supercritical fluid pressure, σ is the cavitation bubble surface tension coefficient, and ρ is the fluid density.

② Other Enhancement Effects

In addition to the cavitation effect, the roles of ultrasound and media can be categorized from the physical point of view, as follows: mechanical effects, thermal effects, and the ability to disrupt cells of raw materials, thereby enhancing the separation process.

The extracted material is generally a non-uniform layered medium. The acoustic impedance of each layer is different so that the pressure and tension and acceleration of the ultrasound abdomen and nodes are changed. The ultrasonic waves of 20 kHz and 9.74 W cm^{-2} are propagated in a supercritical fluid with a density of 500 kg m^{-3}. The amplitude of sound pressure was 117.44 kPa, and the maximum particle vibration velocity was 1.66 m s^{-1}. For 13.20 μm, the maximum particle acceleration can reach 2.08×10^5 m s^2, which is more than 21,224 times the gravity acceleration. This intense and rapidly changing mechanical movement exerts a positive effect on the separation process. At the same time, the ultrasonic wave is a mechanical vibration wave containing energy. When it propagates in the medium, the ultrasonic energy passes through the medium; the medium particle is periodically compressed; the extractive medium has a different acoustic impedance, and a standing wave is generated, causing intermolecular relative motion produces friction. The friction caused by the movement increases the temperature of the medium, speeds up the movement of the solute molecules, produces the thermal effect of ultrasound, and then contributes to the Ultrasound-enhanced extraction process.

The "turbulence effect" [82] produced by the ultrasonic field causes a large turbulence intensity, which makes the vortex perturbation of the system very strong. This makes the mass transfer boundary layer of solid–liquid interface to be reduced, which can strengthen the mass transfer coefficient. The effect of ultrasonic field on leaching rate during solvent extraction is significantly greater than that of mechanical agitation. The "perturbation effect" generated by the ultrasonic field can effectively reach the inside of a particle, causing the extremely fast-changing particle vibration and ultrasonic force to "wash" and "renew" the surface of a particle's inner surface, and then the supercritical fluid enters the inside of the pore. The fluid undergoes "micro-stirring" to reduce the internal diffusion resistance, to accelerate the internal diffusion, and to enhance the mass transfer within the material. The "growth effect" produced by the ultrasonic field breaks the chemical bonds between the substances, destroys or weakens the binding force of the matrix to the solute, changes the original thermodynamic equilibrium conditions during the general extraction process, and improves the extraction rate. The fragmentation effect of the ultrasonic wave breaks the cell wall of the raw material, impairs its resistance of the active ingredient in the cell to dissolution, and allows the solute to be directly transferred into the solvent.

Under the combined effect of the above ultrasonic effects, they play a different role at different stages of extraction and increase the mass transfer coefficient to enhance the supercritical CO_2 extraction process.

The establishment of ultrasound-enhanced supercritical CO_2 extraction kinetics model [83] is based on three methods: an empirical model, a model based on heat transfer analogy, and a model based on differential mass balance. The establishment of the kinetic model not only helps one to understand the mass transfer characteristics and the change in concentration of the active ingredient in the extraction process over time but also predicts the experimental results, which helps reduce the experimental exploration workload and the economic budget before investment. This approach will effectively reduce investment risks. However, most of the kinetic models are complex, and their application is very narrow. Therefore, many aspects of its dynamic model require further study.

After a plant is crushed, a certain amount of solute comes into contact with the outer surface of the granules due to the rupture of the cell wall, and the remaining solute is entrapped inside the solid granules. The distribution of solute inside the particles can be approximated as the distribution of solute in the porous matrix. The mass transfer of solute from the porous matrix to the supercritical CO_2 fluid can be assumed to take place in this physicochemical mode.

(1) Supercritical CO_2 molecules diffuse to the surface of the plant through an interface layer that binds to the surface of the particles.
(2) The diffusion of supercritical CO_2 molecules into the porous matrix through the diffusion in the pores of the particles, that is, into the interior of the plant.
(3) The solute in supercritical CO_2 fluid in pores is solubilized.
(4) The solubilized solute diffuses into the interfacial layer through the pores of the particles, that is, into the surface of the plant.
(5) The solute diffuses into the external supercritical CO_2 fluid through the interfacial layer of particles and supercritical CO_2 fluid.

In the extraction process, steps (4) and (5) often control the rate, thus called the rate-controlling steps. Therefore, the speed of the extraction process mainly depends on these two steps. For general liquid solvents, the diffusion coefficient is relatively small. In the absence of stirring, the influence of step (5) is relatively large; however, for supercritical CO_2 solvents, the influence of step (5) can be ignored since the diffusion coefficient is two to three orders of magnitude larger than that of general liquid solvents. Step (4) (inward diffusion) is considered the main rate-controlling step, which is more easily satisfied in the case of ultrasonic strengthening. Both the general liquid solvent and the supercritical CO_2 solvent can ignore the influence of step (5) but only for supercritical. The CO_2 solvent will be efficient even if it ignores the influence of step (5).

Compared with the ultrasonic field enhancement process, the mechanism of ultrasonic field enhancement process is still its infancy stage, mostly in the stage of qualitative discussion. Hu Aijun and others proposed theoretically that application of the ultrasonic waves in the supercritical fluid can produce cavitation; the result is discussed in the supercritical fluid cannot produce cavitation, because to make the supercritical fluid in the nucleus cavitation. The required sound pressure amplitude and sound intensity are very large, and thus ultrasound cannot be achieved. To verify whether ultrasound can produce cavitation in supercritical fluid, a cavitating foil was placed in a supercritical fluid and treated with ultrasound for 4 h. The changes in foil before and after ultrasound treatment were compared. The results show that the ultrasonic treatment of the foil does not produce cavitation, rupture, and other phenomena; that is, ultrasound does not produce cavitation in the supercritical fluid. The ultrasonic strengthening effect is suggested to have been caused by "mass mixing." In addition, the transfer of ultrasonic energy can activate the solute, reduce the energy barrier of the process, increase the movement of solute molecules, and accelerate the dissolution. Mei et al. further studied the enhancement of ultrasound and concluded that the thermal effect of ultrasound in supercritical fluid and the intense and rapidly changing mechanical effect contribute to the enhancement of supercritical fluid extraction. Low-frequency ultrasound is better than high-frequency ultrasound. The possible reasons are as follows: First, high-frequency ultrasound quickly consumes energy in the fluid. To obtain the same extraction rate, larger energy is required under high frequencies. Second, frequency f and amplitude A are inversely proportional. The smaller the frequency, the greater the amplitude, and this condition is more conducive to the enhancement of the extraction effect. Moreover, low-frequency ultrasound increases the amplitude of the medium molecule (CO_2), so a low-frequency ultrasound increases the CO_2 molecular polarizability σ compared with a high-frequency ultrasound. For the extraction of polar substances, when the polarity of CO_2 molecules increases, according to the principle of similar compatibility, the solubility of polar substances increases, and the extraction rate increases accordingly.

Balachandran et al. also studied the mechanism of ultrasound-enhanced supercritical CO_2 extraction [3]. The remarkable increase in extraction rate is due to the destruction of cell structure and the increased possibility of solvent infiltration into the particle structure, which increases intra-particle diffusivity. Cell damage may simply be caused by the rapid changes in fluid density induced by ultrasonic pressure waves. Additionally, they consider the possibility of cavitation collapse on the surface of ginger particles as a mechanism for mass transfer enhancement. However, they still cannot prove the existence of this phenomenon, leaving the possible cause of cell damage unknown. The mechanism of ultrasound-enhanced supercritical CO_2 extraction thus remains to be further studied.

Characteristics

In the supercritical fluid technology, it is the supercritical fluid extraction technology that is closely related to the separation of traditional Chinese medicines. CO_2 is the most commonly used supercritical fluid. Currently, supercritical fluid extraction is mainly used for the research and development of traditional Chinese medicine, and it has been paid more and more attention by researchers. The use of supercritical CO_2 extraction technology for traditional Chinese medicine research and development and industrialization has many unique advantages compared with the traditional methods of traditional Chinese medicine.

① The supercritical CO_2 extraction has strong extraction capacity and high extraction rate. Using supercritical CO_2 to extract the active ingredients of traditional Chinese medicines, under the optimal conditions, the components to be extracted can be extracted almost completely, thereby greatly improving product yield and resource utilization. At the same time, with the continuous advancement of supercritical CO_2 extraction technology, the application of ammonium carboxylates of perfluoroether (PFPE), supercritical CO_2 extraction has been extended to aqueous systems to make supercritical CO_2 extraction of highly polar compounds such as proteins that are difficult to extract. It has become possible.
② The size of the extraction capacity depends on the density of the fluid and ultimately depends on the temperature and pressure. Changing one of them or simultaneously changing them can change the solubility and can selectively separate a variety of substances from traditional Chinese medicine, thereby reducing impurities and making the traditional Chinese medicine. The active ingredients are highly enriched, which facilitates dose and quality control, and product appearance is greatly improved.
③ The critical temperature of supercritical CO_2 is low, the operating temperature is low, and the effective components of traditional Chinese medicine can be well preserved without being destroyed. Therefore, it is particularly suitable for the extraction of components that are sensitive to heat and easy to oxidatively decompose and destroy, because the traditional methods of steaming, boiling, and frying easily cause the destruction of biologically active substances.
④ The extraction time is fast, and the production cycle is short. The separation begins as soon as the supercritical CO_2 extraction (dynamic) cycle begins and the components will be separated and precipitated after 10 minutes of extraction, and the components will be separated and precipitated after 10 minutes of extraction, about 2–4 h can be completely extracted. At the same time, it does not require steps such as concentration, and even if the entrainer is added, it can be removed by a separation function or simply concentrated.
⑤ Supercritical CO_2 extraction is easy to control the operating parameters, therefore, the active ingredient and product quality are stable.
⑥ Supercritical CO_2 also has antioxidant and sterilizing effects. Therefore, if supercritical fluid extraction technology is used to sap active components, the oxygen-consuming microorganisms in the extract will be completely killed and the biologically active substances will not be destroyed. It can help ensure and improve product quality.
⑦ The supercritical CO_2 extraction process is simple and easy to operate, saves labor and a large amount of organic solvents, and reduces the pollution of three wastes. This undoubtedly provides a new extraction, separation, preparation, and concentration method for the modernization of traditional Chinese medicine. In

the past, traditional Chinese medicine extractants used organic solvents and were harmful to humans and the environment. They also caused fire. The products needed to be purified in order to remove trace amounts of solvents. Supercritical CO_2 is now used instead of these organic solvents to avoid this problem.

Application

Ultrasound-enhanced extraction has been applied in the food, medicine, and chemical industries. Li et al. used extraction rate as a reference indicator to explore the effects of ultrasonication on the supercritical fluid extraction of total anthraquinones from rhubarb under stability parameters. They also compared the extraction efficiencies of the ultrasound, supercritical fluid, and ultrasound-enhanced supercritical fluid extraction method [84]. The results shown in Table 11.8 illustrate that ultrasonication can enhance the supercritical fluid extraction of total anthraquinones from rhubarb.

Liang et al. subjected soybean slurries to ultrasonication for protein and solid extraction. They found that treating soybean slurries and bean dregs with low-frequency ultrasound improved extraction rates [85]. Zhang et al. preliminarily studied the supercritical CO_2 extraction of pyrethrins under ultrasonication and showed that supercritical CO_2 extraction in an ultrasound field effectively improved pyrethroid extraction rate and prevented pyrethroid degradation [86]. The Natural Products Sound Chemistry Research Group of School of Light Industry and Food Engineering in South China University of Technology used an Ultrasound-enhanced supercritical fluid extraction device with a capacity of 1 L to extract EPA and DHA from seaweed [87], oil and huiqinren ester from coix seed [88], and flavonoids from *Toona sinensis* leaves [89]. Extraction system pressure, extraction temperature, entrainer amount, and extraction time decreased, and extraction rate increased by 30% when Ultrasound-enhanced supercritical fluid extraction was used relative to those when supercritical fluid alone was used. Lin used Ultrasound-enhanced supercritical CO_2 reverse micro-emulsion extraction equipment, an AOT-based surfactant,

Table 11.8 Effect of different extraction methods on the extraction of total anthraquinones from rhubarb [84]

Method	Extraction yield (%)	Extracts of total anthraquinones (%)	Total anthraquinone extraction rate (%)
Ultrasonic extraction	18.89	2.6	32.85
Supercritical extraction	A little	A little	–
Supercritical entrainer extraction	3.11	26.85	52.35
Ultrasound-enhanced supercritical extraction	2.28	26.01	39.67

ethanol cosurfactant, and ultrasound to study the supercritical CO_2 reversed-phase micro-emulsion extraction of ginsenosides [90]. They found that under the extraction pressure, temperature, and time of 24 MPa, 45 °C, and 4 h, respectively, and CO_2 flow rate of 2 L/h, ultrasound-enhanced AOT/ethanol/supercritical CO_2 reverse micro-emulsion provided a ginsenoside extraction rate of 7.42, which was 2.30 times that provided by supercritical CO_2 inverse micro-emulsion extraction method with ethanol as the entrainer.

Sethuraman performed the ultrasound-enhanced supercritical fluid extraction of capsaicin from pepper and obtained high extraction rates and extraction vessel load [91]. Ricra et al. reported that 20 kHz ultrasound positively affected the mass transfer of supercritical fluid, accelerated the mass transfer of almond oil, improved the extraction rate of almond oil by approximately 20%, and shortened extraction time by approximately 30% [92]. Balachandran studied the effect of ultrasonication on the supercritical extraction of freeze-dried ginger particles. Under ultrasonication, the final extraction rate of ginger phenol, the compound that confers ginger's spicy taste, increased by 30% [3]. Table 11.9 summarizes some of the applications of ultrasound-enhanced supercritical extraction in the extraction of compounds from Chinese medicinal plants.

Summary and Prospect

From the perspective of the scope of research, supercritical CO_2 extraction technology has been applied to the extraction and separation of active ingredients of Chinese medicine including alkaloids, combinations, coumarins, Huangli, and volatile oils. However, due to the nonpolar and low molecular weight characteristics of CO_2, this technology is difficult to effectively extract many strong polar and molecular weight components. Although the effect of extraction can be improved by the addition of entrainers, the advantages of this method when compared to conventional extraction methods may not be as pronounced or even inferior to traditional extraction methods. Therefore, the application of new technologies should be comprehensively considered according to their suitable conditions and should not be blindly new.

Nowadays, although the laboratory has conducted a lot of research and accumulated certain experience and data, basic research on supercritical fluids is still relatively weak; in order to amplify the preliminary results of the experiment to the industrialization of large-scale production, there are a lot of basic research and chemical engineering work needs to be resolved.

Experimental studies have shown that ultrasound enhances supercritical CO_2 extraction. Ultrasound not only reduces the pressure of the extraction system, extraction temperature, and the amount of entrainer and extraction time but also improves the extraction rate, while the structure of the extraction object does not change. The possible mechanisms of ultrasound enhancement are attributed to the mechanical fluctuations in ultrasound and thermal effects. The application of Ultrasound-enhanced supercritical CO_2 extraction technology is currently focused on the research on the enlargement design of Ultrasound-enhanced extraction device: the scale-

Table 11.9 Applications of ultrasound-enhanced supercritical extraction in the extraction of compounds from Chinese medicinal plants

Extractive	Traditional Chinese medicine plant	Extraction conditions	Extraction rate
Proanthocyanidins	Larch bark	60% ethanol solution as the solvent, using ultrasound of 320 W for 20 min, two times	10.44% [93]
Cyclohexanecarboxylic acid	*Cynara scolymus* L.	70– 95% ethanol solution as the solvent, using ultrasound of 200–800 W for 0.5–3 h	More than 8%
Synephrine	Citrus dropt fruits	Ultrasonic power 250 W, extraction temperature 45 °C, extraction time 4.5 h	43.9% [94]
Diterpenes and polysaccharides	*Salvia officinalis* L.	The ultrasound-assisted extraction was performed at 40 °C using ethanol–water mixture (70:30) and 18 mass% yield of extract after 45 min	7.3% [95]
α-Amyrin and β-amyrin	*Eugenia pyriformis* C.	30% amplitude, 60 °C temperature, and 1:20 ratio	1.79% [96]
Passion fruit seed oil	*Passiflora edulis*	Ultrasound was applied at 16 MPa and 40 °C temperature	29% [97]

up of Ultrasound-enhanced extraction process and Ultrasound-enhanced extraction mechanism. With the deepening of research, Ultrasound-enhanced supercritical CO_2 extraction technology will offer a very wide range of applications in the chemical, food, biotechnology, environmental engineering, and other technical fields.

11.2.3.2 Ultrasound-Enhanced Subcritical Water Extraction Technology

Subcritical water is high-temperature liquid water with a temperature above 100 °C and a critical temperature of 374 °C under certain pressure conditions. Its physical and chemical properties are quite different from the normal temperature and pressure water, and the polarity of the latter is stronger. The polarity of the former is subcritical and decreases with increasing temperature. The properties are similar to organic solvents. Using this feature, it can realize the continuous extraction of water-soluble and fat-soluble active ingredients from natural products, and it has the advantage of

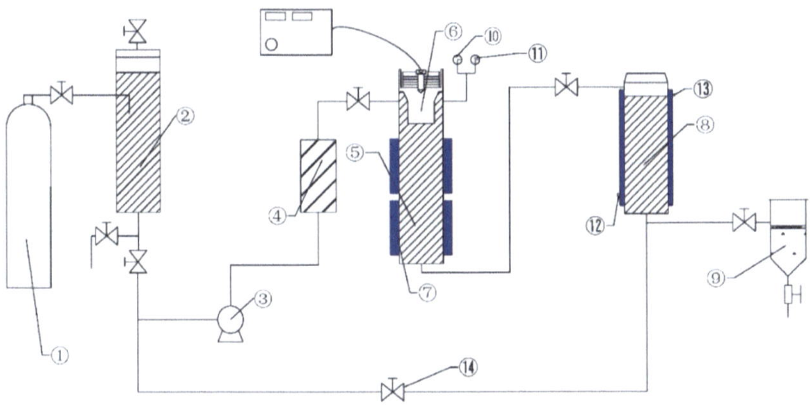

①Nitrogen container ②Reservoir ③Force pump ④Preformed heating device ⑤Extraction kettle ⑥Ultrasonic intensifier ⑦Thermostat ⑧Cooler ⑨Collector ⑩Pressure tester ⑪Temperature measuring device ⑫ Cooling water inlet ⑬ Cooling water outlet ⑭ Control valve

Fig. 11.2 Experiment flowchart of ultrasound-enhanced subcritical water extraction

being cheap, easy to obtain, and green, and has become one of the research hotspots in the current extraction and separation field. However, the temperature of subcritical water is high, and when the extraction time is long, it will easily cause irreversible changes in the heat-sensitive material. When some nonpolar and macromolecular substances are extracted, the conditions are demanding, which affects the extraction efficiency and limits its application and development.

In order to overcome these drawbacks, ultrasonic probes were introduced on the basis of a subcritical water extraction device, and the subcritical water extraction process was enhanced by physical means. The experiment flowchart of ultrasound-enhanced subcritical water extraction is shown in Fig. 11.2.

Principle

Enhancement of a subcritical water system by the ultrasonic waves is mainly manifested in the cavitation effect that occurs when ultrasound propagates in a subcritical water liquid medium, destroying plant cell walls, thereby allowing easy extraction of active substances from plants; vibration, micro-jet, micro-acoustic flow, and other multipolar effects accelerate the diffusion of solutes, making the complete extraction of active compounds from plants possible [98]. The following explains the mechanism of ultrasonic strengthening from the perspective of the cavitation and mechanical effects generated by ultrasound.

Ultrasonic cavitation is the dynamic process involving the oscillation, growth, contraction, and collapse of tiny nucleus in a liquid under the action of the ultrasonic waves. Researchers studying ultrasonic cavitation theory generally use the Rayleigh–Plesset equation (abbreviated as R-P equation) to describe the morphological change in the radius of cavitation bubble with time; this equation was gradually modified to describe the changes in the radius of cavitation bubble with time [99].

Considering the effects of viscous losses, surface tension coefficients, and acoustic radiation in subcritical water, Rayleigh–Plesset equations can be described by the following forms of nonlinear equations:

$$\rho\left(R\ddot{R}+\frac{3}{2}\dot{R}^2\right)=P_g(R,t)-P_a(t)-P_l+\frac{R}{c}\frac{d}{dt}\left[P_g(R,t)-P_a(t)\right]-4\eta\frac{\dot{R}}{R}-\frac{2\sigma}{R}$$

If the subcritical water compressibility and the internal saturated vapor pressure are also considered, bubble motion can be described by the improved Rayleigh–Plesset equation:

$$R\ddot{R}\left(1-\frac{\dot{R}}{c}\right)+\frac{3}{2}\dot{R}^2\left(1-\frac{\dot{R}}{3c}\right)=\frac{1}{\rho}\left(1+\frac{\dot{R}}{c}\right)\left[1+t_g(R,t)-P_a(t)-P_l-4\eta\frac{\dot{R}}{R}-\frac{2\sigma}{R}\right]$$
$$+\frac{R}{\rho c}\frac{d}{dt}\left[P_g(R,t)-P_a(t)\right]$$

The formula shows that:

① R is the radius of the cavitation bubble, the velocity and acceleration of the motion of the cavitation bubble wall are represented by and, respectively, is the density of the subcritical water, $P_g(R,t)$ is the gas pressure in the cavitation bubble, $P_a(t)$ is the ultrasonic pressure exerted on the cavitation bubble wall, P_l is the pressure of the subcritical water, c is the speed of sound in the subcritical water, η is the kinematic viscosity coefficient of the subcritical water, σ is the surface tension coefficient of the subcritical water, and P_v is the saturated vapor pressure in the cavitation bubble.

② Normal power ultrasound, which acts on cavitation bubbles, is a positive and a cosine sound wave. Let this be $P_a(t) = -P_A \sin\omega t$
where P_A is the sound pressure amplitude, $\omega = 2\pi f$ is the angular frequency of sound waves, and f is the frequency of sound waves.

③ Cavitation bubble gas is considered an ideal gas, and heat exchange between gas and water is not considered in the calculation and is obtained according to the adiabatic state equation:

$$P_g(R,t)=P_g(R_0,t_0)\left(\frac{V_0}{V}\right)^r=P_g(R_0,t_0)\left(\frac{\frac{4}{3}\pi R_0^3}{\frac{4}{3}\pi R^3}\right)^r=P_g(R_0,t_0)\frac{R_0}{R}^r$$

where R_0 is the initial radius of the cavitation bubble, $\gamma = \frac{C_p}{C_v}$ is the specific heat ratio, also called the adiabatic coefficient, C_p is the isobaric specific heat, C_v is the equal volume heat, and γ takes 4/3. $P_g(R_0,t_0)$ is the initial pressure of the gas in the cavitation bubble, and it is equal to the initial pressure outside the bubble, $P_g(R_0,t_0) = P_l + \frac{2\sigma}{R_0} - P_V \cdot V(t) = \frac{4}{3}\pi R(t)^3$ is the cavitation bubble in the volume at time t, and $V_0 = \frac{4}{3}\pi R_0^3$ is the volume of the cavitation bubble at the initial time t_0.

The initial conditions in the formula are: $R(t = 0) = R_o$, $\dot{R}(t = 0) = 0$.

The modified Rayleigh–Plesset equation is an ordinary differential equation with respect to the radius R of the cavitation bubble. Obtaining a general solution is difficult. Only the numerical solution is used. For this purpose, we used the fourth–fifth-order Runge–Kutta method (ode45) provided by MATLAB and the second–third-order Runge–Kutta method (ode23) to perform the numerical solution.

Let $R = y_1$, and $\dot{R} = y_2$, then the equation can be organized as the following ordinary differential equations:

$$\dot{y}_1 = y_2$$

$$\dot{y}_1 = \left\{ \begin{array}{l} -\frac{3y_1^2}{2}\left(1 - \frac{y_2}{3c}\right) + \frac{1}{\rho}\left(1+\frac{y_2}{c}\right)\left[\left(P_1+\frac{2\sigma}{R_0} - P_v\right)\frac{R_0^{3r}}{y_1^{3r}} + P_A\sin\omega t - P_1 + P_v - \frac{2\sigma}{y_1} - \frac{4\eta y_2}{y_1}\right] \\ +\frac{y_1}{\rho c}\left[\left(P_1 + \frac{2\sigma}{R_0} - P_v\right)\frac{-3r R_0^{3r} y_2}{y_1^{3r+1}} + P_A\omega\cos\omega t\right] \end{array} \right\}$$

$$y_1(t= 0)=R_0,\ y_2(t= 0)= 0$$

If P_1, P_A, R_0, σ, ρ, c, η, P_V, and f are known, then the relationship of R with time can be solved.

The main strengthening mechanism of ultrasound-enhanced subcritical water extraction is the cavitation effect of the ultrasonic waves. Combined with the characteristics of subcritical water fluids at high temperature and high pressure, the kinematics equation of the ultrasonic waves generated in cavitation bubbles in subcritical waters is established, indicating that ultrasound is produced in subcritical water. Cavitation yield increases with increasing temperature and decreases with increasing pressure. This result shows that the cavitation effect of the ultrasonic waves can enhance the extraction of subcritical water.

Figure 11.3 shows the effect of ultrasound on tin foil materials under different conditions. From the figure, we can see that in the subcritical water of 300 K and 7.20 MPa, there are more tiny holes under the ultrasonic effect of the tin foil, which indicates that under this condition, the cavitation effect can be generated by ultrasound, which is consistent with the calculation result of cavitation enthalpy. In the subcritical water of 300 K and 21.00 MPa, only a few tiny holes appeared under the ultrasonic effect of the foil, which indicates that under this condition, ultrasound can generate the cavitation effects. However, under the same sound pressure, with the increase of pressure, the external pressure resistance that the cavitation bubble needs to overcome when it expands also increases. As a result, cavitation is more difficult to generate, resulting in fewer voids. In the cavitation threshold calculation, the cavitation enthalpy value under this condition far exceeds the maximum sound intensity that can be provided by the ultrasonic generator, and ultrasonic cavitation is not possible. However, contrary to the experimental results, the mechanism of its

a) Without ultrasound b) 300K,7.2MPa c) 300K,21.00MPa d) 288K,21.00MPa e) Ultrasound in water

Fig. 11.3 Effect of ultrasonic on tin foil materials under different conditions [100]

development needs further study. In the subcritical water of 288 K, 21.00 MPa, no holes were found in the foils under the ultrasonic effect, which indicates that under this condition, the cavitation effect was not achieved by ultrasound, which is consistent with the calculation result of the enthalpy of cavitation. Compared with the ultrasonic cavitation effect in subcritical water of 300 K, 21.00 MPa, it can be seen that temperature rise is more conducive to the occurrence of ultrasonic cavitation.

Ultrasonic cavitation is one of the major factors that enhance subcritical water mass transfer. Huang measured the cavitation yield under different extraction conditions by using iodometry. The results showed that cavitation yield increases with increasing temperature and pressure. The prolongation of action time, the increase in ultrasonic output power, and the decrease in frequency of action show an increasing trend, illustrating that the ultrasonic cavitation effect is indeed one of the main mechanisms affecting mass transfer in subcritical water. Huang discussed the mechanical effects of the ultrasonic waves in media propagation by calculating the physical parameters related to particle vibration under subcritical water extraction conditions. The results show that ultrasound can cause severe particle concentration in local space in a subcritical water system. The effects of vibration, performance of the maximum particle vibration velocity, and maximum particle acceleration are remarkable, so considerable and rapid changes in these parameters greatly contribute to the enhancement of subcritical mass transfer. This microscopic form of strenuous exercise can cause some damages to the microstructure of *Lithospermum* granule. The ultrastructure of *Lithospermum granulosum* extracted by subcritical water before and after ultrasound was compared by scanning electron microscopy. After a certain period, the cell with a regular structure displayed numerous irregular grooves. The dense structure of plant material becomes loose, and the permeability remarkably increases, leading to the increase in swelling index and the decrease in mass transfer resistance. Therefore, the vibration effect of an ultrasonic machinery is another important strengthening mechanism. Moreover, Huang compared the extraction rate of a volatile oil in *Lithospermum* using an extraction solvent with ultrasound treatment before and after subcritical water enhancement. Furthermore, Huang discussed the effect of ultrasonic field on mass transfer rate, mass transfer, and diffusion coefficient in subcritical water system. It shows that ultrasound can significantly increase the mass transfer rate, which can promote not only diffusion and mass transfer but also the internal diffusion through the effect of perturbation.

Characteristics

Subcritical water extraction has been recently applied in the extraction and analysis of organic pollutants in environmental samples. Subcritical water extraction has only been applied in extraction of active ingredients in plants in recent years; thus, this technology is a new extraction method in this field. The technology is cheap, non-toxic, non-polluting as it uses water as an extractant under moderate pressure. A subcritical water is a water that is heated to a high temperature, which is above 100 °C but below the critical temperature of 374 °C, and it remains in liquid state. Subcritical water extraction offers the following advantages: ① The operating pressure is very low, only 5–6 MPa, which is considerably lower than the pressure required by the supercritical fluid extraction technology; thus, the required equipment and maintenance are cheaper; ② extraction time is short; ③ water polarity can be changed by adjusting the temperature to selectively extract nonpolar, weakly polar, and polar organic compounds from a sample; ④ pure water is used as extractant, and no or very little organic solvent is used so the technology causes little or no pollution to the environment.

Subcritical water extraction is a promising and efficient extraction method. Moreover, it is a potential alternative for both conventional technologies (steam distillation and solvent extraction) and new technologies (supercritical fluid extraction), thus showing great and decisive advantage [101].

However, current research on subcritical water extraction shows that the amount of raw materials and solvent (water) is fixed, the water is at rest, and the mass transfer efficiency is relatively low; thus, the extraction rate still cannot reflect the advantages of subcritical water extraction. With the application of a constant pressure and the constant input of new subcritical water, raw materials are constantly in contact with fresh extractant, thereby improving the efficiency of mass transfer; as a result, extraction of the active components of raw materials is maximized and thus the extraction rate is increased. However, dynamic extraction uses more extracting solution than static extraction, the extract concentration is low, and therefore, the workload of subsequent processing is increased obviously.

Application and Prospect

Ultrasound subcritical water coupling has the advantages of good extraction uniformity, reducing the extraction time and increasing the extraction rate [102]. By changing the extraction temperature, solvent polarity can be easily changed, and solutes can be continuously extracted at varying degrees depending on the polarity. The Laboratory of South China University of Technology is conducting research on "Ultrasound-enhanced subcritical water extraction technology," the technology applied to extract plant-based flavors; when the ultrasonic wave propagation in a liquid medium (which creates a unique cavitation effect) is coupled with ultrasonic propagation of mechanical vibration, micro-jet, micro-flow, and other multipolar effect, effective materials are more easily extracted at lower temperatures; moreover, plant flavor nutrients can be obtained from water-soluble and fat-soluble plant components by changing the extraction parameters (temperature, pressure, and time); selective extraction, high extraction rate, short extraction time, low energy consumption, and environmental protection are the advantages of this approach [103].

In recent years, subcritical water extraction has been considered as a very promising extraction technique. Table 11.10 lists the application of subcritical water extraction of volatile oil and other major components of plants [104]. The improved subcritical water extraction method was used to separate the laurel oil. Compared to other conventional extraction methods, the extraction time was shorter, the product oil quality was better, and the oil was cheaper and cleaner [105]. The extraction of fennel from subcritical water extraction, steam distillation, and dichloromethane extraction showed that the subcritical water extraction was more rapid and cleaner, and the concentration and yield of quinone oxides were higher [106]. Jim et al. used subcritical water to extract the smashed ground marjoram leaves at a water pressure of 5.0 Mpa, a temperature of 150 °C, and a flow rate of 2 mL/min for 15 min, and the amount of oil obtained is equivalent to steam distillation for 3 h [107].

The researchers reported the use of subcritical water extraction (SWE) volatile oils compared with conventional steam distillation (SD), supercritical fluid extraction (SFE), and Soxhlet extraction (SE) [108, 109]; see Table 11.11.

From these studies, it can be found that the application of subcritical water shows a very important advantage over steam distillation. The subcritical water extraction time is shorter, the extracted volatile oil is of good quality, the amount of oxidizing compound substances is high, and the concentration of terpenic compounds is low. Therefore, the subcritical water extracted volatile oil is more aromatic and natural. Most importantly, the use of subcritical water extraction is more economical than steam distillation. Subcritical water extraction consumes about 10 times more water than steam distillation, but it saves far less energy than the value of these waters [110]. Therefore, subcritical water extraction technology is a promising alternative to steam distillation.

Subcritical water extraction and supercritical CO_2 extraction techniques have also been reported. The main drawback of subcritical water extraction for volatile oil is that it needs to be performed at high temperatures; therefore, it is necessary to study the stability of the extracted volatile oil at this temperature. Since water

Table 11.10 Application of subcritical water extraction of flavors and fragrances from plant material

Extractive	Medicinal herbs	Extraction temperature/°C	Extraction time
Volatile oil	Sweet bay	150	35
Volatile oil	Fennel	150	50
Volatile oil	Marjoram	150	15
Volatile oil	*Rosmarinus officinalis*	125–175	50
Volatile oil	*Mentha spicata* and peppermint oil	100–150	
Essential oil	Peppermint oil	125 or 150	
Kavalactones	*Piper methysticum* Forst	175	40
Volatile oil	*Origanum vulgare*	125	
Volatile oil	*Hyptis suaveolens*	150	30

Table 11.11 Comparison of subcritical water extraction with other extraction methods of essential oil [108]

	SD	SE	SFE	SWE
Extraction time	Long	Long	Short	Shorter
Drying step	Need not	Need	Need	Need not
Environmentally friendly	General	Poor	Good	Better
Effectiveness	Low	Low	High	High
Volatile oil quality	General	General	Good	Better
Extraction conditions	Intense	Intense	High pressure	High temperature
Cost	High	High	High	Low

is insoluble in CO_2, it is often necessary to pre-dry the medicinal material in the supercritical fluid extraction process. Drying requires additional costs, and there is a risk of loss of aromatic compounds, which can be avoided by subcritical water extraction. Subcritical water extraction usually has a much lower operating pressure than supercritical fluid extraction and is cheaper to install and maintain.

At present, there are few researches on ultrasound-enhanced subcritical water extraction, and there is great potential for its development in the extraction of Chinese herbal medicines. It is called a revolutionary technology with a green and promising perspective.

11.2.3.3 Summary and Prospect

Studies have recently reported on the ultrasonic extraction and separation of active ingredients in natural products. The raw materials used in these studies include plants, herbs, foodstuffs, environmental samples, biological products, and raw chemical materials, and through the combination of several other enhanced separation methods, these technologies broaden the scope of application and enhance the effect of enhancement. In terms of the current research trends, the mechanism and model of the ultrasonic wave, the automation of ultrasonic equipment, the online combination of ultrasound and other new separation technologies, and the application of ultrasound in emerging fields will be the future development trends and research hotspots.

Compared with the traditional methods, the current extraction methods and equipment have made a great progress, and the advanced science and technology and equipment, such as ultrasonic extraction, supercritical CO_2 extraction, and subcritical water extraction, have been widely used in plant extraction processes. With the continuous development of various extraction technologies and the increasing automation of equipment, selective coupling and of these technologies and taking into consideration the characteristics of the extracted objects will greatly improve the extraction efficiency and will ensure that high-quality extracts are obtained, thereby broadening the range of industrial applications of power ultrasound extraction technology.

11.2.4 Pulsed Electric Field Technology

High-voltage pulsed electric field (PEF) technology was originally used for sterilization. In recent years, many researchers at home and abroad have applied it to the extraction of components, showing a great advantage.

The high-voltage pulsed electric field is a method in which a liquid food is placed as an electrolyte in a container, and two discharge electrodes on the edge of the container are passed through a high-voltage current to generate electric pulses. PEF can disrupt the cell wall and membrane potential of the treated cells in an instant, change their permeability, and even break down the cell wall and cell membrane, causing irreversible damage, resulting in metabolic disorder of the cells and the outflow of essential components in the cells. It is an ideal way to recover useful substances from cells。

11.2.4.1 Principle

The main equipment of PEF technology is high-voltage pulsed electric field equipment. In general, the PEF processing system is mainly composed of a high-voltage pulse power supply, a processing chamber, a liquid supply system, and a measurement and protection system. High-voltage pulsed power supplies and processing chambers are the two most important subsystems. The schematic diagram of PEF system is shown in Fig. 11.4.

Studies on PEF sterilization have shown that irreversible cell membrane damage directly causes cell inactivation. Theoretical hypotheses on the mechanism of cell membrane damage have been presented, including the dielectric breakdown theory of membranes, electroporation theory, electromechanical compression and instability theory, permeability imbalance theory, viscoelastic model, lipophilic hydrophilic pore transformation theory, and transformation theory [111]. Each of the hypotheses has its own characteristics and is supported by research results. Among these theories, the dielectric breakdown and electroporation theories are universally accepted. The specific mechanism presented in these theories is very similar to that presented in other hypotheses.

In the absence of an applied electric field, the potential difference between the inside and the outside of the cell membrane is approximately 70 mV to maintain the normal physiological functions of cells. Cell membrane electroporation is the result of the physical interaction between the electric field and the two materials with different dielectric constants K: (1) phospholipid Kmem = 2–3; (2) electrolyte Ke = 70–80 = Kw [112]. Due to the difference in permittivity, the external electric field mainly acts on the cell membrane, and the cell membrane acts as an electric field concentrator. When an external electric field is applied, the electric field inside the cell membrane is approximately two orders of magnitude of the applied field strength. The macromolecules and some charged groups in the membrane are affected by the transmembrane potential, and they are very sensitive to the electric field. Under the action of the transmembrane potential, an irreversible conformational change occurs, and even if the electrical pulse acts for a short time, the damage observed in the organism is considerable [113].

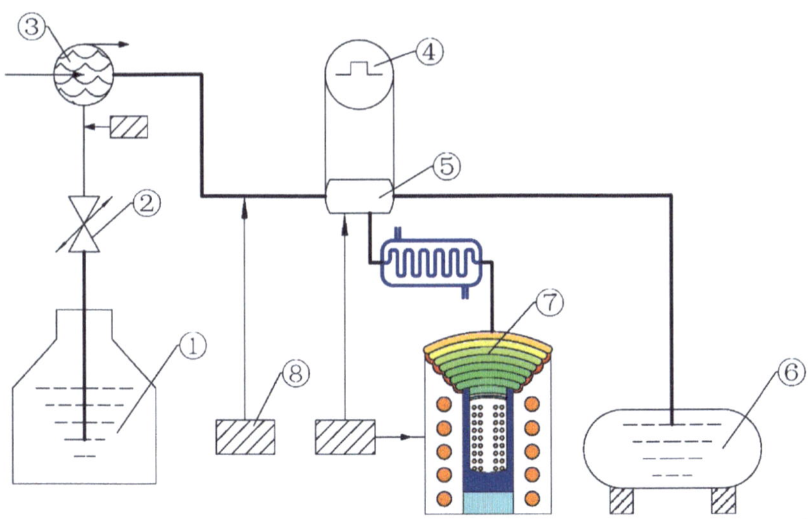

①Sample ②Degasser ③Pump ④High voltage impulse generator ⑤Process chamber ⑥Storage tanks ⑦ Cooling installation ⑧Sensor

Fig. 11.4 Schematic diagram of PEF systems

According to the H. P. Schwan equation, the transmembrane voltage is reached when the cells in the $E(t)$ field are electrostatically balanced.

$$\Delta u(t) = 1.5\, E(t)\, R_{\text{cell}}\, \cos\theta$$

where R_{cell} is the cell radius and θ is the angle between the applied electric field $E(t)$ and the measured normal vector of the cell membrane [110].

For biological cells with a diameter of approximately 10 μm, the maximum transmembrane potential is 600–750 mV when the field strength of the external electric field is 80–100 V/mm. Proteins, TPS molecules, or lipids are uncharged when they flow slowly. However, the ionophores in these molecules are charged and form dipoles. The strong electric field inside the film causes charge separation, which can cause the molecules to turn or recombine. Some covalent bonds cause holes in the bilayer lipid membrane or induce opening of protein channels. Simultaneous changes in the transmembrane potential result in changes in ion concentration on both sides of the membrane [114]. As a result, the membrane permeability increases sharply. Under normal circumstances, some macromolecules that do not easily enter the cell can be introduced into the cell. When the transmembrane potential reaches 500–1000 mV, electroporation occurs [115].

"Membrane perforation" theory considers the bilayer membrane structure of a cell as an equivalent capacitance, and the distribution of the charge on the membrane causes a change in differential pressure. When external electric field is applied on both ends of a cell, the transmembrane potential increases. On the opposite sides of the membrane, anisotropic ions generate mutual attraction forces. This force is equivalent to the force experienced by a membrane that is subjected to compressive forces on both sides. As the electric field intensity or the treatment time increases, the transmembrane potential and pressure continuously increase. At this time, the viscoelastic restoring force of the cell membrane resists the compression force on both sides of the membrane. Since the extrusion force grows faster than the viscoelastic recovery force, the cell membrane thickness continuously decreases; when the transmembrane potential reaches 1 V, the cell membrane is partially destroyed, creating micropores. However, cells can heal themselves. The term "reversible breakdown" is used when pores can self-repair; if the applied electric field increases beyond the critical electric field strength (that is, the electric field already causes a long-term effect), a large number of large pores will form, and the "reversible wearing" will become "irreversible breakdown;" the cell membranes are mechanically destroyed, leading to cell death [116]. Scanning electron microscopy was used to observe the appearance of depressions on the surface of yeast cells after application of PEF; this phenomenon illustrates the theory of "membrane perforation." The yeast cells subjected under PEF remained intact. However, large and small depressions were observed in dead cells, and macromolecular substances, such as proteins and nucleic acids, were extravasated (the number of high-voltage PEF pulses was 1000, pulse width was 11 μs, and the applied pulse square wave is 5–9 kV/cm). The interaction of the PEF with cells is based on cell membrane charging. When an external electric field charges a cell membrane, opposite charges exist on both sides of the membrane, the electric force in the membrane increases, and the cell membrane becomes thin. The degree of thinning depends on the elastic modulus of the membrane. When the cell membrane was 0.607 times thicker than the original membrane, the cell membrane is disrupted due to instability. To keep the pores open, the lower limit of the pulse required for cell membrane electroporation must be longer than the charge time of the membrane and the elastic recovery time of the membrane. Generally, the time required for cell membrane perforation is approximately 5 μs [117]. The experiment on cell membrane electroporation using a number of pulses shows that PEF exerts a cumulative effect on the cell membrane. In cell membranes, membrane potential is dominated by the total amount of charge. The cell membrane can be regarded as a self-discharge-type capacitor. Under the action of the PEF, the charge and discharge time of the cell membrane is an important parameter. The cumulative effect of electroporation is closely related to the charge time of the cell membrane, namely the pulse width tw, the time required for cell membrane discharge tr, and the pulse interval ti. When tr»tw, ti, the cell membrane just begins to discharge when the next pulse arrives, and the electroporation effect is completely accumulated; when tr < ti, the cell membrane has been partially discharged when the next pulse arrives, and the electroporation effect is partially accumulated; when tr < ti, the cell membrane discharge has ended when the next pulse arrives, and the electroporation effect is not accumulated.

11.2.4.2 Influencing Factors

The key factors influencing the destructive effect of PEF can be divided into three categories: processing equipment parameters, media conditions to be processed, and biomaterial properties [118].

(1) PEF equipment parameters

① Electric field strength

The electric field strength is one of the most important factors influencing the destructive effect. When the critical value of the transmembrane potential was reached, the electric field strength further increased, and the survival rate of the target bacteria decreased significantly. In the high-voltage PEF sterilization process, the intensity of the PEF is a major parameter that leads to the inactivation of microorganisms. When the number of pulses is fixed, the killing rate of the cells will increase as the electric field strength increases. These observations are consistent with the electroplating (permeation) theory because the induced electromotive force on both sides of the cell membrane is proportional to the electric field strength, and higher electric field strength can kill more bacteria. The high-voltage PEF for sterilization is generally 5–100 kV/cm [119].

② Pulse number

Under the conditions of fixed electric field strength and pulse width, the inactivation rate of the cells is determined by the number of pulses applied. When the number of pulses increases, the curve of the relationship between the electric field intensity and the fluorescence value moves upward and the electroporation rate of the cells increases accordingly. These results indicate that the number of survived *Escherichia coli* in skimmed milk is a function of the number of pulses. The electric field strength of the applied pulse square wave is 50 kV/cm, and the pulse width is 2 μs. The first inoculated *E. coli* in skimmed milk was approximately 80×10^7 cfu/mL, and the curve considerably changed when the number of pulses was five.

③ Processing time

The processing time is the product of the number of pulses and the pulse width (μs); increasing the pulse width or the number of pulses increases the destruction rate to some extent. Sun et al. treated *Saccharomyces cerevisiae* and *E. coli* with a high-voltage PEF. With the increase in electric field intensity, pulse width, and the number of pulses, the survival rate of the microorganisms decreases significantly [120]. The killing effect is closely related to the PEF parameters, such as processing time, pulse width, and pulse frequency.

④ Pulse width

The pulse width is a factor that influences the PEF effect. The result obtained when a pulse width of 2 μs was applied is not significantly different from that obtained at a pulse width of 6 μs but quite different from that obtained at a pulse width of l μs. The highest possible value for pulse width should be used [121].

⑤ Pulse frequency

Increasing the pulse frequency increases the bactericidal effect. The reason behind is that after the frequency is increased, there is more pulse number corresponding to each capacitor discharge, and thus the decrease in exponential decay curve is reduced. This phenomenon ensures a longer sterilization time [122]. The high-frequency PEF frequency for sterilization is 1–100 Hz, and the discharge frequency is 1–20 Hz [119].

⑥ Energy of the electric field

The most suitable pulsed field energy is 100 J/mL [123]

$$W(\text{J/mL}) = \sigma \cdot \tau \cdot E_2,\ Q = n \cdot W$$

where σ is the conductivity (S/cm), T is the pulse length (μs), E is the electric field strength (U/μm), n is the number of pulses, and Q is the input energy.

When the electric field intensity and the pulse length are constant, the same number of pulses is processed, and the sample having a large electric conductivity will generate more energy. The low-conductivity solvent demonstrates a good destructive capacity when the input energy is constant [124]. The presence of nutrients, such as proteins and carbohydrates, provides a certain degree of protection against microorganisms. The higher the energy obtained per volume of treatment medium, the better the treatment effect [125].

⑦ Pulse waveform

In waveforms commonly used in PEF, pulses can be used in three forms: square wave, exponential, and alternating wave; the best effect is observed in square wave, followed by that of exponential wave and then that of the alternating wave. For pulse polarity, a double polar wave is more effective than a unipolar wave. The exponential pulse is commonly used at present, and it is generated by a circuit composed of a resistor and a capacitor; such a circuit has a relatively simple structure, is relatively cheap, and thus is suitable for industrial applications. When the capacitor is charged, the energy on the capacitor is released to the processing chamber at the instant of closing of the high-speed electronic switch, forming an exponential pulse; therefore, the high-power charging power supply and the high-speed electronic switch are the keys to the design of the high-voltage pulse generator [126]. The design of the treatment device is very important in treatment of PEF-induced damage. The intensity of the field strength and total processing time increases the damage effect; moreover, the waveform, pulse frequency, and treatment chamber design all affect the results.

With the development of PEF research, a growing number of studies have tended to study the specific behavior of the processing unit in the process and the specific influence law of various factors. In order to make the processing conditions explicit and accurate, the damage effect can be controlled.

(2) Influence of the media condition being processed

The state of the medium to be treated [physical and chemical properties of the medium, especially conductivity, pH, ionic strength, temperature, viscosity, and water activity (Aw)] greatly influences the treatment effect of PEF. Low-conductivity liquid foods require low power, resulting in low heating; therefore, treatment with PEF is more effective.

① Ionic strength of the medium

Jayaram et al. studied the effect of *Lactobacillus brevis* in different liquid media. When the conductivity of the liquid increases, the resistance of the process chamber decreases; as a result, the pulse width and destruction efficiency decrease. Jayaram believes that ion movement consumes cellular energy and causes the cell membrane to be easily disrupted [127]. Vega et al. investigated the effects of pH and ionic strength on simulated milk ultrafiltrate in PEF treatment. When the ionic strength was reduced from 168 to 28 mM, the destruction efficiency of *E. coli* was the highest. This potency of ionic strength may be related to membrane perforation rate and physical damage, whereas pH may be related to changes in cytoplasmic conditions caused by permeation imbalance resulting from membrane perforation. However, divalent ions reduce the lethal rate of PEF treatment because their presence exerts a protective effect on cell integrity [128].

② Dielectric conductivity

Pulse frequency increases as the conductivity of the medium increases, and the pulse width decreases. Therefore, when the capacitor discharges, the number of pulses does not change; that is, the total destruction time decreases, and the damage effect accordingly decreases. Medium conductivity affects the pulse intensity and the number of pulses during discharge. Increasing the pulse frequency increases the damage effect. This phenomenon is observed because after the frequency is increased, the pulses per discharge of the capacitor increase; therefore, the decay of the exponential decay curve slows down, thus ensuring a longer destruction process [129].

③ pH of the medium

The pH of the medium is one of the most investigated factors, and its influence on PEF inactivation is the object of discussion [118]. pH changes affect the strength of PEF. The pH of a solvent determines the degree of influence of a biological cell material treated with PEF; thus, pH is an important factor that must be considered. Vega-Mercado found that when pH drops (<7.0), destruction rate increases. When cells undergo electroporation, changes in pH affect the physiological function of the protoplasm, which in turn destroys the structure of the biological cells, resulting in decreased survival. The activation mechanism varies with pH and biomaterials. The manner in which PEF acts on cell membrane is a function of the pH of the solvent, and this property corresponds to the relative PEF resistance of various biological species [130]. Regardless of the shape of their cell wall, bacteria are generally more resistant in acidic conditions than in acidic conditions under heat and pressure [131].

④ AW

AW decreased from greater than 0.99–0.93, indicating that the destructive effect of PEF decreased.

⑤ Biomaterial concentration

A study subjected samples containing a high and low number of bacteria to the pulse intensity for a certain period; the logarithm of the number of bacteria decreased at a greater degree in the former than the latter. For example, the effect of a sterilized fresh juice is better than that of a concentrated juice.

⑥ Processing temperature

Jayaram et al. believe that temperature (−50 to 50 °C) and PEF sterilization exert a synergistic effect. With field strength held constant, the survival rate of the target bacteria decreases as the temperature rises. As the treatment temperature rises (within the range of 240–600 °C), the bactericidal effect improves generally by up to 10 times. The treatment temperature greatly influences the inactivating effect of PEF treatment [132]. The higher the processing temperature, the higher the non-activity of the cell.

⑦ Mixing and stirring

Macroscopically speaking, the synergistic effect mentioned above acts as a solvent that disperses PEF-treated intracellular components into the surroundings, and such a phenomenon most likely increases cell membrane permeability, causing cell death to occur more easily. Specifically, mixing and stirring facilitate change in position of the cells located on the front surface of an electrode so that the limited membrane surface can be effectively electroporated by the pulse voltage of the electric field. Increasing the cell rotation speed and the permeation plane is necessary to promote the effect of PEF [133]. That is, the smaller the cosine, the higher the damage. The presence of microorganisms in the treatment medium also influences the destruction effect, and the attached bacteria demonstrated a greater survival rate than the free cells. Therefore, the PEF treatment must include a stirring process.

(3) Properties of biomaterials

① Biomaterial resistance and conductivity

The resistance of biomaterials ranges from 0.42 Ω M (high salt and high moisture) to 100 Ω M (high grease). Resistance and conductivity are inversely related. The greater the resistivity of a biological material, that is, the smaller the ionic strength in the biological material, the better the destruction effect. If the resistance is too large, current will pass through the low-resistance liquid medium and thus cannot act on the biological material. Some biomaterials consist of several substances (non-homogeneous), and the conductivity of each component is different; the electrical characteristics of non-homogeneous biomaterials change after pulse treatment, and these biomaterials are unsuitable for PEF treatment.

② Species of organism

The cell walls and cell membranes of different types of organisms have different compositional structures, and they display different sensitivities to electric fields. Therefore, the effect of high-frequency PEF sterilization also differs. From the perspective of lethality: yeast > *E. coli*, that is, yeast is more sensitive to electric fields than *E. coli*, consistent with the results of Qin et al. [134]. Under the action of an external electric field, a cell produces a transmembrane potential. When this potential reaches approximately 1 V, the cell membrane is disrupted. This value is called the breakdown potential. According to the mathematical model $U(t) = 3/2rE\cos\theta$ established by Neumann and Zimmermann, for a cell with a radius r in a uniform field strength E, the larger the radius, the easier it is to reach the breakdown potential of the cell membrane and die.

③ Size of biological materials

Knorr [135] used a PEF with a field strength of 50 kV/cm to act on crushed carrots, which is squeezed at room temperature for 10 min at 10 MPa; a yield of 76.1% (not treated with PEF) of a finely processed fruit juice sample (particle size of 1.5 mm) was obtained. The sample yield was 51.3%; the yield of the crude treatment sample (particle size is 5 mm) was 70.3% (the yield of the sample not treated with PEF was 30.0%). Moreover, the yeast cells were large (1–5 μm × 5–30 μm); *E. coli* cells were small (0.85 × 3 μm). The yeast is more sensitive to the electric field than the *E. coli*. In addition, the composition and thickness of cell wall and cell membrane of yeast and *E. coli* differ. Yeast cell wall mainly consists of glucan and mannan. The cell wall of *E. coli* is tougher than that of yeast, and this observation is attributed to peptidoglycan, outer membrane, outer membrane proteins, and periplasmic space composition. Thus, *E. coli* is more tolerant from electric field. The smaller the cells, the greater the resistance because the PEF causes a larger potential impedance. That is, the smaller the cell, the higher its PEF impedance.

④ Growth stage

The bacteria in the logarithmic phase of the growth cycle were more likely to be killed than those in the stable and declining phases. Gaskova reported that the mortality of yeast cells was 30% higher in log phase than in the stationary phase [136]. This phenomenon is observed because most of the logarithmic phase of bacteria is in a dividing stage, and the cell membrane is sensitive to the effect of electric field.

⑤ Cell shape

The PEF intensity required to inactivate rod-shaped cells is fivefold higher than that required by round cells of the same surface area. Cell shape has a higher PEF impedance than cell size [131]. Large, round cells are more sensitive to PEF than small, long cells. Research shows that the biological species, growth stage, size, and shape all exert an impact on the destruction efficiency of PEF. Generally, microorganisms in the exponential growth phase are most sensitive to the effects of PEF [137].

11.2.4.3 Application and Prospect

Numerous physiologically active compounds from Chinese medicinal plants are easily degraded and destroyed once they leave the body given their instability. The physiological activities of these compounds must be carefully protected during production. Therefore, the extraction of physiologically active substances from Chinese medicinal plants often requires mild conditions and is performed at low temperatures in a clean environment. Thus, the use of PEF technology in TCM production has emerged.

Fan et al. used the PEF method to extract various active ingredients from ginseng, such as ginsenosides, ginseng polysaccharides, AF, and AFG. They drastically improved the extraction rate of active ingredients by optimizing extraction conditions. PEF requires mild operating conditions and produces non-toxic waste products. Moreover, its operating parameters can be easily controlled. It has the characteristics of uniform and fast, and can completely destroy cell organization and fully dissolve effective substances while minimizing damage to the active ingredient of interest.

Table 11.12 Polysaccharide, ginsenoside, AF, and AFG extraction ratios of different extraction methods [138]

Method of extraction	Ginseng polysaccharide	Ginsenoside	AF/%	AFG/%
Method of PEF	7.20	0.98	0.19	0.11
Method of decoction	1.94	0.22	0.01	0.02
Ultrasonic method	5.42	0.58	0.03	0.01

Table 11.13 Applications of high-voltage pulsed electric field extraction in the extraction of active constituents from medicinal plants

Extractive	Traditional Chinese medicine plant	Extraction conditions	Extraction rate
Flavones and tea polyphenols	Bamboo leaves	Electric field intensity is 22 kV/cm, and the number of pulses is 9	Flavone extraction rate is 2.49%, and the extraction rate of tea polyphenols is 1.26% [140]
Ginsenoside, ginseng polysaccharide, AF, and AFG	Ginseng	Electric field intensity is 25 kV/cm, and the number of pulses is 10	Extraction rates of ginsenoside, ginseng polysaccharide, AF, and AFG in the aqueous extracts of ginseng were 0.98, 7.20, 0.19, and 0.11%, respectively [138]
Polysaccharide	Corn bran	Electric field intensity is 25 kV/cm, and the frequency of electric field is 2080 Hz	15.36% [141]
Guanfubase A	*Aconitum coreanum*	Electric field intensity is 20 kV/cm, and the number of pulses is 10	0.391% [142]

Compared with the decoction and ultrasonic methods, the high-voltage PEF method generates less heat energy and is cheaper, simpler, safer, and more time efficient and reliable [138] (Table 11.12).

Dai et al. used the high-voltage PEF method to extract geranium flavonoids and optimized puerarin extraction parameters to retain the biological activity of flavonoids. Puerarin extraction through the PEF method provides the advantages of low energy consumption and rapidity while minimizing damage to heat-sensitive substances [139]. Table 11.13 lists some applications of the PEF method in the extraction of active ingredients from medicinal plants.

The high-voltage PEF extraction method can rapidly and efficiently extract material components by destroying plant cell membranes and promoting intracellular substance dissolution. Compared with traditional extraction methods, the PEF method can better retain active ingredients in extracts. The development of industrial TCM production will inevitably require the use of advanced technology with low energy consumption and low operating costs given the intensifying market competition and increasing operating costs. The in-depth development of PEF technology will allow its industrial application in the near future.

11.3 Conclusion and Perspectives

Abundant TCM resources exist, and extraction technologies for the preparation of TCM have a long history of development. However, the promotion and popularization of TCM remain insufficient, and the cultivation and development of TCM resources have not yet been industrialized. The TCM industry remains in its infancy and lacks the complete production chain of procurement, storage, extraction, and pharmaceutical development. The extraction technologies used to produce TCM seem to be arrested in the exploration stage. Given this situation, we must constantly improve our own extraction technology capabilities and relevant system management details. At the same time, to improve extraction technology, we require our closely related departments and staff to receive professional training and to adhere to strict quality requirements.

Given the vigorous promotion of TCM in recent years, the use of extraction technology to prepare TCM products has received increasing attention from industrial enterprises and the public. The development of extraction technology is the key to safe, efficient, stable, and controllable TCM production. The modernization and standardization of TCM production require simple, rapid, efficient, non-polluting extraction technologies that enable the effective separation of the components of interest from the plant matrix, as well as the optimization of analytical sensitivity and accuracy. The above TCM extraction technologies are emerging as research hotspots and must be further developed given their broad application prospects.

Novel extraction technologies provide more advantages than conventional methods for the extraction of TCM components, such as high extraction efficiency, low solvent residue, time-saving, and easy operation. These technologies play crucial roles in the development, modernization, industrialization, refinement, and standardization of TCM production. The promotion and extensive application of these new extraction technologies will enhance the competitiveness of TCM products in the international market.

References

1. Yan Y, Huang X, Xiao G et al (2012) Study on the application of new technology in Chinese traditional medicine extraction. Chem Ind Times 26(1):49-53 (in Chinese)
2. Wang Y, Zhao X (2015) Green catalytic process and technology. Chemical Industry Press (in Chinese)
3. Balachandran S, Kentish SE, Mawson R et al (2006) Ultrasonic enhancement of the supercritical extraction from ginger. Ultrason Sonochem 13(6):471–479
4. Chen W (2000) The principle and application of supercritical fluid extraction. Chemical Industry Press (in Chinese)
5. Li C (2015) Supercritical CO_2 extraction of volatile oil from compound recipe of clove and cinnamon. Huanan Ligong Daxue Xuebao. J South China Univ Technol 36(7):67–71
6. Zhang H,Yao Y, Yan X (2016) Application of supercritical fluid extraction and separation technology. J Capital Norm Univ 37(6):50–53 (in Chinese)
7. Zhu B (2001) Catalytic reaction engineering. China Petrochemical Press (in Chinese)
8. Pourmortazavi SM, Rahimi-Nasrabadi M, Mirsadeghi S (2014) Supercritical fluid extraction in plant analysis. Wiley
9. Berche B, Henkel M, Kenna R (2009) Critical phenomena: 150 years since Cagniard de la Tour. Rev Bras Ensino Fís 13(3):2602.2601–2602.2604
10. Li C (2002) Extraction technique of supercritical fluid and its application. Pet Geol Recovery Effi 9(4):31–32 (in Chinese)
11. Yu C (2007) Research advances of supercritical fluid technology application. Chem World 48(2):118–120 (in Chinese)
12. Xu Z (2004) Brief introduction of artemisinin drug effect study. Anhui J Prev Med 181–182 (in Chinese)
13. National Pharmacopoeia Commission (2010) Pharmacopoeia of the People's Republic of China: 2010 edition. China Medical Science and Technology Press (in Chinese)
14. Ling Z, Chen B, Wei J et al (2008) The history of the development of pepper efficacy. J Yunnan Univ Tradit Chin Med 31(4):49-50 (in Chinese)
15. Chen Z, Xu C, Xie L (2001) Study on chemical constituents of the essential oil from *Zanthoxylum bungeanum* Maxim by supercritical CO_2 fluid extraction. China J Chin Mater Med 26(10):687–688 (in Chinese)
16. Xia K, Zhu M (1991) Application of CO_2 supercritical fluid extraction. Chin Tradit Herbal Drugs 202–204 (in Chinese)
17. Yuan H, Zhang C (1999) Determination of rhein, emodin and emodin in radix polygoni multiflori by supercritical fluid extraction-HPLC. Chin Tradit Herbal Drugs 30(4):258–260 (in Chinese)
18. Bian J, Cai D, Gu M et al (1997) Study on extraction of strychnine from strychnos piersiana A.W. Hill by CO_2 SFE. Chin J Pharm 5–8 (in Chinese)
19. Bian J, Cai D, Gu M et al (1999) A RP-HPLC method in the separation and determination scopolamine and strychnine was developed. Chinese Pharmaceutical Association Academic Annual Meeting (in Chinese)
20. Jiang J, Xin G (1997) Studies on supercritical CO_2 fluid extraction of colchicine from Guangguzi. Chin Tradit Herbal Drugs 147–149 (in Chinese)
21. Li X, Yuan H, Gou K et al (2000) Study of piperine in Fructus piperis Longi by supercritical fluid extraction. Chin J Hosp Pharm 20(10):597–599 (in Chinese)
22. Liao Z, Jiang J, Wang H et al (1998) Extraction of total saponins and polysaccharides from *Ganoderma lucidum* with supercritical CO_2. Chin Tradit Herbal Drugs (9):601–602 (in Chinese)
23. Zhang J, Wu X, Cui W (1996) Study on the extraction of ginsenosides from ginseng leaves by supercritical CO_2 Extraction. National Supercritical Fluid Technology Academic and Application Seminar (in Chinese)

24. Shi Q, Hui G (1998) Study on the extraction of diosgenin from *Dioscorea panthaica* by supercritical CO_2. National Supercritical Fluid Technology Academic and Application Seminar (in Chinese)
25. Cai Q, Yang S, Ma X et al (2001) Supercritical extraction of coumarins from Radix Angelicae Pubescentis. Liaoning J Tradit Chin Med 28(9):559–560 (in Chinese)
26. Fang Y, Huang J, Zhang K (2001) Extraction of paeonol from cortex moutan by supercritical CO_2. Fine Chem 18(4):201–202 (in Chinese)
27. Deng Q, Gao Y (1999) A study on effective components from ginkgo leaves by second supercritical fluid extraction. Chin Tradit Herbal Drugs 30(6):419–423 (in Chinese)
28. Li J, Feng Y (1996) A study on extraction of tea cateciuns with supercritical CO_2. Nat Prod Res Dev 42–47 (in Chinese)
29. Li X, Wang D, He R et al (1999) The formulation principle of surfactants and their applications in the printing and dyeing of textiles china surfactant. Deterg Cosmet 17–18 (in Chinese)
30. Yu N (2011) Theory and application of supercritical fluid extraction. Chem Intermed
31. Chemat F, Rombaut N, Sicaire AG et al (2017) Ultrasound assisted extraction of food and natural products. Mechanisms, techniques, combinations, protocols and applications. A review. Ultrason Sonochem 34:540
32. Zhenmei LI, Jilie LI, Xiao Z et al (2016) Advance in application of subcritical water extraction technology in oil plants. China Oils Fats
33. Nerín C, Salafranca J, Aznar M et al (2009) Critical review on recent developments in solventless techniques for extraction of analytes. Anal Bioanal Chem 393(3):809–833
34. Huang P (2010) Ultrasound enhanced subcritical water extraction of active ingredients of different polarity from *Lithosperrnum* erytlrrarhizon. South China Univ Technol (in Chinese)
35. Curren MS, King JW (2001) Solubility of triazine pesticides in pure and modified subcritical water. Anal Chem 73(4):740
36. Ong ES, Cheong JS, Goh D (2006) Pressurized hot water extraction of bioactive or marker compounds in botanicals and medicinal plant materials. J Chromatogr A 1112(1):92–102
37. Rodil R, Popp P (2006) Development of pressurized subcritical water extraction combined with stir bar sorptive extraction for the analysis of organochlorine pesticides and chlorobenzenes in soils. J Chromatogr A 1124(1–2):82–90
38. Latawiec AE, Reid BJ (2010) Sequential extraction of polycyclic aromatic hydrocarbons using subcritical water. Chemosphere 78(8):1042–1048
39. Kronholm J, Hartonen K, Riekkola ML (2007) Analytical extractions with water at elevated temperatures and pressures. Trends Anal Chem 26(5):396–412
40. Guo J, Qiu T, Yang R et al (2009) Subcritical water extraction of onion essential oil. J South China Univ Technol (Nat Sci Ed) 37(4):143–148 (in Chinese)
41. Khajenoori M, Asl AH, Hormozi F et al (2009) Subcritical water extraction of essential oils from *Zataria multiflora* Boiss. J Food Process Eng 32(6):804–816
42. Khajenoori, Haghighi, Hormozi (2009) Proposed models for subcritical water extraction of essential oils. Chin J Chem Eng 17(3):359–365
43. Eikani MH, Golmohammad F, Rowshanzamir S (2007) Subcritical water extraction of essential oils from coriander seeds (*Coriandrum sativum* L.). J Food Eng 80(2):735–740
44. Prommuak C, De-Eknamkul W, Shotipruk A (2008) Extraction of flavonoids and carotenoids from Thai silk waste and antioxidant activity of extracts. Sep Purif Technol 62(2):444–448
45. Güçlü-Üstündağ Ö, Balsevich J, Mazza G (2007) Pressurized low polarity water extraction of saponins from cow cockle seed. J Food Eng 80(2):619–630
46. Ong ES, Len SM (2003) Pressurized hot water extraction of berberine, baicalein and glycyrrhizin in medicinal plants. Anal Chim Acta 482(1):81–89
47. Ong ES, binte Apandi SN (2001) Determination of berberine and strychnine in medicinal plants and herbal preparations by pressurized liquid extraction with capillary zone electrophoresis. Electrophoresis 22(13):2723–2729
48. Herrero M, Arráezromán D, Segura A et al (2005) Pressurized liquid extraction-capillary electrophoresis-mass spectrometry for the analysis of polar antioxidants in rosemary extracts. J Chromatogr A 1084(1–2):54–62

49. Shalmashi A, Golmohammad F, Eikani MH (2008) Subcritical water extraction of caffeine from black tea leaf of Iran. J Food Process Eng 31(3):330–338
50. Kulkarni A, Yokota TSS, Etoh H (2008) Subcritical water extraction of barley to produce a functional drink. Biosci Biotechnol Biochem 72(1):236–239
51. Karacabey E, Mazza G, Bayındırlı L et al (2012) Extraction of bioactive compounds from milled grape canes (*Vitis vinifera*) using a pressurized low-polarity water extractor. Food Bioprocess Technol 5(1):359–371
52. García-Marino M, Rivas-Gonzalo JC, Ibáñez E et al (2006) Recovery of catechins and proanthocyanidins from winery by-products using subcritical water extraction. Anal Chim Acta 563(1):44–50
53. Choi MP, Chan KK, Leung HW et al (2003) Pressurized liquid extraction of active ingredients (ginsenosides) from medicinal plants using non-ionic surfactant solutions. J Chromatogr A 983(1–2):153
54. Deng C, Xu X, Yao N et al (2006) Rapid determination of essential oil compounds in *Artemisia Selengensis* Turcz by gas chromatography-mass spectrometry with microwave distillation and simultaneous solid-phase microextraction. Anal Chim Acta 556(2):289–294
55. Deng C, Yao N, Wang B et al (2006) Development of microwave-assisted extraction followed by headspace single-drop microextraction for fast determination of paeonol in traditional Chinese medicines. J Chromatogr A 1103(1):15–21
56. Tollbäck J, Bigatá MB, Crescenzi C et al (2008) Feasibility of analyzing fine particulate matter in air using solid-phase extraction membranes and dynamic subcritical water extraction. Anal Chem 80(9):3159
57. Kim WJ, Kim J, Veriansyah B et al (2009) Extraction of bioactive components from *Centella asiatica* using subcritical water. J Supercrit Fluids 48(3):211–216
58. Richter P, Toral MI, Toledo C (2006) Subcritical water extraction and determination of nifedipine in pharmaceutical formulations. J AOAC Int 89(2)
59. Ong ES, Len SM (2015) Evaluation of surfactant-assisted pressurized hot water extraction for marker compounds in *Radix Codonopsis pilosula* using liquid chromatography and liquid chromatography/electrospray ionization mass spectrometry. J Sep Sci 26(17):1533–1540
60. Ong ES, Len SM (2004) Evaluation of pressurized liquid extraction and pressurized hot water extraction for tanshinone I and IIA in *Salvia miltiorrhiza* using LC and LC-ESI-MS. J Chromatogr Sci 42(4):211–216
61. Chen Y, Tian JK, Cheng YY (2006) Novel technique of extracting essential oil from Chinese herb and medicine-subcritical water extraction. Chem Eng 34(8):59–62
62. Hageman KJ, Mazeas L, Grabanski CB et al (1996) Coupled subcritical water extraction with solid-phase microextraction for determining semivolatile organics in environmental solids. Anal Chem 68(22):3892–3898
63. Yang Y, And SBH, Miller DJ (1997) Class-selective extraction of polar, moderately polar, and nonpolar organics from hydrocarbon wastes using subcritical water. Environ Sci Technol 31(2):430–437
64. Yang Y, Belghazi M, Lagadec A et al (1998) Elution of organic solutes from different polarity sorbents using subcritical water. J Chromatogr A 810(1–2):149–159
65. Reeve RN, Barnes JD (2011) Environmental analysis: analytical chemistry by open learning
66. Schantz MM (2006) Pressurized liquid extraction in environmental analysis. Anal Bioanal Chem 386(4):1043–1047
67. Herrero M, Cifuentes A, Ibanez E (2006) Sub- and supercritical fluid extraction of functional ingredients from different natural sources: Plants, food-by-products, algae and microalgae: a review. Food Chem 98(1):136–148
68. Mendiola JA, Herrero M, Cifuentes A et al (2007) Use of compressed fluids for sample preparation: food applications. J Chromatogr A 1152(1–2):234–246
69. Indriatmoko SY, Limantara L (2015) Separation of photosynthetic pigments by high-performance liquid chromatography: comparison of column performance, mobile phase, and temperature. Proc Chem 14:202–210

70. Herrero M, Jaime L, Martín-Alvarez PJ et al (2006) Optimization of the extraction of antioxidants from *Dunaliella salina* microalga by pressurized liquids. J Agric Food Chem 54(15):5597
71. Sereewatthanawut I, Prapintip S, Watchiraruji K et al (2008) Extraction of protein and amino acids from deoiled rice bran by subcritical water hydrolysis. Bioresour Technol 99(3):555
72. Liu Y, Shi J, Langrish TAG (2006) Water-based extraction of pectin from flavedo and albedo of orange peels. Chem Eng J 120(3):203–209
73. Turner C, Turner P, Jacobson G et al (2006) Subcritical water extraction and glucosidase-catalyzed hydrolysis of quercetin glycosides in onion waste. Green Chem 8(11):949–959
74. Lee WC, Yusof S, Hamid NSA et al (2006) Optimizing conditions for enzymatic clarification of banana juice using response surface methodology (RSM). J Food Eng 73(1):55–63
75. Sin HN, Yusof S, Hamid NSA et al (2006) Optimization of enzymatic clarification of sapodilla juice using response surface methodology. J Food Eng 73(4):313–319
76. Bogialli S, D'Ascenzo G, Corcia AD et al (2008) A simple and rapid assay based on hot water extraction and liquid chromatography–tandem mass spectrometry for monitoring quinolone residues in bovine milk. Food Chem 108(1):354–360
77. Yang H, Su M, Wang L et al (2006) Subcritical water extraction combined with fast acetylcholinesterase inhibition assay as an improved methodology for rapid detection of methamidophos residues in vegetables. Fresenius Environ Bull 15(10):1266–1270
78. Qu Q, Jiang W, Yao Y et al (2009) Application and trends in ultrasonic-based equipments for pretreatment of analytical samples. Chemistry 72(5):407–411 (in Chinese)
79. Fan X, Qiu T (2007) Power ultrasonic extraction technology and its latest development. Chinese Academy of Acoustics Power Ultrasound Branch 2007 Academic Annual Meeting (in Chinese)
80. Wei MC, Yang YC (2015) Kinetic studies for ultrasound-assisted supercritical carbon dioxide extraction of triterpenic acids from healthy tea ingredient *Hedyotis diffusa* and *Hedyotis corymbosa*. Sep Purif Technol 142:316–325
81. Yang R, Qiu T, Luo D (2005) Threshold investigation of ultrasonic cavitation in supercritical fluid CO_2. J South China Univ Technol (Nat Sci Ed) 33(12):100–104 (in Chinese)
82. Gogate PR (2017) Intensification of chemical processing applications using ultrasonic and microwave irradiations. Curr Opin Chem Eng 17:9–14
83. Luo D, Xu B, Qiu R et al (2013) Comparison and dynamic models of supercritical CO_2 extraction enhanced by different technologies. Mod Food Sci Technol 1921–1925 (in Chinese)
84. Li W, Wang Z, Liu J et al (2010) Ultrasound-enhanced supercritical extraction of total anthraquinone from rhubarb. Chin J Exp Tradit Med Formulae 16(10):30–32 (in Chinese)
85. Liang H, Yang R, Guo Q (1998) Effect of ultrasonic treatment on raising the extraction rate of proteins and total solids from soybean slurry. Sci Technol Food Ind 7–11 (in Chinese)
86. Zhang C, Li J, Lin J et al (2004) Preliminary study on the effect of ultrasound on supercritical CO_2 extraction of pyrethrins. Chem Eng Equip 22–24 (in Chinese)
87. Qiu T, Ding C, Hu A et al (2004) Analyses of factors affecting the ultrasonically–enhanced supercritical fluid extraction of EPA and DHA from algae. J South China Univ Technol 32(4):43–47
88. Qiu T, Yang R, Hu A et al (2005) Influencing factors in ultrasonic enhanced supercritical fluid extraction for extrating CLSO and CLSE from coix lacryma-jobi seeds. J Chem Eng Chin Univ 19(1):30–35 (in Chinese)
89. Ding C, Qiu T, Lu H (2005) Double-frequency ultrasounds alternately enhanced supercritical fluid extraction of effective components of plants. Chem Eng 33(6):67–70 (in Chinese)
90. Luo D, Qiu T, Fen Y (2006) Effect of ultrasound on ginsenosides extraction by supercritical CO_2 reverse microemulsions. J Jiangsu Univ (Nat Sci Ed) 27(3):202–206 (in Chinese)
91. Sethuraman R (1997) Supercritical fluid extraction of capsaicin from peppers. Texas Tech University
92. Riera E, Golás Y, Blanco A et al (2004) Mass transfer enhancement in supercritical fluids extraction by means of power ultrasound. Ultrason Sonochem 11(3–4):241–244

93. Jiang G, Fang G, Zhang Z et al (2013) Optimization of extraction technology of proanthocyanidins from the larch bark by supercritical CO_2 assisted and the ultrasonic enhanced. Sci Technol Food Ind 34(18):257–261
94. Zhou M, Wang R, Liu Y (2012) Ultrasound-assisted extraction of synephrine from citrus dropt fruits in supercritical CO_2. J Food Sci Biotechnol 31(9):72–81
95. Glisic SB, Ristic M, Skala DU (2011) The combined extraction of sage (*Salvia officinalis* L.): ultrasound followed by supercritical CO_2 extraction. Ultrason Sonochem 18(1):318–326
96. Klein EJ, Santos KA, Palú F et al (2018) Use of supercritical CO_2 and ultrasound-assisted extractions to obtain α/β-amyrin-rich extracts from uvaia leaves (*Eugenia pyriformis* Cambess.). J. Supercrit Fluids
97. Barrales FM, Rezende CA, Martínez J (2015) Supercritical CO_2 extraction of passion fruit (*Passiflora edulis* sp.) seed oil assisted by ultrasound. J Supercrit Fluids 104:183–192
98. Yang R, Qiu T, Fan X et al (2009) Design of ultrasound-enhanced subcritical water extraction apparatus and analysis on the threshold of ultrasonic cavitation. Mod Chem Ind 29(2):70–74 (in Chinese)
99. Zhao C (2014) Ultrasound-enhanced subcritical water extraction of polysaccharides from *Lycium barbarum* L. South China Univ Technol (in Chinese)
100. Qiu T, Li C, Tan W et al (2010) Kinetic model and mechanism of ultrasonic-enhanced subcritical CO_2 extraction. J South China Univ Technol (Nat Sci Ed) 38(12):124–128 (In Chinese)
101. Guo J, Qiu TQ, Yang RF et al (2007) Progress in subcritical water extraction technique for extraction of natural products. Mod Chem Ind
102. Zhang X, Wang R, Ma Y et al (2015) Subcritical water extraction technology and application on natural active ingredients. Food Res Dev 132–136 (in Chinese)
103. Gbashi S, Adebo OA, Piater L et al (2016) Subcritical water extraction of biological materials. Sep Purif Methods 46(1):21–34
104. Chen Y, Tian J, Chen Y (2006) A novel technique of extracting essential oil from Chinese herb and medicine—subcritical water extraction. Chem Eng 34(8):59-62 (in Chinese)
105. Fernández-Pérez V, Jiménez-Carmona MM, Castro MDLD (2000) An approach to the static–dynamic subcritical water extraction of laurel essential oil: comparison with conventional techniques. Analyst 125(3):481–485
106. Gámizgracia L, Castro LD (2000) Continuous subcritical water extraction of medicinal plant essential oil: comparison with conventional techniques. Talanta 51(6):1179
107. Jiménez-Carmona MM, Ubera JL, Castro MDLD (1999) Comparison of continuous subcritical water extraction and hydrodistillation of marjoram essential oil. J Chromatogr A 855(2):625–632
108. Castro MDLD, Jiménez-Carmona MM, Fernández-Pérez V (1999) Towards more rational techniques for the isolation of valuable essential oils from plants. TrAC, Trends Anal Chem 18(11):708–716
109. Basile A, Jimenezcarmona MM, Clifford AA (1998) Extraction of rosemary by superheated water. J Agric Food Chem 46(12):5205–5209
110. Kotnik T, Miklavcic D (2000) Analytical description of transmembrane voltage induced by electric fields on spheroidal cells. Biophys J 79(2):670
111. Saulis G (2010) Electroporation of cell membranes: the fundamental effects of pulsed electric fields in food processing. Food Eng Rev 2(2):52–73
112. Weaver JC (1995) Electroporation in cells and tissues: a biophysical phenomenon due to electromagnetic fields. Radio Sci 30(1):205–221
113. Wang B, Yang J, Guo Q et al (1997) Experimental study and mechanism analysis on bioeffects by nanosecond electromagnetic pulses. Sci China 40(3):301–304
114. Yao X, Chen J, Chuanxiang XU (2003) Analysis for non-thermal effects of pulsed current electromagnetic fields on biological membranes. Adv Technol Electr Eng Energy 22(2):76–80
115. Xiao H, Yan P, Mou Q (2005) The experimental study of electroporation of cell membrane induced by pulsed electric field of high intensity. J Grad Sch Chin Acad Sci 22(4):462-466 (in Chinese)

116. Zimmermann U (1986) Electrical breakdown, electropermeabilization and electrofusion. Rev Physiol Biochem Pharmacol 105(6):176–256
117. Rubinsky B (2013) Mechanisms of abiotic horizontal gene transfer: comment on "Lightning-triggered electroporation and electrofusion as possible contributors to natural horizontal gene transfer" by Tadej Kotnik. Phys Life Rev 10(3):377
118. Wouters PC, Alvarez I, Raso J (2001) Critical factors determining inactivation kinetics by pulsed electric field food processing. Trends Food Sci Technol 12(3–4):112–121
119. Article O (2016) Journal of Food Science and Technology. Ann Libr Inf Stud 54(4)
120. Sun J, Kong F, Zu G et al (2004) The effects of high power pulsed electric field on the survivabitlity of *S. cerevisiae* and *E. coli*. Food Sci 25(2):87–90 (in Chinese)
121. Yin Y, Yin J (1997) Inactivation of bacterial spores treated by high voltage pulsed electric field. Trans Chin Soc Agric Eng 13(1):190–193 (in Chinese)
122. Jayaram S, Castle GS, Margaritis A (1992) Kinetics of sterilization of *Lactobacillus brevis* cells by the application of high voltage pulses. Biotechnol Bioeng 40(11):1412
123. Wu J, Xiao G, Chen W et al (2005) Study on mulberry juice sterilization by pulsed electric field equipment. Acad Periodical Farm Prod Process 41–42 (in Chinese)
124. Zhao W, Yang R, Zhang W et al (2010) The progress on effects of pulsed electric fields on microorganisms, enzymes and constituents in food. Food Mach 26(3):153–157 (in Chinese)
125. Pan D (2011) Study of pulsed electric field processing on microbial popution and quality characteristic of carrot juice. Fujian Agriculture And Forestry University (in Chinese)
126. Ge S (2005) The application of high-voltage pulse electric field in food sterilization. Phys Eng 15(1):42-44 (in Chinese)
127. Vega-Mercado H, Pothakamury UR, Chang FJ et al (1996) Inactivation of *Escherichia coli* by combining pH, ionic strength and pulsed electric fields hurdles. Food Res Int 29(2):117–121
128. Da HJ, Bo YJ, Park DH (2013) Effects of an electric pulse on variation of bacterial community and metabolite production in kimchi-making culture. Biotechnol Bioprocess Eng 18(5):909–917
129. Ravishankar S (2015) Pulsed electric fields. Food Sci Technol Int 14(5):360–427
130. García D, Gómez N, Mañas P et al (2007) Pulsed electric fields cause bacterial envelopes permeabilization depending on the treatment intensity, the treatment medium pH and the microorganism investigated. Int J Food Microbiol 113(2):219–227
131. García D, Gómez N, Raso J et al (2005) Bacterial resistance after pulsed electric fields depending on the treatment medium pH. Innov Food Sci Emerg Technol 6(4):388–395
132. Salengke S, Sastry SK, Zhang HQ (2012) Pulsed electric field technology: modeling of electric field and temperature distributions within continuous flow PEF treatment chamber. Int Food Res J 1137–1144
133. Mattar JR, Turk MF, Nonus M et al (2014) Stimulation of *Saccharomyces cerevisiae* cultures by pulsed electric fields. Food Bioprocess Technol 7(11):3328–3335
134. Sato M (2011) Inactivation of microorganisms by pulsed electric field. Proceedings of the Institute of Electrostatics Japan, 27
135. Knorr D, Geulen M, Grahl T et al (1994) Food application of high electric field pulses. Trends Food Sci Technol 5(3):71–75
136. Gášková D, Sigler K, Janderová B et al (1996) Effect of high-voltage electric pulses on yeast cells: factors influencing the killing efficiency. Bioelectrochem Bioenerg 39(2):195–202
137. Dutreux N, Notermans S, Wijtzes T et al (2000) Pulsed electric fields inactivation of attached and free-living *Escherichia coli* and *Listeria innocua* under several conditions. Int J Food Microbiol 54(1–2):91–98
138. Fan C, Xu L, Sun C et al (2017) Optimization of high voltage pulse electric field assisted extraction of ginseng aqueous extracts. J Northwest A&F Univ (Nat Sci Ed) 45(6):193–198 (in Chinese)
139. Dai M, You L, Wen L et al (2016) Technology optimization of total flavonoids extraction from *Pueraria lobata* by the high-intensity pulsed electric fields assistance. Food Res Dev 37(23):94-98 (in Chinese)

140. Tang S, Wang W, Zhang Y et al (2014) Extracting technology of flavonoids and polyphenols in bamboo leaves by high pressure pulse electric intensity. Food Nutr China 50–53 (in Chinese)
141. Liu X, Guo X, Lin S (2016) Improvement of extraction rate of corn bran polysaccharide by high-voltage pulsed electric field. J Food Saf Qual 7(6):2419–2425 (in Chinese)
142. Bai Y, Li C, Zhan J et al (2013) Extracting technology of guanfubase A from *Aconitum coreanum* by high-voltage pulsed electric field method. Chin Tradit Herbal Drugs 44(10):1267–1270 (in Chinese)

Chapter 12
Vegetable Oils and Fats: Extraction, Composition and Applications

Zahia Ghouila, Moussa Sehailia and Smain Chemat

Abstract Lipids contain a broad category of non-polar molecules that are barely soluble or completely insoluble in water, but soluble in chloroform, hexane, methanol and diethyl ether. The main sources of edible lipids are from the agricultural origin, mainly extracted from fruits, seeds or fish. They are found in the form of free or bound fatty acids, phospholipids, waxes, sterols, tocopherols, carotenoids, cholesterol and similar compounds. Oilseeds and fats production requires several unit operations, starting with a pre-treatment stage that includes washing, drying, heating and flaking followed by a processing stage, dependent on the type of raw material and the target product, using different techniques alone or combined such as rendering, pressing and solvent extraction. Crude fat and any other soluble material can be co-extracted in addition to lipids. These may include residual moisture, pigments, carotenes, urea and other compounds. The final stage concerns purification the obtained oils via multiple operations which involve refining, bleaching, destearinating or winterizing, hydrogenation and deodorization. In the last decade, oilseeds and fats have witnessed the extension of their application range from edible food towards new applications. Nowadays, they are included in many preparations of a large range of cosmetics, functional ingredients or nutraceuticals and recently have made the breakthrough as synthons and greener solvent options as described later in this chapter.

Keywords Edible oil · Fats · Extraction · Refining · Green solvents · Oilcake · Fatty acids · Olive oil · Sunflower oil · Rapessed oil

---

Z. Ghouila · M. Sehailia · S. Chemat (✉)
Research Centre in Analytical Chemistry and Physics (C.R.A.P.C), BP 384 Zone Industrielle de Bousmail (Ex. PASNA), Bousmail, Tipaza CP 42004, Algeria
e-mail: chemats@yahoo.fr

Y. Li and F. Chemat (eds.), *Plant Based "Green Chemistry 2.0"*,
Green Chemistry and Sustainable Technology,
https://doi.org/10.1007/978-981-13-3810-6_12

12.1 Introduction

Global oilseeds production reached record levels of 207 MT in 2015 and expected to continue its moderate expansion at a growth rate of 1.6% p.a. far less the last decade's growth of 3.5% p.a which keeps oilseed prices under pressure. Oilseeds production from palm represented in 2015 more than 30% of total oilseeds + fats production, while soybean oil reached 51 MT (25.1%) bumped by production in the USA and South America followed by rapeseed oil (11.3%), sunflower oil (8%) [1]. This is due to the growing demand for protein meal which favours soybeans over other oilseeds. Due to saturation in per capita, food demand in many emerging economies and reduced growth in biodiesel production from vegetable oils, prices are projected to increase less than the assumed inflation rate according to OECD outlook [2]. Vegetable oil includes the oil from crushing oilseeds (around 53%), palm (36%), palm kernel, coconut and cottonseed.

Global production of meals/cakes in 2014/15 would expand strongly and could increase by 9% to 160 million tonnes driven entirely by soy, with incremental world soymeal output estimated at close to 11 million tonnes (expressed in protein equivalent) [3]. Other meals are expected to shrink, except for a small rise in palm kernel meal and stable cottonseed meal.

In developed countries, more than 40% of dietary fat comes from isolated fats and oils, with 60% obtained from basic foods, whereas the dietary fat in less developed countries is obtained from fruits, cereals, vegetables, dairy products and meats, and relatively little is consumed in the form of isolated fat products. Most fruits and vegetables have from 0.1 to 2% fat, with the exception of avocados and olives, which are exceptional in their high-fat content. Cereals range from 1 to 7%, and nuts may contain as much as 70% fat [4].

The main storage form of lipids in plants and animals is in the form of glycerol backbone with three fatty acids (triglycerides) which include fats such as lard, shortening, butter and margarine (solid at room temperature) and oils such as olive oil, peanut oil, soybean oil or sunflower oil (liquid at room temperature). Lipids tend to be solid when these three fatty acids are mostly or all saturated, while it is in liquid form when the fatty acids are mostly or all unsaturated. Above the proteins and lipids present in seeds, the wall surrounding the cell is mainly composed of cellulose, hemicelluloses, lignin and pectin. They are bound with these primary metabolites such as proteins and carbohydrates with van der Waals interactions, electrostatic, hydrogen or covalent binding.

Due to the nature of the matrix, the processing methods are generally neither specific to lipids, nor they ensure 100% recovery of the lipid material. In the case of crude fat, diethyl ether or petroleum ether stands as favourite solvents as they are relatively non-polar and extract most non-polar components (triacylglycerols, sterols and tocopherols). However, they fail to extract polar lipids, such as glycolipids and phospholipids.

Special conditions are required to store oilseeds' raw material to avoid alterations caused by insects or moulds and prevent or delay lipid oxidation, which is known to be favoured by enzymes, temperature, humidity and oxygen. For instance, reduction in the seeds' humidity by means of dryers and storage in silos under controlled atmosphere using adequate ventilation without oxygen are widely used techniques to avoid such deterioration.

12.2 Conventional and Innovative Extraction Techniques and Solvents

Before processing, oilseeds' moisture must not exceed a certain limit to prevent the growth of fungi and the occurring lipase formation, resulting in a free fatty acid increase. In this case, conditioning the raw material is required which includes a combination of drying with heating to induce the hulls surrounding the oilseed to dry and shrink away from the meat part of the seed.

Generally, three processing methods are used to recover fats from oil-bearing tissues: (1) rendering, (2) pressing with mechanical presses and (3) extracting with volatile solvents [4].

12.2.1 Rendering

Fruits and seeds

Rendering oil from oleaginous fruits consists of heaping them in piles, exposing them to the sun and collecting the oil that exudes. This process is still used in the preparation of palm oil where the fresh palm fruits are boiled in water, and the oil is skimmed from the surface. Such processes are only suitable for seeds or fruits (such as olive and palm) containing large quantities of easily released fatty matter.

Animal fats

The rendering process is applied on a large scale to the production of animal fats such as tallow, lard, bone fat and whale oil. It involves cutting or chopping the fatty tissue into small pieces that are boiled in open vats or cooked in steam digesters. The fat is then liberated gradually from the cells and floats to the surface of the water, where it is collected by skimming. Additional fat is obtained from the separation of the greaves from the aqueous layer by pressing using hydraulic or screw presses. The residue is used for animal feed or as a fertilizer.

In parallel, centrifugal separation processes were developed to dislodge the protein tissue from the liquid phase, followed by a second centrifuge separating the fat from the aqueous protein layer. Compared with conventional rendering, the centrifugal methods provide a higher yield of better-quality fat, and the separated protein has great potential as an edible meat product.

12.2.2 Pressing Processes

In this process, the cell walls are broken by grinding, flaking, rolling or pressing under high pressures to liberate the oil. Pressing oilseeds and nuts follows a general sequence:

(1) Removal of any stray bits of metal from seeds using magnetic separators
(2) Removal of the shells or hulls, if necessary
(3) Conversion of the kernels or meats to a coarse meal by grinding them between grooved rollers or with special types of hammer mills and
(4) Pressing using expeller, hydraulic or screw presses with or without preliminary heating, depending on the type of oil-bearing seeds/fruits and the desired quality specification of oil.

Oil expression without heating contains the least amount of impurities; thus, it does not require refining or further processing. Such premium oils are known as cold-drawn, cold-pressed or virgin oils [5].

Using heat during pressing, the coarse meal removes more oil, that is highly coloured than cold-pressed oils, but accompanied with non-glyceride impurities such as phospholipids, carotenoids and unsaponifiable matter. From the cooker, the flack seeds are passed through the expeller, also known as screw press. This process squeezes out about 70% of the oil, which is routed to a settling tank [5]. Residual meals are concentrated sources of high-quality protein and are generally used in animal feeds.

12.2.3 Solvent Extraction

Generally, crude oils are recovered by solvent from the meal comprising also proteins and carbohydrates. During the solvent extraction process, the oilseeds are flaked due to thermal stress which ruptures the cell wall and exposes the oil present inside the cell to the bulk medium and facilitates percolation of the solvent through the matrix. The solvent dissolves the oil, while the protein remains in the meal with the fibres and carbohydrates. In the case of the aqueous extraction process of oil, the soluble components diffuse into the water, and the partially soluble oil forms a separate liquid phase or is partially emulsified with the water [4]. The properties of the oil such as refractive index, density, saponification and free fatty acid values were affected by the type of solvent used, and therefore, the selection of solvent is of high importance to fulfil quality standards.

12.2.3.1 Conventional Extraction

Usually, the seeds undergo a cracking or grinding step before the conditioning process described earlier. The temperature (30–40 °C) and moisture content (~8%) are adjusted before the flaking process which is known as conditioning. After cooking, the seed is passed for flaking and flatten to a thickness of ~0.25 mm [6]. This process helps in the extraction process by rupturing the cells which enhanced the percolation of the solvent. Various solvents have been used commercially to extract lipids from oil-bearing material. Hexane and hexane-based solvents are broadly accepted as the most efficient solvents for oil extraction despite concerns for its flammability, explosiveness, mild toxicity and environment impacts. The petroleum distillate containing about two-thirds n-hexane is normally used in the commercial extraction of soybean oil [7].

The modern solvent extraction process consists of percolating the solvent in a successive countercurrent mechanism through the cracked, ground, flaked or pressed oleaginous materials. Solvent is recovered from the oil by rising film evaporator followed by vacuum distillation.

Extraction of oilseed involves several mechanisms for removing a liquid from a solid including leaching, washing, diffusion and dialysis. Flaking helps in distortion of the cells and rupture of cell wall. It also reduces particle size, thickness and the distance required for transfer of oil into bulk materials. The transfer of oil from the rupture cells is governed by capillary, and the rate of oil transfer is partially dependent upon viscosities of solvent and micelle [8]. Little oil (less than 1%) is left within the cake and is normally removed by osmosis. The solvent is separated from the meal by distillation and recycled for the next process [6].

Lajara (1990) reported the importance of flake surface rather than the extent to which cells have been opened during the flaking operation where heat treatment causes extraction of more phosphatides [9]. During rapeseed dehulling, moistening the seed with steam and drying the moistened seed in a fluidized bed resulted in maximum dehulling efficiency [10].

Another process developed by Strop & Perry comprises an acid degumming treatment step of the oil before it is forced to leave the oilseed [11]. In this case, the resulting crude oil from milling is degummed with water and then treated with an acid to decompose the non-hydratable phosphatides to obtain an oil quality of a low phosphorus content that can be physically refined after bleaching [6]. In the Strop & Perry process (1990), the mixture is heated under vacuum, causing some water to evaporate, and the resulting slurry is then separated in a centrifuge to recover the cooked oil. According to the invention, the cooked oil recovered from the centrifuge only requires washing and filtering before being steam refined, cooled and stored. This process allows oil refinery without a need for degumming, neutralization and bleaching steps [11].

In the case of palm oil, crude oil is obtained after a digestion step followed by a pressing stage. Digestion helps the rupture or breaking down of the oil-bearing cells of the mesocarp, thus releasing the palm oil in the fruit. The digester consists of a steam-heated cylindrical vessel at about 90 °C fitted with a central rotating shaft that carries out a number of beater arms that is responsible to mash the fruits [12]. Then, the mashed fruit is passed through a screw press to release the oil, and a residue called the press cake is obtained which requires further processes to produce palm kernel. Digestion at high temperature helps to reduce the viscosity of the oil, destroys the fruits' exocarp and completes the disruption of the oil cells. However, pressure has to be optimized as high pressure will obviously cause undesirable higher nut breakage, and thus, high broken kernel and subsequently higher kernel loss in encountered. In addition, palm oil will be contaminated by the "kernel oil" [13].

Refining
Crude edible oil extracted from oleaginous commodities and marine catch often require further downstream processing such as degumming, alkali refining, soap removal, bleaching, filtration, fractionation, physical deacidification and deodorization and crystallization. These operations help the removal of impurities and obtaining uniform oil with good appearance and flavour quality and higher shelf life. At the industrial scale, two broad refining categories are practised: chemical refining and physical refining. The main difference between them resides in the removal of free fatty acids (FFA) either by chemical addition or by physical removal through distillation.

In case of the chemical or wet refining process, the impurities are removed by contacting the oil with certain chemicals. It involves several stages including degumming, deacidification, bleaching and deodorization. Crude oils usually possess a number of desirable and undesirable compounds, while acylglycerols, tocopherols, carotenoids, phytosterols and polyphenols are some of the essential components of oil [14]. Since consumers expect bland and odourless oil with high oxidative stability, the refining stages are designed to discard the major class of impurities that reduce the oil quality like phospholipids, FFA, metal ions, oxidation products and other volatiles. In parallel, physical refining uses stripping with steam to remove under vacuum the FFA, impurities and unsaponifiable matters from the oil.

Bleaching increases the taste of the oil by removing other pigments and residual soap while other remaining off-flavours are removed during the deodorization step by high vacuum (2–4 mmHg) distillation process in which oil is treated with steam (2–4%) at high temperature (260–265 °C). To avoid the presence of waxes and fully saturated triglycerides, winterization is performed by treating the deodorized oil to a cooling period followed by filtration of the precipitate.

For crude palm oil, refining is necessary to separate palm oil (35–45%) from water (45–55%) and fibrous materials [14]. It consists of pumping the crude oil into a horizontal or vertical clarification tank. Then, steam is injected into the tank that is maintained at around 90 °C to create an oil–water mixture which is left to settle for a few hours. Upon settling, the clean oil undergoes centrifugation to discard impurities and is subsequently passed through a vacuum drier to reduce moisture before it is sent to the storage tanks.

12.2.3.2 Microwave Extraction

Microwaves have been proved to accelerate extraction rates of lipids from seeds and tissue lipids. Using a laboratory microwave oven, maximal recoveries of glycerolipids are reported to be similar to the classical Folch's procedure [15]. Dielectric heating (both high-frequency field and microwave field) has also been used for pre-treating oilseeds using 2:1 (*v/v*) chloroform/methanol mixture instead of hexane in order to extract the lipid fraction from the beans. The use of microwave irradiation to enrich olive oil was also found to be effective under conditions of 0.1–5 W/g of oil and plant material mixture [16]. Such approach could be exploited in a number of industries to obtain special extracts in the form of scented extracts and other active ingredients.

Chiavaro et al. [17] indicated the oxidative deterioration under microwave power of 720 W for 1.5–15 min of refined peanut, high-oleic sunflower and canola oil. It was found that canola was more extensively oxidized by microwave heating. Both peanuts and canola oil experienced changes in the heating profiles, free fatty acids composition and peroxide value. Under microwave irradiation at 2450 MHz at varying power ranging from 3 kW to 5 kW, quality indicators of oilseeds (free fatty acids, phospholipids as well as iodine and acid values) from soybean and rice bran showed acceptable standards [18].

12.2.3.3 Ultrasound Extraction

A growing trend is registered for application of ultrasound in extraction of oils and fats by enhancing rates, improving quality and/or safety and reducing processing time. The benefits of ultrasound are attributed to acoustic cavitation: when a mixture is subjected to ultrasound, micro-bubbles will grow and oscillate quickly before collapsing due to pressure changes. These violent implosions are believed to fragment or disrupt at the surface of the solid matrix, therefore enhancing mass transfer and accelerating diffusion. Ultrasound-assisted extraction can be used during extraction to enhance the enrichment of edible oils with carotenoids from pomaces. A combination of microwave and ultrasound was also found to maximize aromatization of olive oil when mixed with vegetables, herbs, spices or fruits [16]. Notable are Sincaire et al. studies [19] that investigated ultrasound-assisted extraction for rapeseed oil recovery where a multivariate study enabled them to obtain optimal parameters (7.7 W/cm^2 for ultrasonic power intensity, 40 °C for processing temperature and a solid/liquid ratio of 1/15) in order to maximize lipid yield while reducing solvent consumption and extraction time. Luque-Garc´ıa and Luque de Castro concluded that the fatty acid composition of oilseeds such as soybean, sunflower and rapeseeds is hardly affected by the application of ultrasound [20]. Using ultrasound to extract flaxseed oil enabled to reduce processing time using an ultrasonic power: 50 W at 30 °C and a liquid-to-solid ratio of 6:1 (v/w) [21].

Several systems have been developed at an industrial scale, most of them used for extraction of phytochemicals but also in aromatization of vegetable oils, i.e. extraction of carotenoids and antioxidants from medicinal plants into edible oils. Figure 12.1 shows an ultrasound-assisted system able to process up to 200 L in a single run.

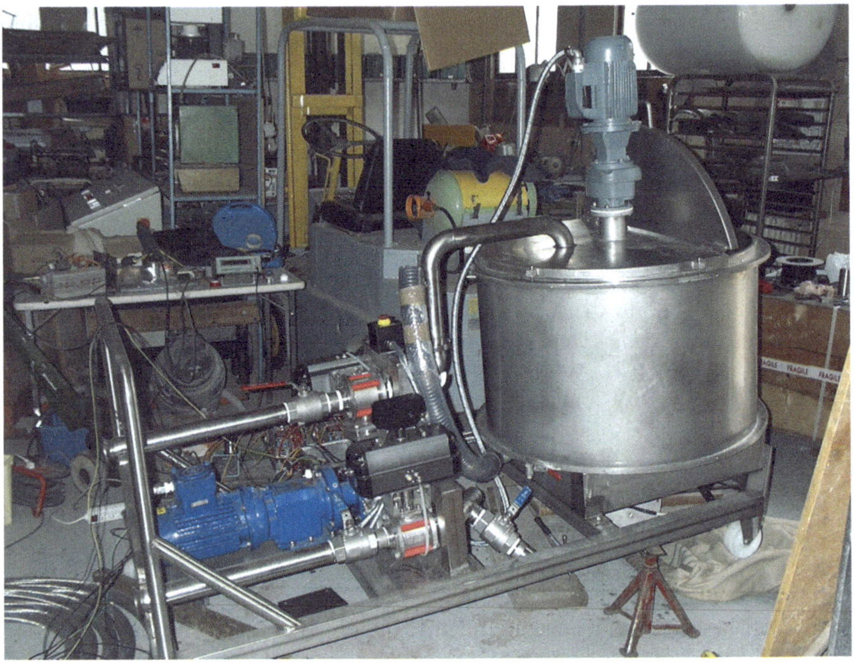

Fig. 12.1 200 L ultrasound-assisted extraction system (courtesy of Charles Gantz (REUS, France))

12.2.3.4 Green Solvent Options for Oilseeds Extraction

A solvent having low solubility at ambient temperature and high solubility at higher temperature is desirable to profit from oil separation without evaporation. It should be non-toxic to human and animal as well as safe for the environment. The solvent should able to extract selectively only triglycerides leaving phosphates, free fatty acids and pigments and should easily be recovered from the meal and oil. As far as safety and environment issues are concerned, petroleum-derived solvents should be replaced by greener options that are non-flammable or low flammable within a narrow range of explosions such as azeotrope solvents, solvents based on aldehydes and ketones, aqueous-based solvents and ionic solvents. In this case, solvents have to be stable in heat, in light and in water while being able to withstand repeated cycles of heating, vaporizing and cooling. They should be non-corrosive to equipment and non-reactive with oil, meal and even with the equipment.

Azeotropic mixture

The alcohol/water mixture azeotropes of ethanol and isopropanol have been considered to be most commercially potential. As water content increases, the solvent becomes more polar, and the solubility decreases towards oils, but capacity to extract non-lipid substances (phosphatides, pigment and sugars) increases. Interestingly, both the azeotropes of ethanol/water and isopropanol/water mixtures had shown to be effective in removing the aflatoxins during the extraction [22]. Compared to the absolute non-protic solvents, the protic organic solvents such as ethanol, methanol, isopropanol and ethyl acetate, mixture of ethanol and ethyl acetate are found to have more advantages in the oil extraction due to the presence of less organic impurities. Azeotrope solvents show higher critical solution temperature and indicate better extraction of oil from the seeds [23].

The loss of solvent is minimum in the azeotrope solvents because of the lower boiling point of azeotrope compared to the pure solvent which reduces the cost of extraction. The properties of the oil extracted by azeotrope solvent show less density and refractive index value as compared to the n-hexane with fewer impurities.

Aldehydes and ketones

Aldehyde solvents, such as furfural (an aldehyde solvent) and furfural alcohol which can be made from agricultural residues, have an excellent solubilizing property for oils at warm temperatures, and the solvent is recovered from crude oil by either chilling the miscella or adding water. Commercially, it is available. However, oils extracted with furfural have high quantity of phosphatides [8].

Eaves et al. showed that acetone and butanone had the ability to extract oil from cottonseed flakes which are similar to hexane. Although, free fatty acid is soluble in acetone, phosphatides are insoluble. A process known as Vaccarino process was developed to extract cottonseed meal with acetone commercially [24].

Aqueous oil extraction

The aqueous extraction process (AEP) uses water alone as an extracting medium in which oil and protein (as a concentrate or an isolated form) can be recovered from the oil-bearing material simultaneously. Extraction of oil by AEP is based on the insolubility of the oil in water rather than the dissolution of oil [8]. During this process, water-soluble components diffuse in the water which leads to the release of oil from the previously bound structure, which can be enhanced by tuning pH, increasing temperature or using enzymes to hydrolyze the cell walls [25]. In fact, mixing of ground and dehulled oilseeds in the vats of hot water is done during which the oil rises on the top of the surface and it is skimmed out [26]. In modern AEP, gravity separation is replaced by centrifugation. Basic unit operations of AEP include grinding, solid–liquid separation, centrifugation, demulsification and drying. Centrifugation separates the oil from the bulk material so that protein and other materials are obtained as concentrate in solid phase or isolates in aqueous phase depending upon the pH and extracting media. Oil is recovered by breaking the emulsion, whereas the final proteins can be obtained by isoelectric precipitation. The oil obtained after the demulsion process has high quality and is required to remove the residual water for the final refining step [27]. Membrane isolation process (MIP) and ultrafiltration (UF) can be associated to recover the protein directly from the aqueous oilseed extract and avoids the generation of whey. Reverse osmosis (RO) can be employed to recover the secondary products and make the water reusable for the next cycle [27].

Supercritical CO_2 (SC-CO_2)
Being non-flammable with relatively low toxicity, CO_2 became the supercritical fluid of choice because it is naturally abundant and hence inexpensive, and its critical temperature is readily accessible (31 °C). Unfortunately, CO_2 has low solubility for high molecular weight and/or polar compounds, but strategies have been employed to overcome its poor solvency power by the use of co-solvents (known as entrainers) or surfactants and chelating agents. Edible oils contain non-polar characteristics since they include a broad range of fatty acids, carotenoids, terpenoids, terpenes and/or tocotrienols, making SC-CO_2 suitable for their extraction from solid matrices. Besides, CO_2 is easily removable from the products by decompression at room pressure, making the edible oils solvent-free. SC-CO_2 was combined that alcohols such as methanol (typically at levels less than 10 mol%) to improve CO_2's ability to solvate polar solutes.

The application of CO_2 in SFE of edible oils typically uses conditions ranging from 40 to 80 °C and 20–50 MPa [28]. Density of CO_2 can be tuned by changing the system pressure or temperature, mainly in the region close to its critical pressure. For example, in SFE-CO_2 from olive husk, increasing pressure from 25 to 35 MPa at constant temperature increased carotenoids concentration and decreased tocopherols concentration in olive oil [29]. The low viscosity and the high diffusion coefficient (10^{-3} cm^2 s^{-1}) favour obtaining high extraction rates when using supercritical fluids as solvents. In fact, CO_2 is a gas at 40 °C and 0.1 MPa with density of 2 kg m^{-3}; whereas it is in a liquid status at 27 °C and 50 MPa with density of 1029 kg m^{-3}; and at 40 °C and 10 MPa, it is a supercritical fluid with density of 632 kg m^{-3} and viscosity as low as 17 μ Pa s, which is about 100-fold lower than that of liquids permitting easier penetration and higher extraction rates [30].

12.3 Major and Minor Constituents of Fats and Oils

Edible fats and oils play significant role in diet, thanks to the energy intake provided as needed by the human body. International authorities like World Health Organization (WHO), Food and Agriculture Organization (FAO), the European Food Safety Authority (EFSA) and many others recommend a total fat intake ranging from 20 to 35% of total calories [31–33]. This fat intake ensures an adequate consumption of essential fatty acids and fat-soluble vitamins useful for the proper functioning of the human body. The lowest level of high-density lipoprotein (HDL) of 20% is necessary to prevent atherogenic dyslipidemia as a case of low high-density lipoprotein cholesterol (HDL-C) and high triglyceride-rich lipoproteins [33]. The upper level (35%) is to limit the saturated fat consumption leading to gain weight and increases the risk of coronary heart disease [34, 35]. Edible fats and oils are composed of major and minor constituents. The major constituents represented by the fatty acids family comprises saturated, monounsaturated and polyunsaturated fatty acids (SFA, MUFA and PUFA, respectively) depending upon the presence or absence of double bonds. These are commonly called triglycerides and reach around 96% of total fatty acids

[36]. Free fatty acids (FFA), monoacylglycerols, diacylglycerols, phospholipids, free and esterified phytosterols, polyphenols, triterpene alcohols, tocopherols, pigments (carotene, chlorophylls), squalene and trace metal ions (e.g. iron, sulphur or copper) are considered as minor constituents [37].

12.3.1 Major Constituents

12.3.1.1 Saturated Fatty Acids (SFA)

Saturated fatty acids, relative to compounds with only single carbon-to-carbon bonds, are principally found under solid form in animal fats at ambient temperature. They are relatively present in high amounts in many meats and dairy products, including whole milk, cheese, butter and cream. They can also be found in coconut and palm kernel oils, which contain more saturated fatty acids than lard, beef tallow and butterfat [38]. Among the SFA, lauric acid (C12:0), myristic acid (C14:0), palmitic acid (C16:0) and stearic acid (C18:0) are the most abundant compounds in both fats and oils. Lauric acid less abundant in animal fats and only present in butterfat with 3% of total fatty acids reached high percentages in coconut and palm kernel oil (47% and 48% of total fatty acids, respectively) [39]. Similarly, myristic acid (C14:0) which is a characteristic compound of animal fats can also be found at significant levels in coconut and palm kernel oil. SFA with shorter carbon chain (less than C10:0) is a characteristic of fats (butter) but can be found at trace levels in certain oils like coconut, palm kernel and Indian ghee oils [39–41]. The palmitic acid (C16:0) and stearic acid (C18:0) are the most abundant SFA and are omnipresent in the majority of fats and oils at variable level, and the important level is reached by palmitic acid (40–47% of total fatty acids) in palm oil [39, 42]. Arachidic (eicosenoic) acid (C20:0) has been found at trace level (<1% of total fatty acids) in conventional oils such as safflower (0.17–0.2%), canola (0.2–1.2%) and soybean (0.1–0.6%) oils [39], whereas it reached high amounts in mustard oil (5.48 $\pm$ 0.36%) from India and pracaxi oil (12.3 $\pm$ 0.12%) [41]. The same observation is noted for behinic acid (C22:0) rarely present in common oils, and if it is present, it does not exceed 1% of total fatty acids. The exception is the pracaxi oil which recorded an important level of 22.6 $\pm$ 0.0% of total fatty acids [43]. The presence of SFA at high amount in certain fats and oils was appreciable especially by dietary industry for its resistance to oxidative reactions, property which makes stability to the food products. However, the SFA consumption was not advising since it was responsible for LDL-cholesterol-raising potential causes of cardiovascular disease. Many health authorities suggest keeping saturated fat intake as low as possible (<10% of total fatty acids) [44].

12.3.1.2 Monounsaturated Fatty Acids (MUFA)

Oleic acid (C18:1) is the most important of MUFA, and it is an acid with double carbon-to-carbon bonds found at high level in almost fats and oils except for butterfat, coconut and palm kernel oil which are rich in SFA. Oleic acid level ranging from 5 to 55% of total fatty acids is a characteristic of vegetable oils as for olive oils with the highest level of 55%. Palmitoleic acid (C16:1) is found at low level (<3% of total fatty acids) in olive oils when the majority of vegetable oils have none. Contrary to the SFA, oleic acid reduces effectively at lowest total serum cholesterol and LDL-C levels, and oils rich in oleic acid are advised in feeding diet [45].

12.3.1.3 Polyunsaturated Fatty Acids (PUFAs)

PUFAs are acids with more than one double carbon-to-carbon bonds. Two important compounds are present in fats and oils, linoleic (C18:2) and linolenic (C18:3) acid known as omega 6 and omega 3, respectively. These acids not synthesized by the body are vital for its functionality. The intake needed was provided by the fats and oils diet. Oils contain high level of linoleic acid ranging from 1 to 48% of total fatty acids, while lower levels not exceeding 1% are contained by fats [39]. High amounts of omega 6 have been recorded for safflower (76.58%), sunflower (62.69%) and soybean (54.17%) oils [46]. Certain usual vegetable oils like olive oil rich in oleic acid registered level of linoleic acid ranging from 3.5 to 21.0% of total fatty acids. Others oils, especially Amazonian oils which high level of oleic acid of 65.9 $\pm$ 0.45%, have been reached in passion fruit oil, and their fats also contain more important level than common fats as for cupuassu fat (5.2 $\pm$ 0.0% of total fatty acids) [43]. The omega 6 is also found in oils newly introduced in food such as grape seed oil which contains between 58 and 78% of total fatty acids. Belonging to omega 3 fatty acid class, linolenic (C18:3) or α-linolenic acid (ALA) is considered as essential because it cannot be synthesized by human organism and believed to be precursor of others omega 3 like EPA and DHA. Plants are important sources of such ALA but can also be found in fats, exceptionally in butter and margarine made from vegetable oils. For animal fats, around 0.98–1.2% of total fatty acids have been recorded in butter and lard, and higher level has been registered for vegetable fats such as margarine with 2.4% [47]. However, it has been found beyond these concentrations in no common fats like cupuassu fat in which omega 3 reached 11.5 $\pm$ 0.45% of total fatty acids [43]. The intake in omega 3 from vegetable oils widely consumed in our diet varies between less than 1 and 14% of total fatty acids such as olive, safflower and sunflower oils (<1%), soybean (4.5–11.0%) and canola (5.0–14.0%) oils [39]. The highest levels of omega 3 have been found in oil from oilseed, with level up to 50% of fatty acids, as in the case of linseed oil ranging from 54 to 71% of total fatty acids [47].

12.3.2 Minor Constituents

12.3.2.1 Monoacylglycerols (MAGs) and Diacylglycerols (DAGs)

Mono- and diglycerides are mono- and diesters of fatty acids and glycerol, frequently used as emulsifiers in food industry. They are naturally formed in the intestinal tract as a result of the normal digestion of triglycerides and commercially prepared by the esterification of glycerol and fatty acids or by simple reaction of glycerol and triglycerides. They are the major components of the food additive E471 which can be of animal or vegetal origin [48]. The amounts of MAGs and DAGs in vegetable oils vary in the range of 3.0–7.6% as reported by O'Brien et al. [48].

Catalano et al. established the remarkable differences in total DAGs, i.e. 1,2(2,3)- and 1,3-diacylglycerols as characteristic parameters of virgin olive oil quality [49]. In extra virgin oils, the total DAGs were less than 2.0, 97.7% of the oils had less than 1.5% of 1,2 DAG and less than 0.4% of 1,3 DAG. Should there be a need to know the quality of the oil of origin, it may be more appropriate to include the DAG content as a complementary parameter. Generally, freshly extracted oils have a low DAG content of 2.3–4%. However, the conditions prevailing in the industry such as harvesting conditions, transportation of fruits to factories result in commercial oils having DAG contents of 4.0–7.5%.

12.3.2.2 Phosphatides or Phospholipids

Phosphatides are present at trace level (0.1–3%) in crude oils. They are derived from the association of glycerol with fatty acids and a phosphate ester and are composed essentially by phosphatidylethanolamines, phosphatidylinositides and other phospholipids. Soybean oil, rapeseed oil and sunflower oil are the main sources. They are used as emulsifiers in food industry under the name of lecithin formed only of phosphatidylcholines which is present in high quantity in sunflower and soybean oil (52% and 30% of total phosphatides, respectively) [50]. Adding also their property as an antioxidant often highlighted in many oils [51]. The majority of phosphatides naturally present in oils are removed from oil during the degumming and refining processes which are recovered and treated to yield a variety of lecithin products. Among the oils with important level of phosphatides, soybean oil is at the top of the list with 2.2 ± 1.0%, followed by canola oil with 2.0 ± 1.0%, corn oil (25 ± 0.25%), cottonseed oil (0.8 ± 0.1%), sunflower oil (0.7 ± 0.2%) and safflower oil (0.5 ± 0.1%). Olive oil and other commons oils contain less than 0.1% of phosphatides [48].

12.3.2.3 Sterols

Sterols are found in both animal fats and vegetable oils. The cholesterol is the animal fat forms as for butter which can contain about 215 mg/100 g. This kind of sterols is rarely found in vegetable fats and oils. Phytosterols, the current form found in vegetable fats and oils, can be found free or esterified with other molecules such as fatty acids. The profile of phytosterols allows to detect oil adulteration or to identify the oils and fats in a mixture. The most common concentrations found in vegetable oils are of the order of 1000–5000 mg/Kg. Different types of phytosterols are present in oils as for brassicasterol, campesterol, stigmasterol and β-sitosterol, and campesterol and β-sitosterol are the most abundant and found in the major of oils at high concentrations as for canola oil (1904.4 and 3608.7 mg/Kg), soybean oil (1191.40 and 2060.80 mg/Kg) and sunflower oil (266.50 and 2300.10 mg/Kg). Brassicasterol stands as a characteristic parameter of canola oil with level of 14.3% of total phytosterols reaching a concentration of 986.70 mg/Kg (Table 12.1). Corn oil has recorded 15,050 mg/Kg of sterols and its concentration can vary between 8000 and 22,000 mg/Kg, while olive oil is considered as the least rich in sterols with 100 mg/Kg [48].

12.3.2.4 Tocopherols/Tocotrienols

The tocopherols and tocotrienols are important minor constituents of most vegetable fats and oils. They are recognized as natural fat-soluble antioxidants which permit to retard rancidity. The common types of tocopherols and tocotrienols are alpha (α), beta (β), gamma (γ) and delta (δ), depending on the number and position of methyl groups on a chromanol ring. They are also an excellent source of vitamin E. Alpha-tocopherol has the highest vitamin E activity and the lowest antioxidant activity compared to the others tocopherols and considered as a standard (100%) for vitamin E activity. Nutritionally, α-tocopherol used as additive (E 307) is subjected to health recommendations with daily intake between 0 and 2 mg/Kg of body mass/day. Vegetable oils recorded concentrations of total tocopherols ranging from 50 to 3500 mg/Kg. Canola and sunflower oils contain, respectively, 566 mg/Kg and 898.8 mg/Kg of tocopherols, while the highest concentration of 1146.4 mg/Kg has been recorded in soybean oil (Table 12.1). Others oils with higher concentrations have been registered in corn and wheat germ oil with concentration range of 1100–1850 mg/Kg and 3000–3500 mg/Kg, respectively [53]. Tocotrienols rarely present in conventional oils are found in palm oil in concentration range of 300–1000 mg/Kg.

12.3.2.5 Pigments

Carotenoids and chlorophyll are the most important pigments which can be found in vegetable oils. The carotenoids as for carotenes, lycopene, xanthophylls and lutein are considered as precursors of vitamin A (retinol) and are present in important quantity

Table 12.1 Minor compounds in canola, soybean and sunflower oils (adapted from Gordon et al. [52])

	Oils		
	Canola	Soybean	Sunflower
Phosphatides (%)	2.0 ± 1.0	2.2 ± 1.0	0.7 ± 0.2
Sterols (%)			
Cholesterol	0.65	0.96	0.74
Phytosterol			
Brassicasterol	14.3	Trace	Trace
Campesterol	27.6	25.9	6.5
Stigmasterol	0	16.8	7.8
β-Sitosterol	52.3	44.8	56.1
$^{5}\Delta$-Avenasterol	2.5	6.5	14.8
$^{7}\Delta$-Avenasterol	0.9	2.8	5.9
$^{7}\Delta$-Stigmasterol	2.3	1.4	7.1
Total phytosterol (mg/Kg)	6900	4600	4100
Esterified sterols (mg/Kg)	4231	576	2069
Tocopherol (mg/Kg)			
α-Tocopherol	197	179.3	206.5
β-Tocopherol	–	–	–
ϒ-Tocopherol	369	636.8	239
δ-Tocopherol	–	330.3	453.3
Total Tocopherols	566	1146.4	898.8

in palm oil in range of (1000–2000 mg/Kg). Among the palm oil carotenoids, α, β, γ, δ, ζ carotene are present, respectively, in amount of 2519, 379.9, 2, 5.4 and 4.7 mg/Kg as well as lycopene with 8.7 mg/Kg [49]. Xanthophylls represent 90% of total carotenoid concentration (117 mg/Kg) in crude oil followed by 10% of carotenes (13 mg/Kg) [29]. Chlorophylls are the green pigments found in almost vegetable oils, and they act on photosynthesis of plants. Chlorophylls are at the origin of the green tinge of the oils which sometimes at important levels the oils are denatured, and during oil process (refining) these levels are drastically reduced to ppb levels. Among common vegetable oils, crude canola oil registered the highest amount of chlorophylls range in 4–30 mg/Kg as well as crude olive oil (1–20 mg/Kg) [29]. This latter concentration can exceed 80 mg/Kg for oil obtained from olives in early maturity stage to finally reach 2 mg/Kg when the fruit is ripe [29].

12.3.2.6 Polyphenols

Vegetable oils like crude canola oil have a high amount of polyphenols (about 130 mg gallic acid/Kg oil) which provide antioxidation and stability properties [54]. During degumming and traditional neutralization steps, polyphenol percentage decreases dramatically to 2.1 mg gallic acid/Kg oil. Vinylsyringol, sinapic acid and sinapine are the most important polyphenols found in canola oils [55]. Others phenolic compounds in olive oil as for o- and p-coumaric, cinnamic, caffeic, ferulic, gallic, synaptic and chlorogenic acids have been identified.

12.4 Examples of Fat and Oils

12.4.1 Olive Oil

Olive oil is extracted from olive fruit (*Olea europaea* L.) by several extraction processes. These fruits with different shapes, sizes (e.g. 2–3 cm width and length) and pulp per stone ratios (3.0–6.5) contain between 36 and 53% water (w/w), whereas oil ranges between 18 and 27%. The last 20% are attributed to phenols, dietary fibre (DF), free sugars, nitrogenous compounds and minerals [16]. In attempt to protect the olive growing sector, the International Olive Oil Council (IOC) has promulgated four definitions for olive oil [6]:

1. *Virgin olive oil*: the oil from the fruit of the olive tree obtained by mechanical or other physical means under conditions, particularly thermal, which do not lead to alteration of the oil. Virgin olive oil is suitable for consumption as is and can be designated as "natural" is further defined as:
    (a) *Extra virgin olive oil*—oil that has a flavour rating of 6.5 or better and a 1.0 max free acidity.
    (b) *Fine virgin olive oil*—oil that has a flavour rating of 5.5 or better and a 1.5 max free acidity.
    (c) *Semifine virgin olive oil*—oil that has a flavour rating of 3.5 or more and a 3.3 max free acidity.
2. *Virgin olive oil* with an organoleptic rating of less than 3.5 or a 3.3 g/100 g free fatty acid is considered not fit for human consumption. It is used to produce refined olive oil or for non-food uses.
3. *Refined olive oil* is the oil obtained from virgin olive oil by refining methods which do not lead to alterations in the initial triglyceride structure.
4. Olive oil can consist of a blend of refined olive oil and virgin olive oil in various proportions.

To further improve extra virgin olive oil stability during storage, inert gases (such as nitrogen or argon) were used as conditioner gas in the head-space of the container, to test their ability to remove oxygen and prevent oxidation [56].

12.4.1.1 Production and Localization

The worldwide production of olive oil has considerably increased; it has risen from 1032 MT in 1958/1959 to around 3 MT in 2009/2010, then to 3.4 MT in 2011/2012. This production represents approximately 3% of the world's output of vegetable fats and oils. Expansion in olive orchard plantings around the world has increased gradually, and it rises from 4 Mha of land to 9.5 Mha in only six years with production range in 9–15 MT of olive fruit [1, 2]. The major production part is localized in Southern Europe, North Africa and the Middle East with 95% of the world's olive oil. Europe alone represents more than 67.31% of the world production. According to the latest letter of IOC, world production of olive oil for the 2017/18 season is estimated at 2,894,000 T with 14% higher than the previous season [57]. Spain remains in the lead with an estimated output of 1,090,500 T, followed by Italy with 320,000 T, Greece with 300,000 T and Portugal with 78,800 T. The main increases occurred in Turkey with 287,000 T (+62%), followed by Tunisia with production of 220 000 T (+120%), Morocco with 140,000 T (+27%), Algeria with 80,000 T (+27%), Argentina with 37,500 T (+74%) [57].

12.4.1.2 Extraction Process

Before oil extraction, olive receipt follows four steps in order to preserve olive oil quality, (1) olive storage; (2) removing foreigner bodies; (3) washing, then (4) processing.

Olive oil is generally obtained from olive fruit by mechanical procedures, which involves one of the following extraction processes:

(i) discontinuous process in traditional mills (press olive oil extraction),
(ii) three-phase centrifugal olive oil extraction,
(iii) two-phase centrifugal olive oil extraction.

Extraction starts with the crushing step which aims at the release of oil droplets from the cells of olive flesh. This operation is performed by metallic crushers, such as mobile or fixed hammers, toothed discs, cones, cylindrical or rollers crushers. Functioning at high-speed rotations, these crushers project violently the olives against a fixed or mobile metal grating, which produces an olive paste. It is important to note that metal crushers' type as well as the crushing intensity affects the size of the stone fragments, the extraction yields and the organoleptic properties of virgin olive oil (VOO). In fact, metal or granite crushers influenced the organoleptic characteristics of VOO, with higher content of total phenols (435 mg/Kg), have been obtained with metal (hammer) crushers than granite (millstones) crushers (186 mg/Kg), and the oil is more stable with a more bitter taste [58]. Over time the crushers have evolved considerably, and crushers with new metallic crusher at mobile knives, crushers with granite stone mill and with double olive crushing are nowadays frequency used.

The crushing step is followed by a malaxation phase where a slow stirring of olive paste takes place to obtain larger drops by promoting contact between oil droplets. These larger drops are separated in a continuous oily phase, thus avoiding the formation of emulsions which affects extraction yield [6]. During malaxation, the olive paste is heated by hot water in which addition of water enables paste dilution. Generally, the malaxation phase is performed under certain conditions: (1) limited mixing time 20–30 min, (2) temperature of the paste should not exceed 22–25 °C and (3) paste/water ratio has to be kept between 1:0.4 and 1:0.7. The mixers automatically spread the olive paste onto mats (filtering diaphragms of nylon fibre), which are then stacked for pressing.

After that, there is a choice of three procedures to recover oil from the paste using: hydraulic press, continuous centrifuges or adhesion filtering. Hydraulic press applies high pressure to the paste placed between two metallic discs of trolley with a perforation in central spike which permits the must (mixture olive oil + vegetation + water) to flow. The oily must is then sent to a vertical centrifuge, rotating at 6500 rpm, to separate the VOO from the aqueous phase (vegetation water) [46, 56].

Three fractions are separated from the olive paste: (1) oil, (2) wastewater and (3) husks or residue. The husks are dried, and the remaining oil extracted with solvent; therefore, two oil types are obtained from olives: (1) olive oil, which is pressed without further processing (other than washing, decantation, centrifugation and filtration) and contains less than 3.5% FFA; and (2) pomace oil, which is obtained by solvent extraction of the husks and does not qualify as olive oil.

12.4.2 Palm Oil

The oil palm tree has the appearance of a date palm with a large head of pinnate feathery fronds growing from a sturdy trunk. The most important palm species are *Elaeis guineensis* from Africa and *E. melanococca* from America. The fruit grows in bunches weighing 10–50 kg and each containing 800–2000 individual fruits. The fruit consists of an outer pulp, which is the source of the crude palm oil; an inner shell, which is used for fuel; and two or three kernels, which are the source of another oil type: palm kernel [59]. Palm plant grows under specific conditions as for land below 400 m attitude, moisture more than 80%, moderate temperature between 22 and 33 °C, adequate light not less than 5–6 h daily sunshine and solar radiation of 16–17 MJ/m^2 per day, high annual rainfall between 2500 and 4000 mm and for low wind speed. The most oil palm commercialized is from *E. guineensis* species which has spread to most parts of the tropical and subtropical zones of the world but particularly Malaysia and Indonesia [60].

12.4.2.1 Production and Localization

Oil palm represented 34% of vegetable oils production with 54.385 mill tonne [1, 3]. Palm Oil has the highest oil yield crop with productivity between 4 and 5 metric tonne per hectare [61] on top of the favourable climate conditions and to their highly supportive governmental policy to support long-term strategic planning and sustainable growth for palm oil industry, and Malaysia and Indonesia are the main producers with more 90% of world palm oil production [2, 3]. The rest of world production (10%) is shared with countries like Thaïland (3.6%), Nigeria (1.8%) and Colombia (1.7%).

12.4.2.2 Chemical Composition of Palm Oil

Palm oil has specific fatty acid composition with equal portions of saturated and unsaturated fatty acids and a significant amount of the saturated fatty acids (10–16%) in sn-2 position of triglycerides. Palm oil is semisolid at room temperature with high amounts of triglycerides, palmitic and oleic fatty acids [62]. Crude palm oil contains high-carotene amounts range from 500 to 700 ppm due essentially to a deep orange-red colour, of which 90% consists of alpha- and beta-carotene, and this latter act as prooxidant which affects palm oil stability. The presence of tocotrienols and tocopherols in the range of 600–1000 ppm gives to palm oil high antioxidant protection [60]. Red palm oil is utilized as a natural colourant for margarine and shortening products.

Others forms of palm oil as for crude, olein (fractionated oil) and stearin forms present globally the same physical characteristics except for iodine value, and olein products have iodine value within a relatively narrow range (56–58), but the stearin fractions exhibit a wide range (25–49). For palm stearin, softening point ranges from 56 to 53 °C for the detergent process, 51–50 °C for the slow-dry process and 49°–46° for the fast-dry process.

Currently, palm oil is widely used in food preparation and manufacturing worldwide. Almost 90% of palm oil production goes for edible consumption, while the remaining 10% of palm oil and its products are used for non-edible applications, mainly in the soap and oleochemical industries [60]. The physical characteristics and oil composition are summarized in Table 12.2.

12.5 Fats and Oils as Solvents

On top of the environmental impact, the utilization of volatile organic compounds (VOCs) to affect chemical reactions and/or extract natural products from plants is sparking an outcry because of their associated health, e.g. irritation, headaches, loss of coordination and nausea. To respond to this challenge, scientists were able to exploit a number of alternatives to the utilization of VOCs such as ionic liquids, supercritical

Table 12.2 Palm oil composition and physical characteristics [48]

Parameter	Typical	Range
Specific gravity 50 °C	–	0.888–0.889
Refractive index, 50 °C	–	1.455–1.456
Iodine value	53.0	46.0–56.0
Saponification number	196	190–202
Unsaponifiable number	0.5	0.15–0.99
Titer (°C)	46.3	40.7–49.0
Melting point (°C) (MDP)	37.5	35.5–45.0
Solidification point (°C)	–	35.0–42.0
Cloud point (°C)	−5.6	–
Cold test (hours)	None	–
Carotene content (ppm)	–	500–700
AOM stability (hours)	54.0	53–60
Tocopherol content (ppm)		
α-Tocopherol	172	129–215
β-Tocopherol	30	22–37
γ-Tocopherol	26	19–39
ζ-Tocopherol	13	10–16
Tocotrienol content (ppm)		
α-Tocotrienol	59	44–73
β-Tocotrienol	59	44–70
γ-Tocotrienol	350	262–437
ζ-Tocotrienol	94	70–117
Fatty acid composition (%)		
Lauric (C-12:0)	0.2	0.1–1.0
Myristic (C-14:0)	1.1	0.9–1.5
Palmitic (C-16:0)	44.0	41.8–46.8
Palmitoleic (C-16:1)	0.12	0.1–0.3
Stearic (C-18:0)	4.5	4.5–5.1
Oleic (C-18:1)	39.2	37.3–40.8
Linoleic (C-18:2)	10.1	9.1–11.0
Linolenic (C-18:3)	0.4	0.4–0.6
Arachidi (C-20.0)	0.4	0.2–0.7
Triglyceride composition (%)		
Trisaturated (GS3)	10.2	4.0–10.5
Disaturated (GS2U)	48.0	41.0–59.0
Monosaturated (GSU2)	34.6	32.0–54.0
Triunsaturated (GU3)	6.8	3.0–12.0
Hydrogenated crystal habit	β′	–

(continued)

Table 12.2 (continued)

Parameter	Typical	Range
Solid fat index (%) at		
10.0 °C	34.5	30.0–39.0
21.1 °C	14.0	11.5–17.0
26.7 °C	11.0	8.0–14.0
33.3 °C	7.4	4.0–11.0
40.0 °C	5.6	2.5–9.0

G—glycerides; *S*—saturated; *U*—unsaturated; *AOM*—active oxygen method, MDF—Mettler dropping point

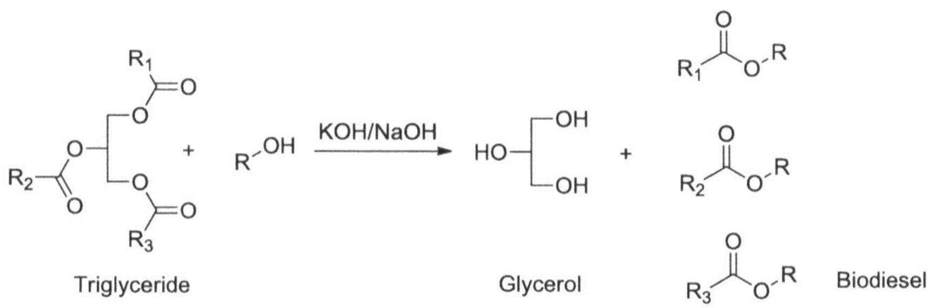

Fig. 12.2 Synthesis of biodiesel and by-product glycerol

fluids and other bio-derived solvents (e.g. 2-MeTHF). Most importantly, the choice of non-VOC solvents in chemical reactions and/or natural product extraction not only relies on the principle of eliminating the health hazard factor posed by VOCs but also needs to respond to a number of green engineering criteria which are necessary to protect our environment and promote growth and sustainability within the chemical industry. For this reason, solvents derived from bio-based fats and oils are seen as potential substitutes.

12.5.1 Solvents in Chemical Reactions

One of the most commonly exploited bio-derived oils used as solvent in chemical reactions is glycerol. The latter is readily accessible from naturally occurring triglycerides via base catalysed trans-esterification reaction in the presence of an alcohol, e.g. methanol. In addition to glycerol, the process is primarily used with the prime objective of making biodiesel (Fig. 12.2) [63, 64].

Given the eco-benign and sustainable source of glycerol, this compound was widely examined as solvent to affect number of metal and non-metal catalysed reactions.

12.5.1.1 Non-metal-Based Catalytic Reactions

A number of pharmaceutically important reactions were successfully synthesized using glycerol as solvent. In this regard, various condensation reactions, nucleophilic additions, electrophilic aromatic substitutions, cycloaddition reactions, Petasis borono–Mannich, Friedel–Crafts alkylations and other multicomponent reactions were successfully achieved using glycerol as solvent. Table 12.3 shows some literature examples where glycerol was used as an efficient solvent for certain type of non-metal catalysed chemical transformations.

12.5.1.2 Metal-Based Catalytic Reactions

Given the importance of metal catalysed reactions in the synthesis of various pharmaceutically important molecules, glycerol was successfully employed as solvent in a number of reactions involving catalysts containing Cu, Zn, Al, Ni, Pd, Ru, Ir and Ru as promoters. Table 12.4 shows a summary of chemical transformations performed using glycerol as solvent.

12.5.2 Fats and Oils as Solvent for the Extraction of Natural Products

Over the past two decades, scientists have successfully developed methods to extract a variety of natural products using widely accessible vegetable oil. These approaches were developed for the sole purpose of overcoming the utilization of VOC-based solvents and to address the various challenges associated with promoting the employment of sustainable and eco-friendly resources in the extraction industry. In addition, it is widely acknowledged in the literature that using vegetable oil for extraction can help stabilize natural products and improve the different safety aspects of the obtained natural products following formulation [89].

Chen et al. [90] successfully used soya oil to extract astaxanthin from crawfish waste. The authors demonstrated that in order to achieve maximum pigment extraction, 1:1 ratio of oil to crawfish waste was needed. Equally, olive oil was used to extract phenolic compounds from rosemary, dry oregano; this process significantly increased the amount of phenolic contents by 1.7 and 3.5 times for rosemary and oregano gourmet oils, respectively [91]. Extraction using vegetable oils was also applied to isolate carotenoids from shrimp waste (*Penaeus indicus*); under optimum conditions (2:1 ratio of oil to shrimp waste), sunflower oil was found to give the highest yield when compared to other used oils [92].

On the other hand, Kang et al. were able to use soybean, corn, grapeseed and olive oils to extract astaxanthin from micro-algae (*Haematococcus pluvialis*) [93]. Under the circumstances, up to 87.5% yield of the desired material was isolated

Table 12.3 Synthesis of useful chemical intermediates using glycerol as solvent

Reactants	Product	Conditions	Ref
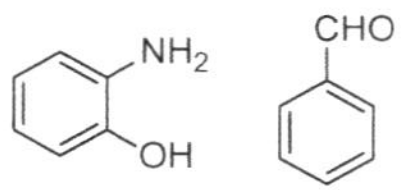	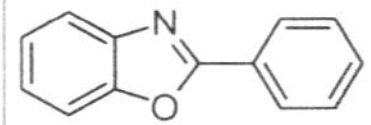	Aq. glycerol, 90 °C	[65]
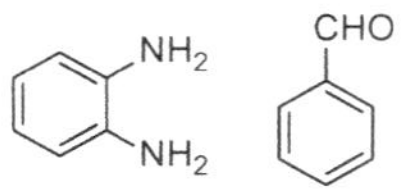	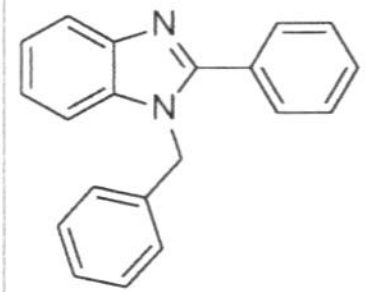	Glycerol, 90 °C	[66]
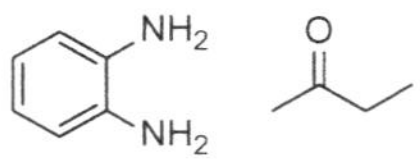	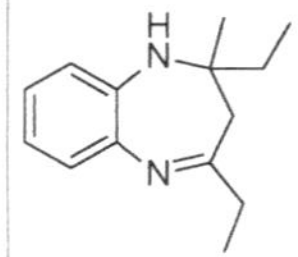	Glycerol, 90 °C	[66]
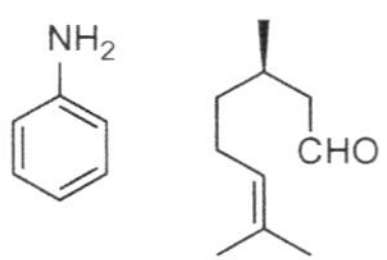	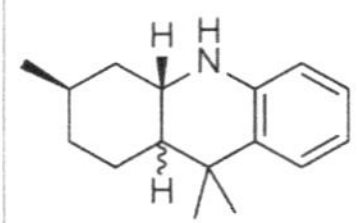	Glycerol, 90 °C	[67]
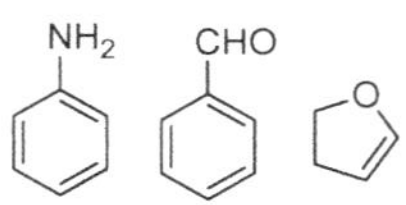	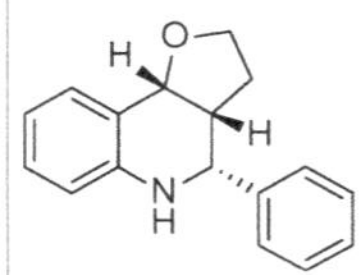	Glycerol, 90 °C	[68]

(continued)

Table 12.3 (continued)

Reactants	Product	Conditions	Ref
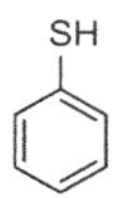	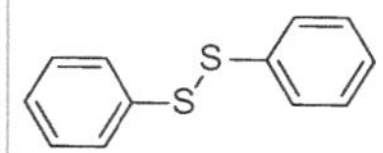	Na_2CO_3, glycerol, 120 °C, microwave	[69]
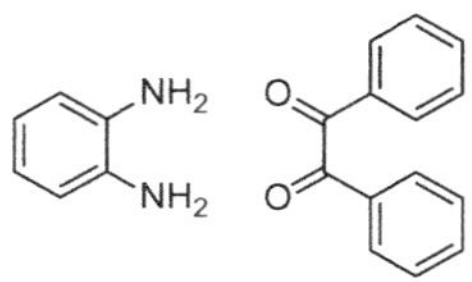	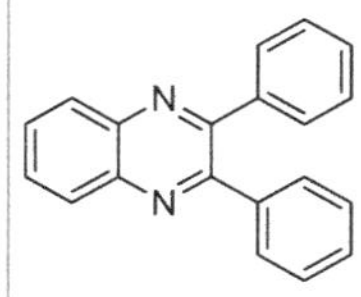	Glycerol, 90 °C, microwave	[70]
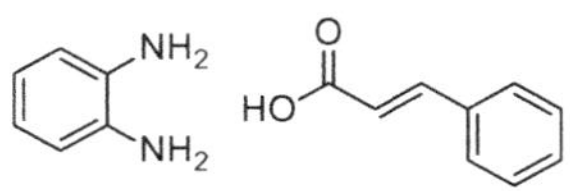	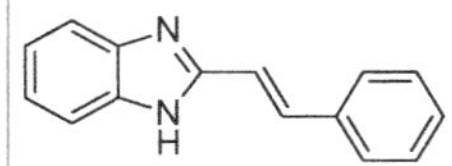	Glycerol, triacetylborate, 160–180 °C	[71]
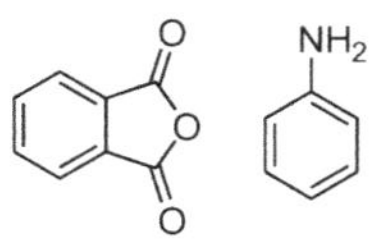	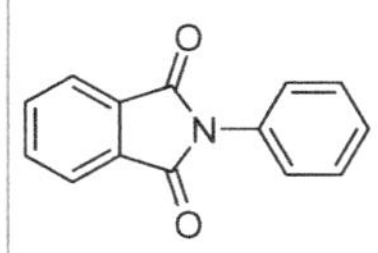	Glycerol, 65 °C	[72]
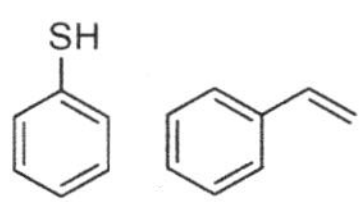	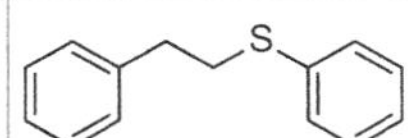	Glycerol, RT	[73]
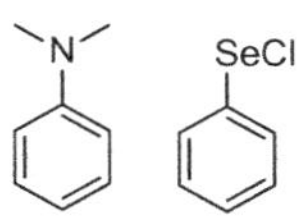	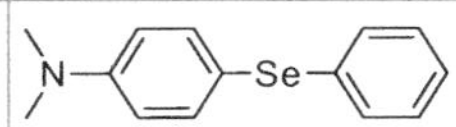	Glycerol, N_2, RT	[74]

(continued)

Table 12.3 (continued)

Reactants	Product	Conditions	Ref
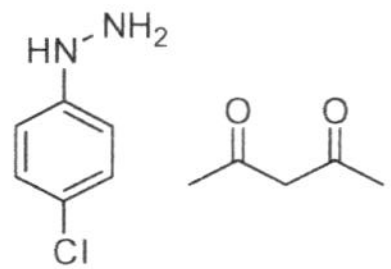	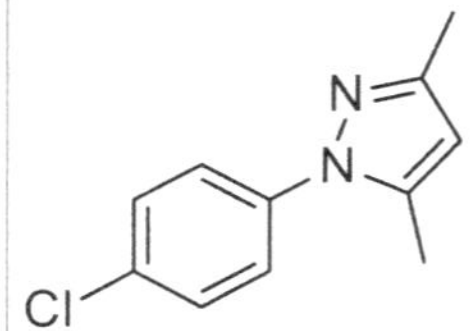	Aq. glycerol, 90 °C	[75]
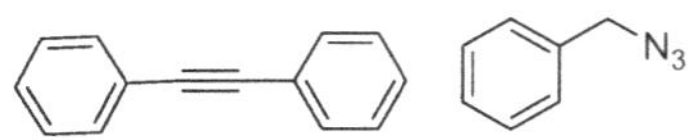	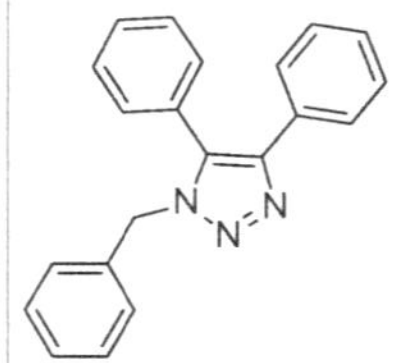	Glycerol, 100 °C, microwave	[76]
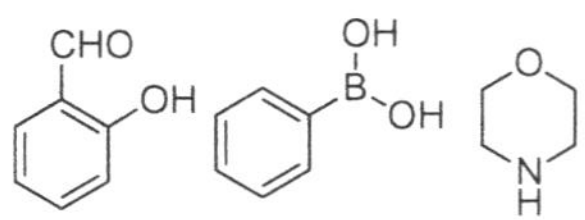	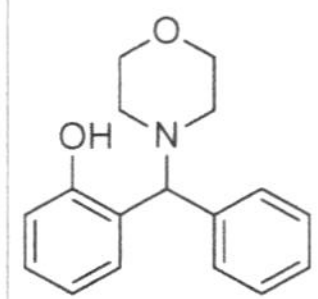	Glycerol, 90 °C	[77]
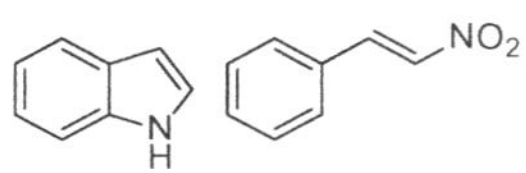	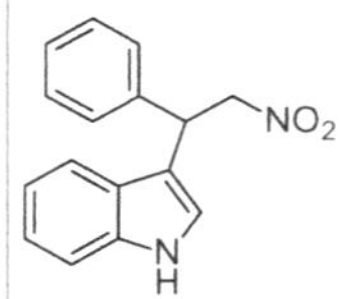	Glycerol, 90 °C	[78]

Table 12.4 Metal catalysed reactions using glycerol as solvent

Reagents	Product	Conditions	Ref.
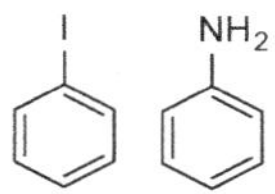	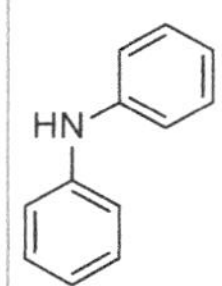	Glycerol, $Cu(OAc)_2$, KOH	[79]
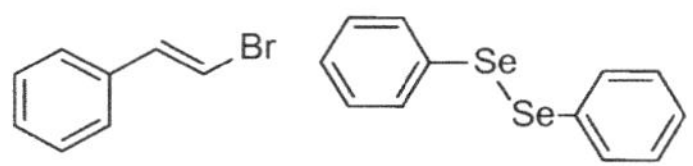	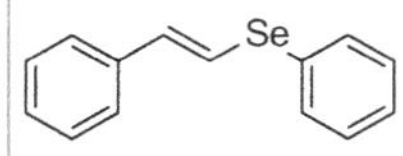	Glycerol, Cu, Zn, 110 °C	[80]
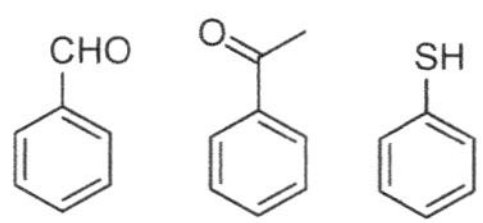	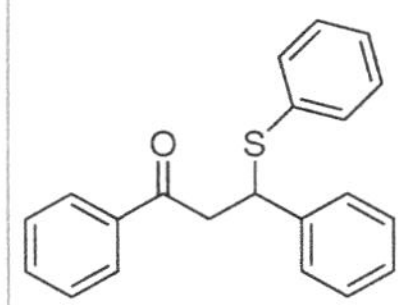	Glycerol, KF, Al_2O_3, 90 °C	[81]
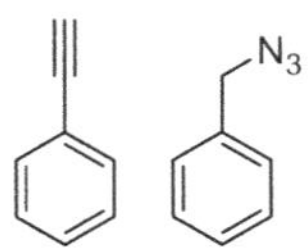	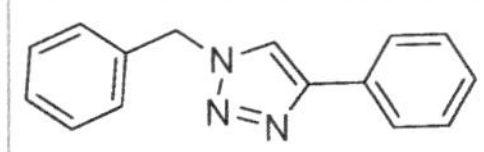	Glycerol, CuI, RT	[82]
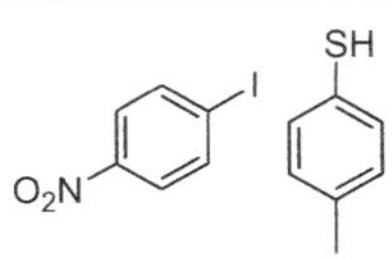	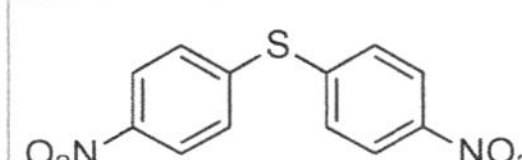	Glycerol, Cu_2ONP, tBuOK, 100 °C	[83]

(continued)

Table 12.4 (continued)

Reagents	Product	Conditions	Ref.
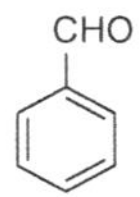	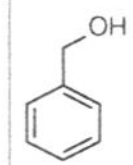	Glycerol, Ru(p-cumene)Cl_2, NaOH + KOH, 60 °C	[84]
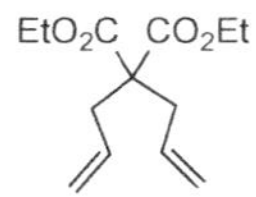	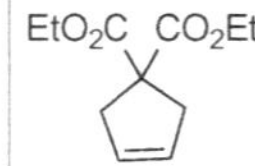	Glycerol, Grubbs I, surfactant, microwave, 50 °C	[85]
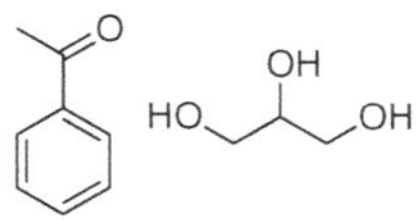	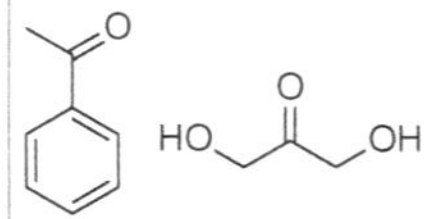	Ir NHC, KOH, 120 °C	[86]
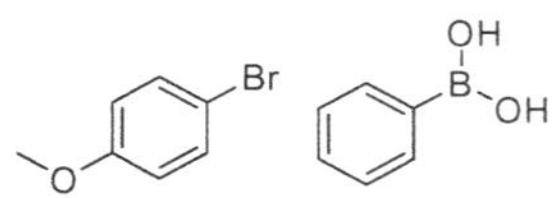	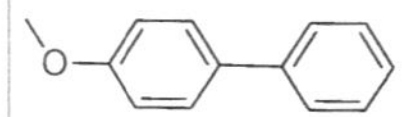	Glycerol, PdNP	[87]
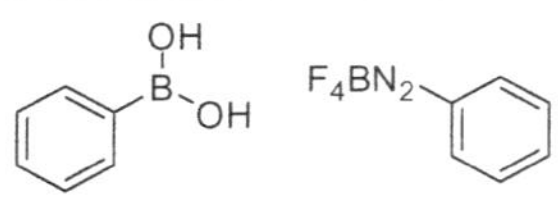	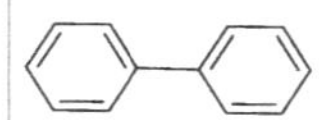	Glycerol, $NiCl_2$·glyme, DMSO, 80 °C	[88]
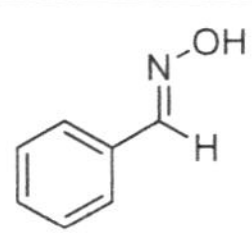	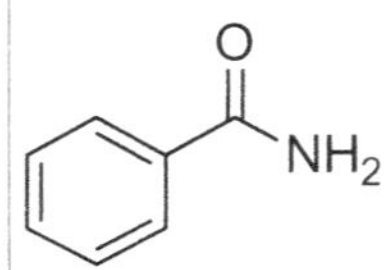	Aq. glycerol, 120 °C	[89]

when 1:1 mixture of oil and micro-algae were enriched at 25 °C. Refined edible oil was equally used to isolate phenolic compounds from olive pomace and leaves [94]. Similarly, phenolic compounds were also isolated using canola frying oil from olive and hazelnut leaves; the extract was enriched with aqueous ethanolic extracts at 200 ppm phenolic equivalence level [95].

Garcia-Martinez et al. were able to investigate the influence of olive variety and the level of orchards on the content of phenolic compounds isolated from Sicilian olive samples from eight different cultivars using Sicilian virgin olive oil (VOO) as solvent [96]. The authors also studied the effect of VOO extracts on human MG-63 cell growth; during the study, it was found that such an effect was strongly dependent on the type of cultivar and the grove altitude.

Corn oil was also used to extract a mixture of phenolic compounds and other antioxidants from thyme dried flowers (*Thymus capitatus*) [97]. The unique antioxidant activity of the thyme oil makes it one of the highly considered and naturally occurring antioxidant sources.

12.6 Fats and Oils as Synthons

The utilization of fatty acids as synthons and/or building blocks in chemical synthesis was widely investigated by the synthetic organic chemistry community worldwide. One of the earliest successful attempts was achieved by Elias J. Corey et al. [98] who successfully synthesised 5-hydroxyeicosatetraenoic acid (5-HETE) starting from readily available arachidonic acid (**1**). The latter compound underwent an initial iodolactonization reaction in the presence of KI, I_2 and $KHCO_3$ in THF to afford iodolactone **2**. Further treatment of **2** with 1,5 diazabicyclo[5.4.0]undec-5-ene (DBU) followed by lactone ring opening in the presence of MeOH and triethylamine afforded the desired methyl ester of 5-HETE (**3**); this was then treated with lithium hydroxide in a mixture of dimethoxyethane–water prompted the targeted 5-HETE in quantitative yield (Fig. 12.3).

In a similar approach to Corey et al., Yamamoto et al. were able to synthesise a series of oxidized fatty acids derivatives, namely 5-HEPA, 5-oxoEPA, 6-HOTE and 6-oxoHOTE, starting from readily available fatty acid **4**. Structure–activity relationship (SAR) studies of the obtained oxidized fatty acids were performed for the activation of peroxisome proliferator-activated receptors (PPARs) [99].

Langseter et al. also synthesised marine natural product 1,6Z,9Z,12Z,15Z-octadecapentaen-3-one starting from either docosahexaenoic acid (DHA) in eight steps or eicosapentaenoic acid (EPA) in 14 steps [100]. In a similar work by Jacobsen and co-workers, the authors demonstrated the synthesis of eicosatetraenoic acid (ETA), docosapentaenoic acid (DPA) and stearidonic acid (SDA) from readily available EPA and DHA [101].

Methyl vernolate (**8**) was also used to generate a mixture of azidohydrin and pyrrole. The former was successfully converted to furnish enantiomerically pure aziridine **11** (Fig. 12.4) [101].

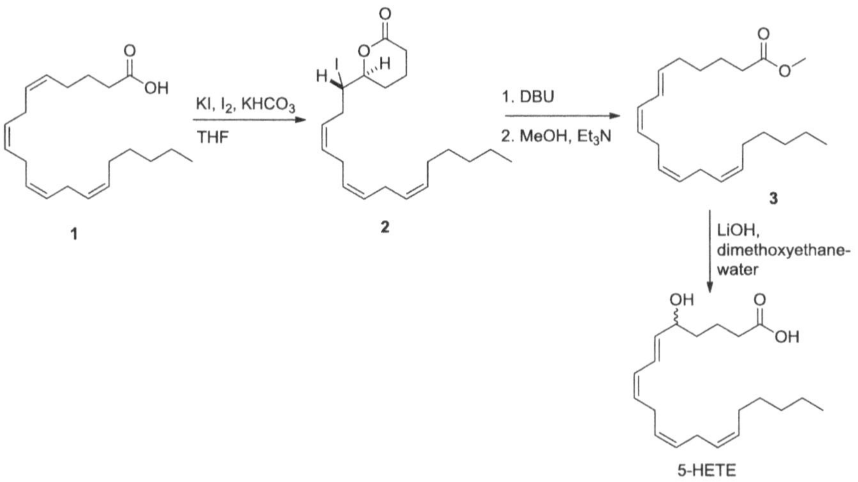

Fig. 12.3 Conversion of arachidonic acid to 5-HETE [98]

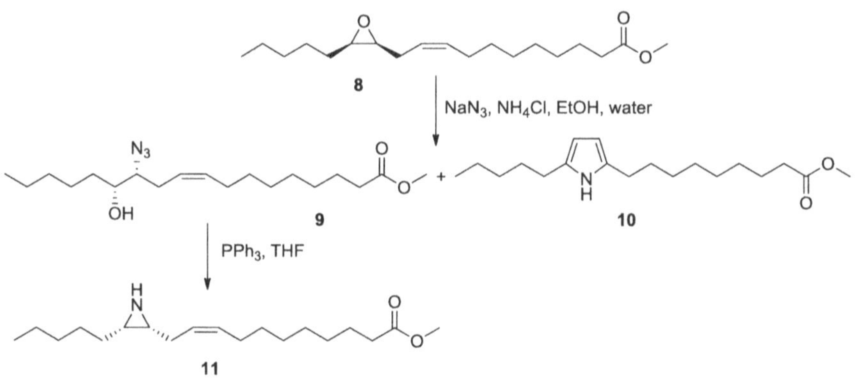

Fig. 12.4 Conversion of methyl vernolate (**8**) to pyrrole (**10**) and azide (**9**) and subsequent generation of aziridine (**11**)

Regio- and stereoselective Diels–Alder reaction was also performed on methyl calandulate (**12**) with maleic anhydride. Under solvent-free conditions, the reaction exclusively furnished the *cis* product **13** with high region and stereoselectivity [102] (Fig. 12.5).

All-cis-4-chlorotetrahydropyran (**14**) was also synthesised from reaction of methyl ester **15** with heptanal in the presence of $AlCl_3$ as catalyst. Under these conditions, the reaction proceeded with high without epimerisation to afford the corresponding product with high enantiomeric excess [103] (Fig. 12.6).

A mixture of δ-stearolactone (**16**) (major) and γ-stearolactone (**17**) (minor) was also obtained from treatment of fatty acid, oleic acid (**18**), with conc. H_2SO_4 in DCM (Fig. 12.7). The mixture contained 15:1 ratio of δ: γ stearolcatone [103].

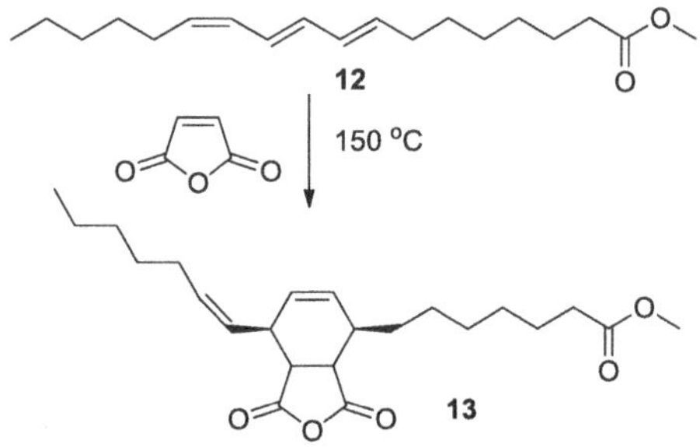

Fig. 12.5 Conversion of methyl calandulate (**12**) to anhydride (**13**)

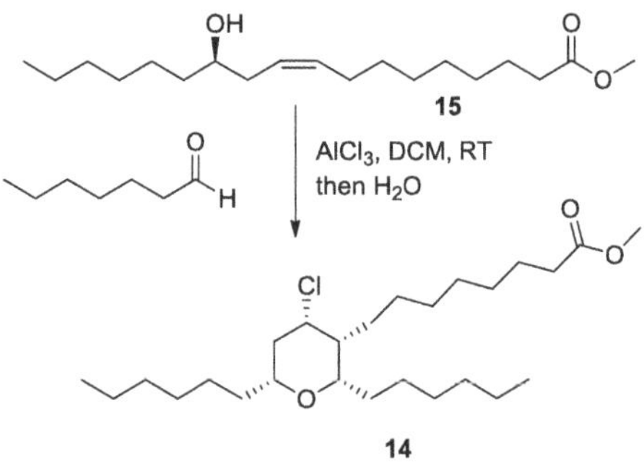

Fig. 12.6 Synthesis of cis-4-chlorotetrahydropyran (**14**) from methyl ester (**15**)

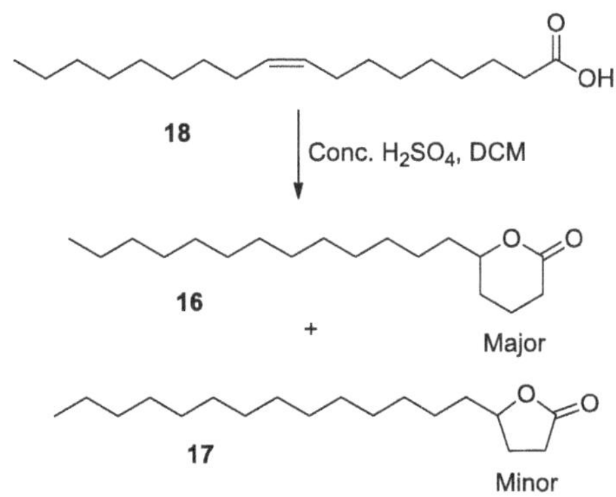

Fig. 12.7 Conversion of oleic acid to stearolactone

Cross-metathesis reactions using derivatives of fatty acids were also exploited; a variety of ruthenium base catalysts were used to affect this type of transformation. For example, polyester monomers **19** and **20** were successfully synthesised from addition of methyl ester **21** to methyl acrylate using Ru-based catalysts **22** or **23** [104] (Fig. 12.8).

The utilization of fatty acid derivatives as building blocks to gain access to various polymers was also envisaged. Second-generation Grubbs catalyst was shown to give highly ordered polymers when high-oleic sunflower oil and triglycerides were subjected to metathesis conditions in the presence of chainstopper (Fig. 12.9). The

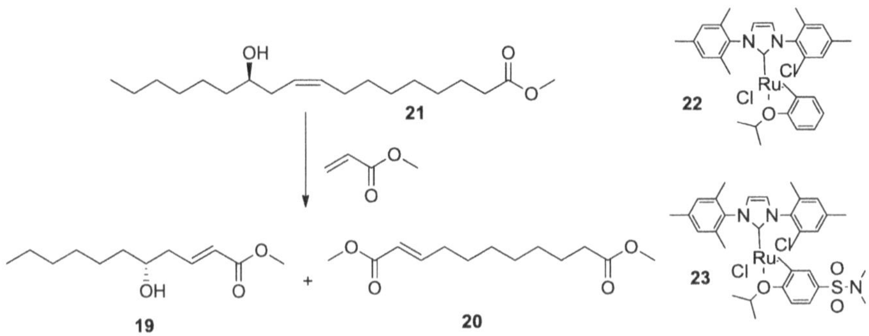

Fig. 12.8 Cross-metathesis reaction between methyl ester 21 and methyl acrylate

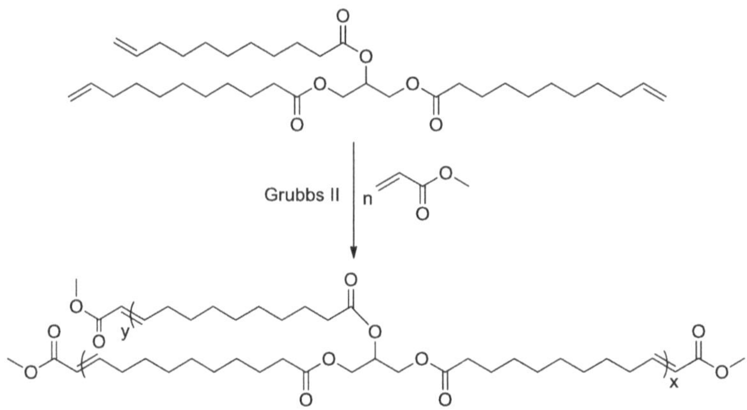

Fig. 12.9 ATMET approach for triglycerides in the presence of chainstopper

Fig. 12.10 Myristyl myristate (**24**) and cetyl ricinoleate (**25**)

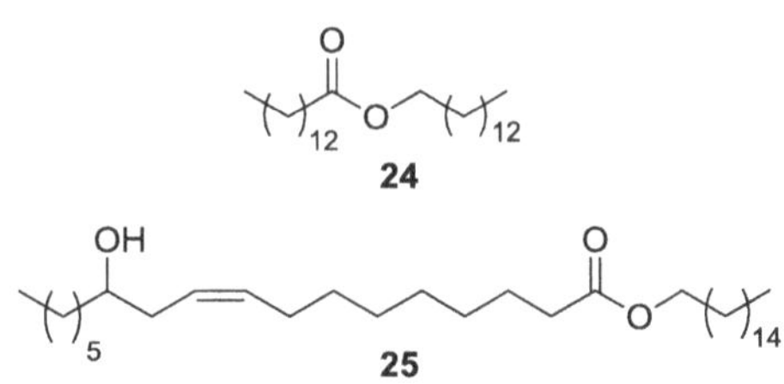

term acyclic triene metathesis (ATMET) was coined to describe the process whereby further cross-linking of the resulting polymer is prevented [105].

Other type of transformations of fatty acid derivatives involves the utilization of enzymes as catalysts. In this regard, lipases are widely used industrially as catalysts to produce cosmeceutical ingredients in the form of emollient esters such as myristyl myristate (**24**) and cetyl ricinoleate (**25**) from mannitol or sorbitol in the presence of fatty acids vinyl esters as donors of acyl groups (Fig. 12.10) [106, 107].

12.7 Future Trends: Sustainability

Prices of oilseeds are expected to increase over the medium term due to rising demand for vegetable oil and protein meal but are not expected to attain any exponential highs. The demand for protein meal is driven mainly by the growth in non-ruminant and milk production and a greater incorporation rate of protein in feed rations in developing countries. Vegetable oil consumption is driven mainly by food demand in developing countries.

According to the OECD report [2], the issues and uncertainties will stay common to most commodities (e.g. macroeconomic environment, crude oil prices and weather conditions) in addition to each sector has its specific supply and demand sensitivities.

Questions over the sustainability of soybeans and palm oil production will be critical as they stem from the high share of soybean production that comes from genetically modified seeds and the expansion of oil palm plantations into rain forests. However, certification schemes and labelling might bost demand for non-GMO modified sources, thus further area expansion is expected and subsequently resulting in more supply growth.

Petroleum oil prices and biofuel policies in the USA, European Union and Indonesia will play a key role in bumping uncertainties in the vegetable oil sector because of increasing share of vegetable oil production devoted to produce biofuels in these countries. Since biodiesel is considered an advanced biofuel in the US Renewable Fuel Standard mandates, all the uncertainties related to that policy have to be taken into consideration. The proposal by the European Commission to reduce the amount of first-generation biofuels that can be counted towards the renewable energy targets from 10 to 5% remains an unachievable target.

References

1. Mielke T (2016) World supply, demand and price outlook for oils and fats. ISTA (International Statistical Agricultural Information Mielke GmbH). http://storage.unitedwebnetwork.com/files/23/351cd4c1229e5a782a3a5acb9f25e819.pdf. Accessed 4 Sept 2018
2. OECD/FAO (2015) Oilseeds and oilseed products. In: OECD-FAO Agricultural Outlook 2015. OECD Publishing, Paris. https://doi.org/10.1787/agr_outlook-2015-8-en
3. Peter T (2016) FAO food outlook May 2015. FAO—Trade and Markets Division 2016
4. Singh RP, Baldwin AR, Formo MW (1998) Fat and oil processing. Encyclopædia Britannica Inc. https://www.britannica.com/science/fat-processing. Accessed 25 Sept 2018
5. Walisundera MN, Daun KJ (2009) Chemical composition of canola and rapeseed oils. In: Gunstone F (ed) Rapeseed and Canola oil: production, processing, properties and uses. Wiley-Blackwell, London
6. Dijkstra AJR (2013) Edible oil processing from a patent perspective. Springer Science + Business Media LLC
7. Ming-Hsun C, Jasreen JK, Sekhon Rosentrater KA, Wang T, Jung S, Johnson LA (2018) Environmental impact assessment of soybean oil production: extruding-expelling process, hexane extraction and aqueous extraction. Food Bioprod Process 108:58–68
8. Johnson LA, Lusas EW (1983) Comparison of alternative solvents for oils extraction. J Am Oil Chem Soc 60:229–242

9. Lajara JR (1990) Solvent extraction of oil from oilseeds: the real basics. In: Erickson DR (ed) Edible fats and oils processing: basic principles and modern practices. AOCS Press, Champaign
10. Thakor NJ, Sokhansanj S, McGregor I, McCurdy S (1995) Dehulling of canola by hydrothermal treatments. JAOCS 72:597–602
11. Strop HR, Perry RR (1990) Vegetable oil extraction process. US Patent 4,944,954 assigned to EPE Incorporated
12. Virgilio P, Tuomas H, Marjukka K, Risto S, Jussi H, Linnanen L (2009) Sustainability of palm oil production and opportunities for finnish technology and know-how transfer. Lappeenranta University of Technology Press
13. Rupani PF, Singh RP, Ibrahim MH, Esa N (2010) Review of current palm oil mill effluent (POME) treatment methods: vermicomposting as a sustainable practice. World Appl Sci J 11(1):70–81
14. Chandrasekar V, Sampath Charanyaa S, Belur PD, Regupathi I (2015) Refining of edible oils: a critical appraisal of current and potential technologies. Int J Food Sci Technol 50(1):13–23
15. Bertrand M, Ludger B (2001) Comparison of different methods for the determination of the oil content in oilseeds. J Am Oil Chem Soc 78(1):95–102
16. Assami K, Chemat S, Meklati BY, Chemat F (2016) Ultrasound-Assisted aromatisation with condiments as an enabling technique for olive oil flavouring and shelf life enhancement. Food Anal Method 9(4):982–990
17. Chiavaro E, Rodriguez-Estrada MT, Vittadini E, Pellegrini N (2010) Microwave heating of different vegetable oils: relation between chemical and thermal parameters. LWT-Sci Technol 43:1104–1112
18. Terigar B, Balasubramanian GS, Sabliov CM, Lima M, Boldor D (2010) Soybean and rice bran oil extraction in a continuous microwave system: from laboratory- to pilot-scale. J Food Eng 104:208–217
19. Sicaire AG, Vian MA, Fine F, Carré P, Tostain S, Chemat F (2016) Ultrasound induced green solvent extraction of oil from oleaginous seeds. Ultrason Sonochem 31:319–329
20. Luque-Garcıa JL, Luque de Castro MD (2004) Ultrasound-assisted Soxhlet extraction, an expeditive approach for solid sample treatment, application to the extraction of total fat from oleaginous seeds. J Chromatogr A 1034:237–242
21. Zhang ZS, Wang LJ, Li D, Jiao SS, Chen XD, Mao ZH (2008) Ultrasound-assisted extraction of oil from flaxseed. Sep Purif Technol 62(1):192–198. https://doi.org/10.1016/j.seppur.2008.01.014
22. Rayner ET, Dollear FG, Codifer LP (1970) Extraction of aflatoxins from cottonseed and peanut meals with ethanol. J Am Oil Chem Soc 47(1):26. https://doi.org/10.1007/BF02680163
23. Rao RK, Arnold LK (1957) Alcoholic extraction of vegetable oils. Part IV. Solubilities of vegetable oils in aqueous 2-propanol. J Am Oil Chem Soc 34:401–404
24. Vaccarino C, Vaccarino S (1961) A new industrial process for cottonseed. J Am Oil Chem Soc 38:143. https://doi.org/10.1007/BF02641236
25. Rosenthal A, Pyle DL, Niranjan K (1996) Aqueous and enzymatic processes for edible oil extraction. Enzyme Microb Technol 19(6):402–420
26. Lusas EW, Jividen GM (1987) Glandless cottonseed: a review of the first 25 years of processing and utilization research. J Am Oil Chem Soc 64(6):839–854. https://doi.org/10.1007/BF02641491
27. Cater CM, Rhee KC, Hagenmaier RD, Mattil KF (1974) Aqueous extraction—an alternative oilseed milling process. J Am Oil Chem Soc 51(4):137–141
28. Prado JM, Veggi PC, Moraes MN, Zabot GL (2017) Supercritical fluid extraction for the recovery of edible oils. In: Chemat S (Ed) Edible oils: extraction, processing and applications. CRC Press (Taylor & Francis), Boca Raton
29. Gracia I, Rodríguez JF, Lucas AD, Fernandez-Ronco MP, García MT (2011) Optimization of supercritical CO_2 process for the concentration of tocopherol, carotenoids and chlorophylls from residual olive husk. J Supercrit Fluid 59:72–77

30. McHugh MA, Krukonis VJ (1994) Supercritical fluid extraction. Butterworth-Heinemann, Boston
31. FAO-WHO (2010) Fats and fatty acids in human nutrition. Rome: FAO Food and nutrition paper # 91. Report of an expert consultation. Geneva, 10–14 Nov 2008
32. EFSA Panel on Dietetic Products, Nutrition, and Allergies (NDA) (2010) Scientific opinion on dietary reference values for fats, including saturated fatty acids, polyunsaturated fatty acids, monounsaturated fatty acids, trans fatty acids, and cholesterol. The EFSA Journal 8:1461
33. U.S. Department of Agriculture, U.S. Department of Health and Human Services (2010) Dietary guidelines for Americans, 7th edn. U.S. Government Printing Office, Washington, DC
34. Report of a Joint WHO/FAO Expert Consultation (2003) Diet, nutrition and the prevention of chronic diseases. WHO Technical Report Series 916. WHO, Geneva
35. Aranceta J, Pérez RC (2012) Recommended dietary reference intakes, nutritional goals and dietary guidelines for fat and fatty acids: a systematic review. Br J Nutr 107(S2):8–22
36. Kostik V, Memeti S, Bauer B (2013) Fatty acid composition of edible oils and fats. J Hygienic Eng Des 4:112–116
37. Ferrari RA, Schulte E, Esteves W, Bruhl L, Mukherjee KD (1996) Minor constituents of vegetable oils during industrial processing. J Am Oil Chem Soc 73:587–592
38. Boyle MA, Anderson SL (2007) Personal Nutrition, 6th edn. Wadsworth, Thomson
39. Data from Codex Alimentarius (2013) Codex standard for named vegetable oils Codex Stan 210-1999. Adopted 1999. Revisions 2001, 2003, 2009. Amendments 2005, 2011, 2013
40. DebMandal M, Mandal S (2011) Coconut (Cocos nucifera L.: Arecaceae) in health promotion and disease prevention. Asian Pac J Trop Med 4(3):241–247
41. Johnson S, Saikia N (2009) Fatty acids profile of edible oils and fats in India. Centre for Science and Environment, New Delhi
42. Mancini A, Imperlini E, Nigro E (2015) Biological and nutritional properties of palm oil and palmitic acid: effects on health. Molecules 20:17339–17361
43. Bezerra CV, Antonio MCR, Pedro DO, Dayala AS, Luiza HMS (2017) Technological properties of amazonian oils and fats and their applications in the food industry. Food Chem 221:1466–1473
44. Siri-Tarino P, Sun Q, Hu F (2010) Saturated fat, carbohydrate, and cardiovascular disease. Am J Clin Nutr 91(3):502–509
45. EFSA Panel on Dietetic Products, Nutrition, and Allergies (NDA) (2010) Scientific opinion on dietary reference values for fats, including saturated fatty acids, polyunsaturated fatty acids, monounsaturated fatty acids, trans fatty acids, and cholesterol. EFSA J 8:1462
46. Boskou D, Blekas G, Tsimidou M (2006) Olive oil composition. In: Boskou D (ed) Olive oil: chemistry and technology, 2nd edn. AOCS Press, Champaign, pp 41–72
47. Odile M (2015) Caractéristiques des huiles de lin et de chanvre. OCL 22(6):D608
48. O'Brien RD (2004) Fats and oils: formulating and processing for applications, 2nd edn. CRC Press, Boca Raton
49. Cboo YM, Yap SC, Ooi CK, Ma AN, Goh SH, Ong ASH (1996) Recovered oil from palm-pressed fiber: a good source of natural carotenoids. Vitamin E and Sterols. J Amer Oil Chem Soc 73:599. https://doi.org/10.1007/bf02518114
50. Campbell E (1983) Sunflower oil. J Amer Oil Chem Soc 60(2):339A. https://doi.org/10.1007/BF02543524
51. King ME, Boyd LC, Sheldon BW (1992) Antioxidant properties of individual phospholipids in a salmon oil model system. J Amer Oil Chem Soc 69:545–551
52. Gordon MH, Miller LAD (1997) Development of steryl ester analysis for the detection of admixtures of vegetable oils. J Amer Oil Chem Soc 74:505–510
53. Morin O, PAGES-XATART-PARES X (2012) Huiles et corps gras végétaux: ressources fonctionnelles et intérêt nutritionnel. OCL 19(2):63–75. https://doi.org/10.1684/ocl.2012.0446
54. Farhoosh R, Einafshar S, Sharayei P (2009) The effect of commercial refining steps on the rancidity of soybean and canola oils. Food Chem 115:933–938

55. Zacchi P, Eggers R (2008) High-temperature pre-conditioning of rapeseed: a polyphenol-enriched oil and the effect of refining. Eur J Lipid Sci Technol 110:111–119
56. Di Giovacchino L, Mucciarella MR, Costantini N, Ferrante ML, Surricchio G (2002) Use of nitrogen to improve stability of virgin olive oil during storage. J Amer Oil Chem Soc 79:339–344
57. Lettre d'information IOC novembre (2017) Les chiffres clés de la production d'huile d'olive. Lettre d'information n°3 du Moulin des Moines- Abbaye Sainte-Madeleine, OIC
58. Angerosa F, Di Giacinto L, Basti C, De Mattia G (1995) Caratterizzazione del difetto di "cascola". Riv Ital Sostanze Grasse 72:61–65
59. Gunstone F (2001) Palm oil supplying much of world demand for fats and oils. INFORM 12:141
60. Ariffin A (1994) Selected readings on palm oil and its uses. Palm Oil Research Institute of Malaysia, Kuala Lumpur
61. Mukherjee I, Sovacool BK (2014) Palm oil-based biofuels and sustainability in southeast Asia: a review of Indonesia, Malaysia, and Thailand. Renew Sust Energ Rev 37:1–12
62. Sarmidi MR, El Enshasy HA, Abdul Hamid M (2009) Oil palm: the rich mine of pharma, food, feed and fuel industries. Am Eurasian J Agric Environ Sci 5:767–776
63. Kemp WH (2006) Biodiesel: basics and beyond. Azlext Press, Ontario
64. Zhou CH, Beltramini JN, Fan YX, Lu CQ (2008) Chemoselective catalytic conversion of glycerol as a biorenewable source to valuable commodity chemicals. Chem Soc Rev 37:527–549
65. Bachhav HM, Bhagat SB, Telvekar VN (2011) Efficient protocol for the synthesis of quinoxaline, benzoxazole and benzimidazole derivatives using glycerol as green solvent. Tetrahedron Lett 52:5697–5701
66. Radatz CS, Silva RB, Perin G (2011) Catalyst-free synthesis of benzodiazepines and benzimidazoles using glycerol as recyclable solvent. Tetrahedron Lett 52:4132–4136
67. Nascimento JER, Barcellos AM, Sachini M et al (2011) Catalyst-free synthesis of octahydroacridines using glycerol as recyclable solvent. Tetrahedron Lett 52:2571–2574. https://doi.org/10.1016/j.tetlet.2011.03.045
68. Somwanshi JL, Shinde ND, Faruqui M (2013) Catalyst-free synthesis of furanoand pyranoquinolines by using glycerol as recyclable solvent. Heterocyclic Lett 3:69–74
69. Cabrera DML, Líbero FM, Alves D et al (2012) Glycerol as a recyclable solvent in a microwave-assisted synthesis of disulfides. Green Chem Lett Rev 5:329–336
70. Zhou WJ, Zhang XZ, Sun XB et al (2013) Microwave-assisted synthesis of quinoxaline derivatives using glycerol as a green solvent. Russ Chem Bull 62:1244–1247
71. Taduri AK, Babu PNK, Devi BR (2014) Glycerol containing triacetylborate mediated syntheses of novel 2-heterostyryl benzimidazole derivatives: a green approach. Org Chem Int 260726
72. Lobo HR, Singh BS, Shankarling GS (2012) Deep eutectic solvents and glycerol: a simple, environmentally benign and efficient catalyst/reaction media for synthesis of *N*-aryl phthalimide derivatives. Green Chem Lett Rev 5:487–533
73. Lenardão EJ, Jacob RG, Mesquita KD et al (2013) Glycerol as a promoting and recyclable medium for catalyst-free synthesis of linear thioethers: new antioxidants from eugenol. Green Chem Lett Rev 6:269–276
74. Thurow S, Penteado F, Perin G et al (2014) Metal and base-free synthesis of arylselanyl anilines using glycerol as a solvent. Green Chem 16:3854–3859
75. Min ZL, Zhang Q, Hong X et al (2015) A green protocol for catalyst-free syntheses of pyrazole in glycerol–water solution. Asian J Chem 27:3205–3207
76. Rodriguez-Rodriguez M, Gras E, Pericas MA, Gomez M (2015) Metal-free intermolecular azide–alkyne cycloaddition promoted by glycerol. Chemistry Eur J 21:18706–18710
77. Rosholm T, Gois PMP, Franzen R, Candeias NR (2015) Glycerol as an efficient medium for the Petasis borono–Mannich reaction. ChemistryOPEN 4:39–46
78. Habib PM, Kavala V, Kuo CW, Yao CF (2008) Catalyst-free aqueous-mediated conjugative addition of indoles to β-nitrostyrenes. Tetrahedron Lett 49:7005–7007

79. Khatri PK, Jain SL (2013) Glycerol ingrained copper: an efficient recyclable catalyst for the N-arylation of amines with aryl halides. Tetrahedron Lett 54:2740–2743
80. Gonçalves LC, Fiss GF, Perin G et al (2010) Glycerol as a promoting medium for cross-coupling reactions of diaryl diselenides with vinyl bromides. Tetrahedron Lett 51:6772–6775
81. Perin G, Mesquita K, Calheiro TP et al (2014) Synthesis of β-aryl-β-sulfanyl ketones by a sequential one-pot reaction using KF/Al_2O_3 in glycerol. Synth Commun 44:49–58
82. Vidal C, García-Álvarez J (2014) Glycerol: a biorenewable solvent for base-free Cu(i)-catalyzed 1,3-dipolar cycloaddition of azides with terminal and 1-iodoalkynes. Highly efficient transformations and catalyst recycling. Green Chem 16:3515–3521
83. Chahdoura F, Pradel C, Gómez M (2014) Copper(i) oxide nanoparticles in glycerol: a convenient catalyst for cross-coupling and azide–alkyne cycloaddition processes. ChemCatChem 6:2929–2936
84. Cravotto G, Orio L, Calcio Gaudino E et al (2011) Efficient synthetic protocols in glycerol under heterogeneous catalysis. Chemsuschem 4:1130–1134
85. Hamel A, Sacco M, Mnasri N et al (2014) Micelles into glycerol solvent: overcoming side reactions of glycerol. ACS Sustain Chem Eng 2:1353–1358
86. Azua A, Mata JA, Peris E (2011) Iridium NHC based catalysts for transfer hydrogenation processes using glycerol as solvent and hydrogen donor. Organometallics 30:5532–5536
87. Chahdoura F, Favier I, Pradel C et al (2015) Palladium nanoparticles stabilised by PTA derivatives in glycerol: synthesis and catalysis in a green wet phase. Catal Commun 63:47–51
88. Bhojane JM, Sarode SA, Nagarkar JM (2016) Nickel–glycerol: an efficient, recyclable catalysis system for Suzuki cross coupling reactions using aryl diazonium salts. New J Chem 40:1564–1570
89. Sehailia M, CHEMAT S (2017) Green extraction of edible oils in edible oils. In: Chemat S (ed) Extraction, processing and applications. CRC Press (Taylor & Francis), Boca Raton
90. Chen HM, Meyers SP (1982) Extraction of astaxanthin pigment from crawfish waste using a soy oil process. J Food Sci 47:892–896
91. Damechki M, Sotiropoulou S, Tsimidou A (2001) Antioxidant and pro-oxidant factors in oregano and rosemary gourmet olive oils. Grasas Aceites 52:207–213
92. Sachindra NM, Mahendrakar NS (2005) Process optimization for extraction of carotenoids from shrimp waste with vegetable oils. Bioresour Technol 96:1195–1200
93. Kang CD, Sim SJ (2008) Direct extraction of astaxanthin from Haematococcus culture using vegetable oils. Biotechnol Lett 20:441–444
94. Sánchez de Medina V, Priego-Capote F, Luque de Castro MD (2012) Characterization of refined edible oils enriched with phenolic extracts from olive leaves and pomace. J Agric Food Chem 60:5866–5873
95. Aydeniz B, Yilmaz E (2012) Enrichment of frying oils with plant phenolic extracts to extend the usage life. Eur J Lipid Sci Technol 114:933–941
96. García-Martínez O, Mazzaglia G, Sánchez-Ortiz A, Ocaña-Peinado FM, Rivas A (2014) Phenolic content of sicilian virgin olive oils and their effect on MG-63 human osteoblastic cell proliferation. Grasas Aceites 65:e032
97. Karoui IJ, Msaada K, Abderrabba M, Marzouk B (2016) Bioactive compounds and antioxidant activities of thyme-enriched refined corn oil. J Agric Sci Technol 18:79–91
98. Corey EJ, Albright JO, Barton AE, Hashimoto S (1980) Chemical and enzymic syntheses of 5-HPETE, a key biological precursor of slow-reacting substance of anaphylaxis (SRS), and 5-HETE. J Am Chem Soc 102(4):1435–1436. https://doi.org/10.1021/ja00524a044
99. Itoh T, Yoshimoto N, Yamamoto K (2010) Synthesis of oxidized fatty acid derivatives via an Iodolactonization reaction. Heterocycles 80(1):637–644. https://doi.org/10.3987/COM-09-S(S)90
100. Langseter AM, Stenstrøm Y, Skattebøl L (2014) Synthesis of All-Z-1,6,9,12,15-octadecapenten-3-one, a vinylketone polyunsaturated marine natural product isolated from *Callyspongа* sp. Molecules 19(3):3804–3812. https://doi.org/10.3390/molecules19033804
101. Gjerde JM, Vik A, Hansen TV (2012) Concise syntheses of three x-3 polyunsaturated fatty acids. Tetrahedron Lett 53(44):5837–5839. https://doi.org/10.1016/j.tetlet.2012.08.009

102. Biermann U, Butte W, Eren T, Haase D, Metzger JO (2007) Regio- and stereoselective diels–alder additions of maleic anhydride to conjugated triene fatty acid methyl esters. Eur J Org Chem 3859–3862. https://doi.org/10.1002/ejoc.200700243
103. Cermak SC, Isbell TA (2000) Synthesis of δ-stearolactone from oleic acid. J Amer Oil Chem Soc 77(3):243–248. https://doi.org/10.1007/s11746-000-0040-6
104. Ho TTT, Meier MAR (2009) A design-of-experiments approach for the optimization and understanding of the cross-metathesis reaction of methyl ricinoleate with methyl acrylate. Chemsuschem 2:749–754. https://doi.org/10.1002/cssc.200900091
105. Fokou PA, Meier MAR (2008) Acyclic triene metathesis (atmet) polymerization with chain-stoppers: molecular weight control in the synthesis of branched polymers. Macromol Rapid Commun 29:1620–1625. https://doi.org/10.1002/marc.200890039
106. Heinrichs V, Thum O (2005) Biocatalysis for the production of care specialties. Lipid Technol 17:82–87
107. Hills G (2003) Industrial use of lipases to produce fatty acid esters. Eur J Lipid Sci Technol 105:601–607. https://doi.org/10.1002/ejlt.200300853

The manufacturer's authorised representative in the EU is Springer Nature Customer Service Centre GmbH, Europaplatz 3, 69115 Heidelberg, Germany. If you have any concerns regarding our products, please contact ProductSafety@springernature.com

Printed and bound by CPI Group (UK) Ltd, Croydon, CR0 4YY
25/09/2026
02222884-0001